OCEAN WAVES AND OSCILLATING SYSTEMS

OCEAN WAVES AND
OSCILLATING SYSTEMS

LINEAR INTERACTIONS INCLUDING
WAVE-ENERGY EXTRACTION

JOHANNES FALNES
Department of Physics
Norwegian University of Science and Technology NTNU

CAMBRIDGE
UNIVERSITY PRESS

CAMBRIDGE UNIVERSITY PRESS
Cambridge, New York, Melbourne, Madrid, Cape Town, Singapore, São Paulo

Cambridge University Press
The Edinburgh Building, Cambridge CB2 2RU, UK

Published in the United States of America by Cambridge University Press, New York

www.cambridge.org
Information on this title: www.cambridge.org/9780521782111

First published 2002
This digitally printed first paperback version 2005

A catalogue record for this publication is available from the British Library

Library of Congress Cataloguing in Publication data

Falnes, Johannes, 1931–
    Ocean waves and oscillating systems : linear interactions including wave-energy
    extraction / Johannes Falnes.
        p.   cm.
    Includes bibliographical references and index.
    ISBN 0-521-78211-2
    1. Ocean waves.   2. Ocean wave power.   I. Title.
    GC211.2 .F35   2002
    551.47′02 – dc21   2001035679

ISBN-13 978-0-521-78211-1 hardback
ISBN-10 0-521-78211-2 hardback

ISBN-13 978-0-521-01749-7 paperback
ISBN-10 0-521-01749-1 paperback

The author has provided a website for updates and corrections. The URL (active at the time of going to press) is <http://www.ntnu.no./~falnes/isbn0521782112corrections/isbn0521782112corrections.htm>

Contents

Preface

This book is intended to provide a thorough consideration of the interaction between waves and oscillating systems (immersed bodies and "oscillating water columns") under conditions where amplitudes are sufficiently small that linear theory is applicable. In practice, this small-wave assumption is reasonably valid for most of the time, during which, for example, a wave-energy converter is generating most of its income. During the rather rare extreme-wave situations, however, non-linear effects may be significant, and such situations influence design loads, and hence the costs, for ships and other installations deployed at sea. This matter is treated in several other books.

The present book is mainly based on lecture notes from a postgraduate university course on water waves and extraction of energy from ocean waves, which I have taught many times since 1979. For the purposes of this book, I have selected those parts of the subject which have more general interest, rather than those parts of my course which pertain to wave-power conversion in particular. I hope that the book is thus of interest to a much wider readership than just the wave-energy community.

Except in 1983, my course has been taught every second year, mainly for doctorate students at the university in Trondheim, but other interested students have also attended. Moreover, a similar two-week course was given in 1986 with participants from Norwegian industry. Another two-week course, with international participation, was held at the Chalmers University of Technology in Gothenburg, Sweden, in 1998.

In February 1980, the lecture notes were issued in a bound volume entitled *Hydrodynamisk teori for bølgjekraftverk* ("Hydrodynamic theory for wave power plants") by L. C. Iversen and me. One hundred copies were published by the University of Trondheim, Division of Experimental Physics. Later I revised the lecture notes, while translating them into English. In 1993 this process resulted in a two-volume work entitled *Theory for extraction of ocean-wave energy*.

I wish to thank the course participant Knut Bønke for his inspiring encouragement to have the lecture notes typed in 1979 and issued in a bound volume, and for his continued encouragement over many years to write a textbook based on the notes. Moreover, I would like to thank Jørgen Hals, a course partcipant from 1997, for working out the subject index of the present book. I am also in debt to my other students for their comments and proof-reading. In this connection I wish to mention, in particular, the following graduate students (the years they completed their doctorate degrees are given): L. C. Iversen (1980), Å. Kyllingstad (1982), O. Malmo (1984), G. Oltedal (1985), A. Brendmo (1995) and H. Eidsmoen (1996). Also my collaborator over many years, P. M. Lillebekken, who attended the course in 1981, has made many valuable comments.

Most of all, I am in debt to my late colleague Kjell Budal (1933–1989), whose initiative inspired my interest in wave-power utilisation at an early stage. During the oil crisis at the end of 1973, we started a new research project aimed at utilising ocean-wave energy. At that time we did not have a research background in hydrodynamics, but Budal had carried out research in acoustics, developing a particular microphone, whereas I had studied waves in electromagnetics and plasma physics. During 1972–4 we jointly authored a (Norwegian) textbook *Bølgjelære* ("Wave Science") for the second-year undergraduate students in physics. This is an interdisciplinary text on waves, with particular emphasis on acoustics and optics.

With this background our approach and attitude towards hydrodynamic waves have perhaps been more interdisciplinary than traditional. In my view, this has influenced our way of thinking and stimulated our contributions to the science of hydrodynamics. This background is also reflected in the present book, notably in Chapter 3, where interaction between oscillations and waves is considered in general; water waves, in particular, are treated in subsequent chapters (Chapters 4–7).

I am also grateful to Elsevier Science for permission to reuse, in Sections 4.9, 5.5 and 7.2, parts of my own contributions to papers in *Applied Ocean Research*. Moreover, I wish to thank Professor J. N. Newman and Dr. Alain Clément for suggesting the use of computer codes WAMIT and AQUADYN, which have been used to compute the numerical results presented in Subsections 5.2.4 and 5.7.3, respectively. I am also grateful to Dr. Stephen Barstow for permission to use Problem 4.4, which he formulated.

Finally, I wish to thank my wife, Dagny Elisabeth, for continuous support during the many years I have worked on this book. Our oldest son, Magne, took the photographs used in composition of the front cover of the book.

4 September 2001   Johannes Falnes

CHAPTER ONE

Introduction

In this book, gravity waves on water and their interaction with oscillating systems (having zero forward speed) are approached from a somewhat interdisciplinary point of view. Before the matter is explored in depth, a comparison is briefly made between different types of waves, including acoustic waves and electromagnetic waves, drawing the reader's attention to some analogies and dissimilarities. Oscillating systems for generating or absorbing waves on water are analogues of loudspeakers or microphones in acoustics, respectively. In electromagnetics the analogues are transmitting or receiving antennae in radio engineering, and light-emitting or light-absorbing atoms in optics.

The discussion of waves is, in this book, almost exclusively limited to waves of sufficiently low amplitudes for linear analysis to be applicable. Several other books (see, e.g., the monographs by Mei,[1] Faltinsen,[2] Sarpkaya and Isaacson[3] or Chakrabarti[4]) treat the subject of large ocean waves and extreme wave loads, which are so important for determining the survival ability of ships, harbours and other ocean structures. In contrast, the purpose of this book is to convey a thorough understanding of the interaction between waves and oscillations, when the amplitudes are low, which is true most of the time. For example, on one hand, for a wave-power plant the income is determined by the annual energy production, which is essentially accrued during most times of the year, when amplitudes are low, that is, when linear interaction is applicable. On the other hand, as with many other types of ocean installations, wave-power plants also have their expenses, to a large extent, determined by the extreme-load design. The technological aspects related to conversion and useful application of wave energy are not covered in the present book. Readers interested in such subjects are referred to other literature.[5-7]

The content of the subsequent chapters is outlined below. At the end of each chapter, except the first, there is a collection of problems.

Chapter 2 gives a mathematical description of free and forced oscillations in the time domain as well as in the frequency domain. An important purpose is to introduce students to the very useful mathematical tool represented by the complex representation of sinusoidal oscillations. The mathematical connection between complex amplitudes and Fourier transforms is treated. Linear systems are discussed in a rather general way, and for a causal linear system in particular, the Kramers-Kronig relations are derived. A simple mechanical oscillating system is analysed to some extent, the concept of mechanical impedance is introduced and a discussion of energy accounting in the system is included to serve as a tool for physical explanation, in subsequent chapters, of the so-called hydrodynamic "added mass".

In Chapter 3 a brief comparison is made of waves on water with other types of waves, in particular with acoustic waves. The concepts of wave dispersion, phase velocity and group velocity are introduced. In addition the transport of energy associated with propagating waves is considered, and the radiated power from a radiation source (wave generator) is mathematically expressed in terms of a phenomenologically defined radiation resistance. The radiation impedance, which is a complex parameter, is also introduced in a phenomenological way. For mechanical waves (such as acoustic waves and waves on water) its imaginary part may be represented by an added mass. Finally in Chapter 3, an analysis is given of the absorption of energy from a mechanical wave by means of a mechanical oscillation system of the simple type considered in Chapter 2. The optimum parameters of this system for maximising the absorbed energy are discussed. The maximum is obtained at resonance.

From Chapter 4 onward, a deeper hydrodynamic discussion of water waves is the main subject. With an assumption of inviscid and incompressible fluid and irrotational fluid motion, the hydrodynamic potential theory is developed. With the linearisation of fluid equations and boundary conditions, the basic equations for low-amplitude waves are derived. In most of the following discussions, either infinite water depth or finite, but constant, water depth is assumed. Dispersion and wave-propagation velocities are studied, and plane and circular waves are discussed in some detail. Also non-propagating, evanescent plane waves are considered. Another studied subject is wave-transported energy and momentum. The spectrum of real sea waves is treated only briefly in the present book. The rather theoretical Sections 4.7 and 4.8, which make extensive use of Green's theorem, may be omitted at the first reading, and then be referred to as needed during the study of the remaining chapters of the book. Whereas most of Chapter 4 is concerned with discussions in the frequency domain, the last section contains discussions in the time domain.

The subject of Chapter 5 is interactions between waves and oscillating bodies, including wave generation by oscillating bodies as well as forces induced by waves on the bodies. Initially six-dimensional generalised vectors are introduced which correspond to the six degrees of freedom for the motion of an immersed (three-dimensional) body. The radiation impedance, known from the phenomenological introduction in Chapter 3, is now defined in a hydrodynamic formulation, and, for a three-dimensional body, extended to a 6×6 matrix. In a later part of the chapter the radiation impedance matrix is extended to the case of a finite number of interacting, radiating, immersed bodies. For this case the generalised excitation force vector is decomposed into two parts, the Froude-Krylov part and the diffraction part, which are particularly discussed in the "small-body" (or "long-wavelength") approximation. From Green's theorem (as mentioned in the summary of Chapter 4) several useful reciprocity theorems are derived, which relate excitation force and radiation resistance to each other or to "far-field coefficients" (or "Kochin functions"). Subsequently these theorems are applied to oscillating systems consisting of concentric axisymmetric bodies or of two-dimensional bodies. The occurrence of singular radiation-resistance matrices is discussed in this connection. Whereas most of Chapter 5 is concerned with discussions in the frequency domain, two sections, Sections 5.3 and 5.9, contain discussions in the time domain. In the latter section motion response is the main subject. In the former section two hydrodynamic impulse-response functions are considered; one of them is causal and, hence, has to obey the Kramers-Kronig relations.

The extraction of wave energy is the subject of Chapter 6, which starts by explaining wave absorption as a wave-interference phenomenon. Toward the end of the chapter (Section 6.4) a study is made of the absorption of wave energy by means of a finite number of bodies oscillating in several (up to six) degrees of freedom. This discussion provides a physical explanation of the quite frequently encountered cases of singular radiation-resistance matrices, as mentioned above (also see Sections 5.7 and 5.8). However, the central part of Chapter 6 is concerned with wave-energy conversion which utilises only a single body oscillating in just one degree of freedom. With the assumption that an external force is applied to the oscillating system, for the purpose of power takeoff and optimum control of the oscillation, this discussion has a different starting point than that given in the last part of Chapter 3. The conditions for maximising the converted power are also studied for the case in which the body oscillation has to be restricted as a result of its designed amplitude limit or because of the installed capacity of the energy-conversion machinery.

Oscillating water columns (OWCs) are mentioned briefly in Chapter 4 and considered in greater detail in Chapter 7, where their interaction with incident waves and radiated waves is the main subject of study. Two kinds of interaction are considered: the radiation problem and the excitation problem. The radiation problem concerns the radiation of waves which is due to an oscillating dynamic air pressure above the internal air-water interfaces of the OWCs. The excitation problem concerns the oscillation which is due to an incident wave when the dynamic air pressure is zero for all OWCs. Comparisons are made with corresponding wave-body interactions. Also, wave-energy extraction by OWCs is discussed. Finally in this chapter, the case is considered in which several OWCs and several oscillating bodies are interacting with waves.

CHAPTER TWO

Mathematical Description of Oscillations

In this chapter, which is a brief introduction to the theory of oscillations, a simple mechanical oscillation system is used to introduce concepts such as free and forced oscillations, state-space analysis and representation of sinusoidally varying physical quantities by their complex amplitudes. In order to be somewhat more general, causal and non-causal linear systems are also looked at and Fourier transform is used to relate the system's transfer function to its impulse response function. With an assumption of sinusoidal (or "harmonic") oscillations, some important relations are derived which involve power and stored energy on one hand, and the parameters of the oscillating system on the other hand. The concepts of resonance and bandwidth are also introduced.

2.1 Free and Forced Oscillations of a Simple Oscillator

Let us consider a simple mechanical oscillator in the form of a mass-spring-damper system. A mass m is suspended through a spring S and a mechanical damper R, as indicated in Figure 2.1. Because of the application of an external force F the mass has a position displacement x from its equilibrium position.

Newton's law gives

$$m\ddot{x} = F + F_R + F_S, \tag{2.1}$$

where the spring force and the damper force are $F_S = -Sx$ and $F_R = -R\dot{x}$, respectively.

If we assume that the spring and the damper have linear characteristics, then the "stiffness" S and the "mechanical resistance" R are coefficients of proportionality, independent of the displacement x and of the velocity $u = \dot{x}$. Then Newton's law gives the following linear differential equation with constant coefficients,

$$m\ddot{x} + R\dot{x} + Sx = F, \tag{2.2}$$

where an overdot is used to denote differentiation with respect to time t.

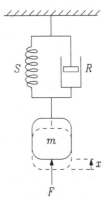

Figure 2.1: Mechanical oscillator composed of a mass-spring-damper system.

2.1.1 Free Oscillation

If the external force is absent, that is, $F = 0$, we may have so-called free oscillation if the system is released at a certain instant $t = 0$, with some initial energy

$$W_0 = W_{p0} + W_{k0} = Sx_0^2/2 + mu_0^2/2, \tag{2.3}$$

written here as a sum of potential and kinetic energy, where x_0 is the initial displacement and u_0 the initial velocity. It is easy to show (see Problem 2.1) that the general solution to Eq. (2.2), when $F = 0$, is

$$x = (C_1 \cos \omega_d t + C_2 \sin \omega_d t) e^{-\delta t}, \tag{2.4}$$

where

$$\delta = R/2m, \quad \omega_0 = \sqrt{S/m}, \quad \omega_d = \sqrt{\omega_0^2 - \delta^2} \tag{2.5}$$

are the damping coefficient, the undamped natural angular frequency and the damped angular frequency, respectively. The integration constants C_1 and C_2 may be determined from the initial conditions as (see Problem 2.1)

$$C_1 = x_0, \quad C_2 = (u_0 + x_0\delta)/\omega_d. \tag{2.6}$$

For the particular case of zero damping force, the oscillation is purely sinusoidal with a period $2\pi/\omega_0$, which is the so-called natural period of the oscillator. The free oscillation as given by Eq. (2.4) is an exponentially damped sinusoidal oscillation with "period" $2\pi/\omega_d$, during which a fraction $1 - \exp(-4\pi\delta/\omega_d)$ of the energy in the system is lost, as a result of power consumption in the damping resistance R. We define the oscillator's *quality factor* Q as the ratio between the stored energy and the average energy loss during a time interval of length $1/\omega_d$:

$$Q = \left(1 - e^{-2\delta/\omega_d}\right)^{-1}. \tag{2.7}$$

If the damping coefficient δ is small, then Q is large. When $\delta/\omega_0 \ll 1$, the following expansions (see Problem 2.2) may be useful:

$$Q = \frac{\omega_0}{2\delta}\left(1 + \frac{\delta}{\omega_0} - \frac{1}{6}\frac{\delta^2}{\omega_0^2} + \mathcal{O}\left\{\frac{\delta^3}{\omega_0^3}\right\}\right)$$

$$\approx \frac{\omega_0}{2\delta} = \frac{\omega_0 m}{R} = \frac{S}{\omega_0 R} = \frac{(Sm)^{1/2}}{R}, \tag{2.8}$$

$$\frac{\delta}{\omega_0} = \frac{1}{2Q}\left(1 + \frac{1}{2Q} + \frac{5}{24Q^2} + \mathcal{O}\{Q^{-3}\}\right) \approx \frac{1}{2Q}. \tag{2.9}$$

As a result of the energy loss, the freely oscillating system comes eventually to rest. The free oscillation is "overdamped" if ω_d is imaginary, that is, if $\delta > \omega_0$ or $R > 2(Sm)^{1/2}$. [The quality factor Q, as defined by Eq. (2.7), is then complex and it loses its physical significance.] Then the general solution of the differential equation (2.2) is a linear combination of two real, decaying exponential functions. The case of "critical damping", that is, when $R = 2(Sm)^{1/2}$ or $\omega_d = 0$, requires special consideration, which we omit here. (See, however, Problem 2.11.)

2.1.2 Forced Oscillation

When the differential equation (2.2) is inhomogeneous, that is, if $F = F(t) \neq 0$, the general solution may be written as a particular solution plus the general solution (2.4) of the corresponding homogeneous equation (corresponding to $F = 0$). Let us now consider the case in which the driving external force $F(t)$ has a sinusoidal time variation with angular frequency $\omega = 2\pi/T$, where T is the period. Let

$$F(t) = F_0\cos(\omega t + \varphi_F), \tag{2.10}$$

where F_0 is the amplitude and φ_F the phase constant for the force. It is convenient to choose a particular solution of the form where

$$x(t) = x_0\cos(\omega t + \varphi_x) \tag{2.11}$$

is the position, and

$$u(t) = \dot{x}(t) = u_0\cos(\omega t + \varphi_u) \tag{2.12}$$

is the corresponding velocity of the mass m. Here the amplitudes are related by $u_0 = \omega x_0$ and the phase constants by $\varphi_u - \varphi_x = \pi/2$. For Eq. (2.11) to be a particular solution of the differential equation (2.2), it is necessary that (see Problem 2.3) the excursion amplitude is

$$x_0 = \frac{u_0}{\omega} = \frac{F_0}{|Z|\omega} \tag{2.13}$$

and that the phase difference

$$\varphi = \varphi_F - \varphi_u = \varphi_F - \varphi_x - \pi/2 \tag{2.14}$$

is an angle which is in quadrant no. 1 or no. 4, and which satisfies

$$\tan \varphi = (\omega m - S/\omega)/R. \tag{2.15}$$

Here

$$|Z| = \{R^2 + (\omega m - S/\omega)^2\}^{1/2} \tag{2.16}$$

is the absolute value (modulus) of the complex mechanical impedance, which is discussed later.

The "forced oscillation", Eq. (2.11) or (2.12), is a response to the driving force, Eq. (2.10). Let us now assume that F_0 is independent of ω, and then discuss the responses $x_0(\omega)$ and $u_0(\omega)$, starting with $u_0(\omega) = F_0/|Z(\omega)|$. Noting that $|Z|_{\min} = R$ for $\omega = \omega_0 = (S/m)^{1/2}$ and that $|Z| \to \infty$ for $\omega = 0$ as well as for $\omega \to \infty$, we see that $(u_0/F_0)_{\max} = 1/R$ for $\omega = \omega_0$ and that $u_0(0) = u_0(\infty) = 0$. We have resonance at $\omega = \omega_0$, where the "reactive" contribution $\omega m - S/\omega$ to the mechanical impedance vanishes. Graphs of the non-dimensionalised velocity response $\sqrt{Sm}\, u_0/F_0$ versus ω/ω_0 are shown in Figure 2.2 for $\sqrt{Sm}/R$ equal to 10 and 0.5. Note that the graphs are symmetric with respect to $\omega = \omega_0$ when the frequency scale is logarithmic. The

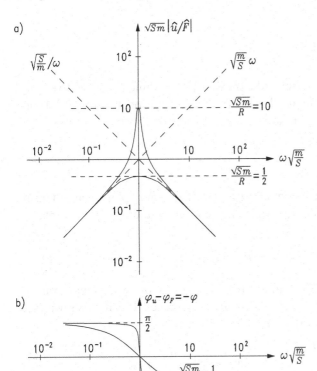

Figure 2.2: Frequency response of relation between velocity u and applied force F in normalised units, for two different values of the damping coefficient. a) Amplitude (modulus) response with both scales logarithmic. b) Phase response with linear scale for the phase difference.

phase difference φ as given by Eq. (2.15) is also shown in Figure 2.2. The graphs of Figure 2.2, where the amplitude response is presented in a double logarithmic diagram and the phase response in a semilogarithmic diagram, are usually called Bode plots or Bode diagrams.[8] Next, we consider the resonance bandwidth, that is, the frequency interval $(\Delta\omega)_{res}$, where

$$\frac{u_0(\omega)}{F_0} > \frac{1}{\sqrt{2}}\left(\frac{u_0}{F_0}\right)_{max} = \frac{1}{R\sqrt{2}}, \tag{2.17}$$

that is, where the kinetic energy exceeds half of the maximum value. At the upper and lower edges of the interval, ω_u and ω_l, the two terms of the radicand in Eq. (2.16) are equally large. Thus, we have

$$\omega_u m - S/\omega_u = R = S/\omega_l - \omega_l m. \tag{2.18}$$

Instead of solving these two equations, we note from the above-mentioned symmetry that $\omega_u \omega_l = \omega_0^2 = S/m$, that is, $S/\omega_u = m\omega_l$ and $S/\omega_l = m\omega_u$. Evidently

$$(\Delta\omega)_{res} = \omega_u - \omega_l = R/m = 2\delta. \tag{2.19}$$

The relative bandwidth is

$$\frac{(\Delta\omega)_{res}}{\omega_0} = \frac{2\delta}{\omega_0} = \frac{R}{\sqrt{Sm}}, \tag{2.20}$$

and it is seen that this is inverse to the maximum non-dimensionalised velocity response $\sqrt{Sm}\, u_0(\omega_0)/F_0 = \sqrt{Sm}/R$ as indicated on the graph in Figure 2.2. From Eq. (2.8) we see that this is approximately equal to the quality factor Q, when this is large ($Q \gg 1$). In the same case,

$$(\Delta\omega)_{res}/\omega_0 \approx 1/Q. \tag{2.21}$$

Next we consider the excursion response, which, in non-dimensionalised form, may be written as $Sx_0(\omega)/F_0$. It equals unity for $\omega = 0$ and zero for $\omega = \infty$. At resonance its value is

$$Sx_0(\omega_0)/F_0 = S/\omega_0 R = (Sm)^{1/2}/R = \omega_0/2\delta, \tag{2.22}$$

as obtained by using Eqs. (2.13) and (2.16). Note that $x_0(\omega)$ has its maximum at a frequency which is lower than the resonance frequency. It can be shown (see Problem 2.4) that, if $R < (2Sm)^{1/2}$ or $\delta < \omega_0/\sqrt{2}$, then

$$\frac{Sx_{0,max}}{F_0} = \frac{\omega_0}{2\delta}\left(1 - \frac{\delta^2}{\omega_0^2}\right)^{-1/2} \quad \text{at} \quad \omega = \omega_0\left(1 - \frac{2\delta^2}{\omega_0^2}\right)^{1/2}, \tag{2.23}$$

and if $R > (2Sm)^{1/2}$ then

$$\frac{Sx_{0,max}}{F_0} = \frac{Sx_0(0)}{F_0} = 1. \tag{2.24}$$

For large values of Q there is only a small difference between $x_0(\omega_0)/F_0$ and

$x_{0,max}/F_0$. Using Eq. (2.9) we find

$$\frac{Sx_0(\omega_0)}{F_0} = Q - \frac{1}{2} + \frac{1}{24Q} + \mathcal{O}\{Q^{-2}\} \tag{2.25}$$

and

$$\frac{Sx_{0,max}}{F_0} = Q - \frac{1}{2} + \frac{1}{6Q} + \mathcal{O}\{Q^{-2}\} \tag{2.26}$$

for

$$\omega = \omega_0 \left(1 - \frac{1}{4Q^2} + \mathcal{O}\{Q^{-3}\}\right). \tag{2.27}$$

2.1.3 Electric Analogue: Remarks on the Quality Factor

For readers with a background in electric circuits, it may be of interest to note that the mechanical system of Figure 2.1 is analogous to the electric circuit shown in Figure 2.3, where an inductance m, a capacitance $1/S$ and an electric resistance R are connected in series. The force F is analogous to the driving voltage, the position x is analogous to the electric charge on the capacitance and the velocity u is analogous to the electric current. If Kirchhoff's law is applied to the circuit, Eq. (2.2) results. The "capacitive reactance" S/ω is related to the capacitance's ability to store electric energy (analogous to potential energy in the spring S of Figure 2.1), and the "inductive reactance" ωm is related to the inductance's ability to store magnetic energy (analogous to kinetic energy in the mass of Figure 2.1). The electric (or potential) energy is zero when $x(t)=0$, and the magnetic (or kinetic) energy is zero when $u(t)=\dot{x}(t)=0$. The instants for $x(t)=0$ and those for $u(t)=0$ are displaced by a quarter of a period $\pi/2\omega_0$, and at resonance the maximum values for the electric and magnetic (or potential and kinetic) energies are equal, $mu_0^2/2 = m\omega_0^2 x_0^2/2 = Sx_0^2/2$, because $m\omega_0 = S/\omega_0$. Thus, at resonance the stored energy is swinging back and forth between the two energy stores, twice every period of the system's forced oscillation.

By Eq. (2.7) we have defined the quality factor Q as the ratio between the stored energy and the average energy loss during a time interval $1/\omega_d$ of the free oscillation. An alternative definition would have resulted if instead the forced oscillation at resonance had been considered, for a time interval $1/\omega_0$ (and not $1/\omega_d$). The stored energy is $u_0^2/2$ and the average lost energy is $Ru_0^2/2\omega_0$ during a time $1/\omega_0$ (as is shown in more detail later, in Section 2.3). Such an alternative quality factor equals the right-hand side of the approximation (2.8), and

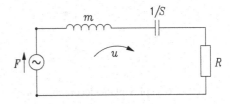

Figure 2.3: Electric analogue of the mechanical system shown in Figure 2.1.

would have been equal to the inverse of the relative bandwidth (2.20), to the non-dimensionalised excursion amplitude (2.22) at resonance (i.e., the ratio between the excursion at resonance and the excursion at zero frequency), and to the ratio of the reactance parts $\omega_0 m$ and S/ω_0 to the damping resistance R. The term quality factor is usually used only when it is large, $Q \gg 1$. In that case the relative difference between the two definitions is of little importance.

2.2 Complex Representation of Harmonic Oscillations

2.2.1 Complex Amplitudes and Phasors

When sinusoidal oscillations are dealt with, it is mathematically convenient to apply the method of complex representation, involving complex amplitudes and phasors. A great advantage with the method is that differentiation with respect to time is simply represented by multiplying with $i\omega$, where i is the imaginary unit ($i = \sqrt{-1}$). We consider again the forced oscillations represented by the excursion response $x(t)$ or velocity response $u(t)$ that is due to an applied external sinusoidal force $F(t)$, as given by Eqs. (2.11), (2.12) and (2.10), respectively.

With the use of Euler's formulas

$$e^{i\psi} = \cos \psi + i \sin \psi, \tag{2.28}$$

or, equivalently,

$$\cos \psi = (e^{i\psi} + e^{-i\psi})/2, \quad \sin \psi = (e^{i\psi} - e^{-i\psi})/2i, \tag{2.29}$$

the oscillating quantity $x(t)$ may be rewritten as

$$\begin{aligned}
x(t) &= x_0 \cos(\omega t + \varphi_x) \\
&= \frac{x_0}{2} e^{i(\omega t + \varphi_x)} + \frac{x_0}{2} e^{-i(\omega t + \varphi_x)} \\
&= \frac{x_0}{2} e^{i\varphi_x} e^{i\omega t} + \frac{x_0}{2} e^{-i\varphi_x} e^{-i\omega t}.
\end{aligned} \tag{2.30}$$

Introducing the *complex amplitude* (see Figure 2.4),

$$\hat{x} = x_0 e^{i\varphi_x} = x_0 \cos \varphi_x + i x_0 \sin \varphi_x, \tag{2.31}$$

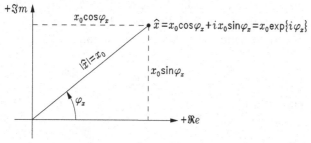

Figure 2.4: Complex-plane decomposition of the complex amplitude $\hat{x}$.

Table 2.1. Various mathematical descriptions of a sinusoidal oscillation, $\hat{x} = x_0 \exp\{i\varphi_x\}$
$x(t) = x_0 \cos(\omega t + \varphi_x)$
$x(t) = \frac{\hat{x}}{2}e^{i\omega t} + \frac{\hat{x}}{2}e^{-i\omega t}$
$x(t) = \frac{\hat{x}}{2}e^{i\omega t} + \text{c.c}$
$x(t) = \text{Re}\{\hat{x}e^{i\omega t}\}$

and the complex conjugate of $\hat{x}$,

$$\hat{x}^* = x_0 e^{-i\varphi_x} = x_0 \cos\varphi_x - ix_0 \sin\varphi_x, \tag{2.32}$$

we have

$$2x(t) = \hat{x}e^{i\omega t} + \hat{x}^* e^{-i\omega t}. \tag{2.33}$$

Note that the sum is real, while the two terms are complex (and conjugate to each other). Another formalism is

$$2x(t) = \hat{x}e^{i\omega t} + \text{c.c.} \tag{2.34}$$

where c.c. denotes complex conjugate of the preceding term.

The complex amplitude (2.31) contains information on two parameters:

1. the (absolute) amplitude $|\hat{x}| = x_0$, which is real and positive.
2. the phase constant $\varphi = \arg \hat{x}$, which is an angle to be given in units of radians (rad) or degrees (°).

A summary of the method is given in Table 2.1. In the above expression the real function $x(t)$ denotes a physical quantity in harmonic oscillation. [Note: frequently one may see it written as $x(t) = \hat{x}e^{i\omega t}$. Then, because $x(t)$ is now complex, it must be implicitly understood that it is the real part of $x(t)$ which represents the physical quantity.]

The instantaneous value

$$x(t) = x_0 \cos(\omega t + \varphi_x) = \text{Re}\{\hat{x}e^{i\omega t}\} \tag{2.35}$$

equals the projection of the rotating complex-plane vector $\hat{x}e^{i\omega t}$ on the real axis (see Figure 2.5). The value $x(t)$ is negative when $\omega t + \varphi_x$ is an angle in the second or third quadrant of the complex plane.

Let us consider position, velocity and acceleration and their interrelations. Assume that the position of an oscillating mass point is

$$x(t) = x_0 \cos(\omega t + \varphi_x) = \frac{\hat{x}}{2}e^{i\omega t} + \text{c.c} = \text{Re}\{\hat{x}e^{i\omega t}\}. \tag{2.36}$$

The velocity is

$$u = \dot{x} = \frac{dx}{dt} = -\omega x_0 \sin(\omega t + \varphi_x)$$

$$= \omega x_0 \cos(\omega t + \varphi_x + \pi/2)$$

$$= \text{Re}\{\omega x_0 \exp\{i(\varphi_x + \pi/2)\} \exp(i\omega t)\}. \tag{2.37}$$

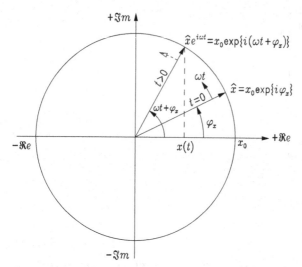

Figure 2.5: Phasor shown at times $t = 0$ and $t > 0$.

Because $e^{i\pi/2} = \cos \pi/2 + i \sin \pi/2 = i$, this gives

$$u(t) = \dot{x} = \text{Re}\{i\omega x_0 \exp(i\varphi_x) \exp(i\omega t)\} = \text{Re}\{i\omega \hat{x} e^{i\omega t}\}. \tag{2.38}$$

Differentiation of

$$x = \text{Re}\{\hat{x} e^{i\omega t}\} = \frac{\hat{x}}{2} e^{i\omega t} + \frac{\hat{x}^*}{2} e^{-i\omega t} \tag{2.39}$$

gives the same result:

$$\begin{aligned} u = \dot{x} &= \frac{\hat{x}}{2} i\omega e^{i\omega t} + \frac{\hat{x}^*}{2}(-i\omega) e^{-i\omega t} \\ &= \frac{i\omega \hat{x}}{2} e^{i\omega t} + \frac{(i\omega \hat{x})^*}{2} e^{-i\omega t} \\ &= \text{Re}\{\hat{u} e^{i\omega t}\} = \frac{\hat{u}}{2} e^{i\omega t} + \text{c.c.} \end{aligned} \tag{2.40}$$

The complex velocity amplitude is

$$\hat{u} = i\omega \hat{x}, \tag{2.41}$$

and its complex conjugate is $\hat{u}^* = (i\omega \hat{x})^* = -i\omega \hat{x}^*$.

Thus, differentiation of a physical quantity (in harmonic oscillation) with respect to time t means to multiply the corresponding complex amplitude by $i\omega$. Formally $i\omega$ is a simpler operator than d/dt. This is one advantage with the complex representation of harmonic oscillations.

Similarly, the acceleration is

$$a(t) = \dot{u} = \frac{du}{dt} = \text{Re}\{i\omega \hat{u} e^{i\omega t}\} = \text{Re}\{\hat{a} e^{i\omega t}\}. \tag{2.42}$$

The complex acceleration amplitude is

$$\hat{a} = i\omega \hat{u} = i\omega \, i\omega \hat{x} = -\omega^2 \hat{x}. \tag{2.43}$$

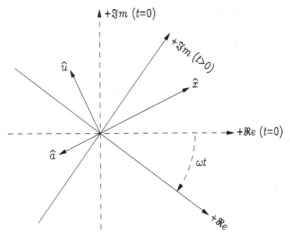

Figure 2.6: Phasor diagram for position x, velocity u and acceleration a.

The complex amplitudes $\hat{x}$, $\hat{u}$ and $\hat{a}$ are illustrated in the phase diagram of Figure 2.6. Instead of letting the complex-plane vectors (the phasors) rotate in the counterclockwise direction, we may envisage the phasors being stationary, while the coordinate system (the real and imaginary axes) rotates in the clockwise direction. Projection of the phasors on the real axis equals the instantaneous values of position $x(t)$, velocity $u(t)$ and acceleration $a(t)$.

In analogy with Eq. (2.31), we may use (2.10) and (2.12) to write the complex amplitudes for the external force and the velocity as

$$\hat{F} = F_0 \exp(i\varphi_F), \quad \hat{u} = u_0 \exp(i\varphi_u), \tag{2.44}$$

respectively. Because

$$\hat{u} = \dot{\hat{x}} = \omega x_0 \exp\{i(\varphi_x + \pi/2)\}, \tag{2.45}$$

we have

$$u_0 = \omega x_0, \quad \varphi_u = \varphi_x + \pi/2. \tag{2.46}$$

Similarly, we have the complex acceleration amplitude

$$\hat{a} = \dot{\hat{u}} = \ddot{\hat{x}} = (i\omega)^2 \hat{x} = -\omega^2 \hat{x}. \tag{2.47}$$

2.2.2 Mechanical Impedance

We shall now see how the dynamic equation (2.2) will be simplified, when we represent sinusoidally varying quantities by their complex amplitudes. Inserting

$$F(t) = \frac{\hat{F}}{2} e^{i\omega t} + \frac{\hat{F}^*}{2} e^{-i\omega t}, \tag{2.48}$$

$$x(t) = \frac{\hat{x}}{2} e^{i\omega t} + \frac{\hat{x}^*}{2} e^{-i\omega t} \tag{2.49}$$

into Eq. (2.2) gives (after multiplication by 2)

$$\hat{F}e^{i\omega t} + \hat{F}^* e^{-i\omega t} = [(i\omega)^2 m + i\omega R + S]\hat{x}e^{i\omega t} + [(-i\omega)^2 m - i\omega R + S]\hat{x}^* e^{-i\omega t}$$

$$= \left(R + i\omega m + \frac{S}{i\omega}\right)\hat{u}e^{i\omega t} + \left(R - i\omega m - \frac{S}{i\omega}\right)\hat{u}^* e^{-i\omega t}$$

(2.50)

or

$$\left[\hat{F} - \left(R + i\omega m + \frac{S}{i\omega}\right)\hat{u}\right]e^{i\omega t} + \left[\hat{F}^* - \left(R - i\omega m - \frac{S}{i\omega}\right)\hat{u}^*\right]e^{-i\omega t} = 0.$$

(2.51)

Now introducing the complex *mechanical impedance*

$$Z = i\omega m + R + S/(i\omega) = R + i(\omega m - S/\omega)$$

(2.52)

gives

$$(\hat{F} - Z\hat{u})e^{i\omega t} + (\hat{F}^* - Z^*\hat{u}^*)e^{-i\omega t} = 0.$$

(2.53)

This equation is satisfied for arbitrary t if $\hat{F} - Z\hat{u} = 0$, giving

$$\hat{u} = \hat{F}/Z$$

(2.54)

or

$$\hat{x} = \hat{u}/(i\omega) = \hat{F}/(i\omega Z).$$

(2.55)

When the absolute value (modulus) is taken on both sides of this equation, Eq. (2.13) results – if Eq. (2.44) is also observed.

An electric impedance is the ratio between complex amplitudes of voltage and current. Analogously, the mechanical impedance is the ratio between the complex force amplitude and the complex velocity amplitude. In SI units mechanical impedance has the dimension

$$[Z] = \frac{N}{m/s} = \frac{Ns}{m} = kg/s.$$

(2.56)

The impedance $Z = R + iX$ is a complex function of ω. Here R is the *mechanical resistance* and $X = \omega m - S/\omega$ is the *mechanical reactance*, which is indicated as a function of ω in Figure 2.7. Note that $X = 0$ for $\omega = \omega_0 = \sqrt{S/m}$. For this frequency, corresponding to resonance, the velocity-amplitude response is maximum:

$$|\hat{u}/\hat{F}|_{max} = |u_0/F_0|_{max} = 1/|Z|_{min} = 1/R.$$

(2.57)

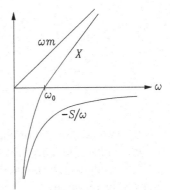

Figure 2.7: Mechanical reactance of the oscillating system versus frequency.

The impedance Z as shown in the complex plane of Figure 2.8 is

$$Z = R + iX = |Z|e^{i\varphi}, \tag{2.58}$$

where

$$|Z| = \sqrt{R^2 + X^2} \tag{2.59}$$

in accordance with Eq. (2.16). Further, we have

$$\cos\varphi = R/|Z|, \quad \sin\varphi = X/|Z|, \tag{2.60}$$

from which Eq. (2.15) follows. Note that because $R > 0$ we have $|\varphi| < \pi/2$.

We have

$$\hat{u} = \frac{\hat{F}}{Z} = \frac{|\hat{F}|e^{i\varphi_F}}{|Z|e^{i\varphi}}$$
$$= |\hat{F}/Z|\exp\{i(\varphi_F - \varphi)\} \tag{2.61}$$

or

$$|\hat{u}| = |\hat{F}|/|Z|, \quad \varphi_u = \varphi_F - \varphi. \tag{2.62}$$

We may choose the time origin $t = 0$ such that $\varphi_F = 0$; that is, $\hat{F}$ is real and positive. Then $\varphi_u = -\varphi$. A case with positive φ is illustrated with phasors in Figure 2.9 and with functions of time in Figure 2.10.

When $\varphi > 0$ (as in Figures 2.9 and 2.10) the velocity u is said to lag in phase, or to have phase lag, relative to the force F. In this case, corresponding to $\omega > \omega_0$, the reactance is dominated by the inertia term ωm. In the opposite case, $\omega_0 > \omega$,

Figure 2.8: Complex-plane decomposition of the mechanical impedance Z.

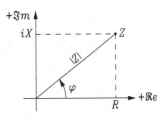

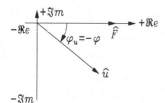

Figure 2.9: Complex amplitudes of applied force F and of resulting velocity u (when $\varphi > 0$).

$\varphi < 0$, the reactance is dominated by the stiffness, and the velocity leads in phase (or has a phase advance) relative to the force.

Frequency-response diagrams for two different values of damping resistance R are shown in Figure 2.2. If $\omega_0 \ll \omega$, we have $Z \approx i\omega m$, that is, $\hat{u} \approx \hat{F}/(i\omega m)$ or $\hat{a} = i\omega\hat{u} \approx \hat{F}/m$. Thus

$$ma(t) \approx F(t). \tag{2.63}$$

In contrast, if $\omega_0 \gg \omega$, we have $Z \approx -iS/\omega$, that is $\hat{u} \approx i\omega\hat{F}/S$ or $\hat{x} = \hat{u}/i\omega \approx \hat{F}/S$. Hence

$$Sx(t) \approx F(t). \tag{2.64}$$

Let us consider the force balance within the mechanical oscillation system. The differential equation (2.2) expresses how the applied external force is balanced against inertia, damping and stiffness forces. The force balance may also be represented by complex amplitude relations or by a force-phasor diagram:

$$m\hat{a} + R\hat{u} + S\hat{x} = \hat{F} \tag{2.65}$$

($\hat{a} = i\omega\hat{u}$, $\hat{x} = \hat{u}/i\omega$), or

$$(R + iX)\hat{u} = Z\hat{u} = \hat{F}. \tag{2.66}$$

Figure 2.11 shows a phasor diagram for the numerical example $R = 4\,\text{kg/s}$, $\omega m = 5\,\text{kg/s}$ and $S/\omega = 2\,\text{kg/s}$. For those values we obtain $X = 3\,\text{kg/s}$, $|Z| = 5\,\text{kg/s}$ and $\varphi = 0.64\,\text{rad} = 37°$. The damping force $R\hat{u}$ is in phase with the velocity $\hat{u}$. The stiffness force $S\hat{x}$ and the inertia force $m\hat{a}$ are in anti-phase. The former has a phase lag of $\pi/2$ and the latter a phase lead of $\pi/2$ relative to the velocity.

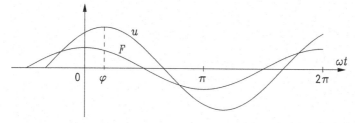

Figure 2.10: Applied force F and resulting velocity u versus time (when $\varphi > 0$). The velocity has a phase lag relative to the force.

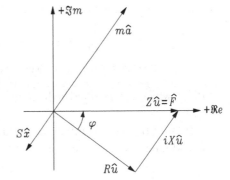

Figure 2.11: Phasor diagram showing balance of forces.

2.3 Power and Energy Relations

We shall now consider energy delivered by the external force to the simple mechanical oscillating system, shown in Figure 2.1, and see how this energy is exchanged with energy consumed in the damping resistance R and with stored potential and kinetic energy in the spring S and mass m. The mechanical power (rate of work) delivered by supplying the external force $F(t)$ is

$$P(t) = F(t) u(t) = F(t) \dot{x}(t), \tag{2.67}$$

where

$$F(t) = F_m(t) - F_R(t) - F_S(t)$$
$$= ma(t) + Ru(t) + Sx(t), \tag{2.68}$$

in accordance with the dynamic equation (2.2). Thus, we have

$$P(t) = P_R(t) + [P_k(t) + P_p(t)], \tag{2.69}$$

where the powers delivered to R, m and S are

$$P_R(t) = - F_R(t) u(t) = Ru^2, \tag{2.70}$$

$$P_k(t) = F_m(t) u(t) = m\dot{u}u = \frac{d}{dt} W_k(t), \tag{2.71}$$

$$P_p(t) = - F_S(t) u(t) = S\dot{x}x = \frac{d}{dt} W_p(t), \tag{2.72}$$

respectively, where

$$W_k(t) = \frac{m}{2} [u(t)]^2 \tag{2.73}$$

is the kinetic energy and

$$W_p(t) = \frac{S}{2} [x(t)]^2 \tag{2.74}$$

is the potential energy stored in spring S. The energy stored in the oscillating system is

$$W(t) = W_k(t) + W_p(t), \tag{2.75}$$

and the corresponding power or rate of change of energy is

$$P_k(t) + P_p(t) = \frac{d}{dt} W(t). \tag{2.76}$$

The delivered power $P(t)$ has two components.

1. $P_R(t)$, which is consumed in the damping resistance R (cf. instantaneous "active power").
2. $P_k(t) + P_p(t)$ (cf. instantaneous "reactive power"), which is exchanged with stored kinetic energy (in mass m) and potential energy (in spring S).

2.3.1 Harmonic Oscillation: Active Power and Reactive Power

With harmonic oscillation we have

$$\begin{aligned}
P(t) &= F(t)\, u(t) \\
&= \left(\frac{\hat{F}}{2} e^{i\omega t} + \frac{\hat{F}^*}{2} e^{-i\omega t} \right) \left(\frac{\hat{u}}{2} e^{i\omega t} + \frac{\hat{u}^*}{2} e^{-i\omega t} \right) \\
&= \tfrac{1}{4} (\hat{F}\hat{u}^* + \hat{F}\hat{u}^* + \hat{F}\hat{u} e^{2i\omega t} + \hat{F}^* \hat{u}^* e^{-2i\omega t}). \tag{2.77}
\end{aligned}$$

The two last terms, being complex conjugate to each other, represent a harmonic oscillation with angular frequency 2ω, and hence their sum has a time-average value equal to zero. The delivered time-average power is

$$P \equiv \overline{P(t)} = \frac{1}{4} \hat{F}\hat{u}^* + \text{c.c} = \frac{1}{2} \text{Re}\{\hat{F}\hat{u}^*\} \tag{2.78}$$

$$= \frac{1}{2} \text{Re}\{Z\hat{u}\hat{u}^*\} = \frac{1}{2} \text{Re}\{Z|\hat{u}|^2\} = \frac{R}{2} |\hat{u}|^2$$

$$= \frac{R}{2} \left| \frac{\hat{F}}{Z} \right|^2 = \frac{R}{2|Z|^2} |\hat{F}|^2 = \frac{R}{2(R^2 + X^2)} |\hat{F}|^2. \tag{2.79}$$

The consumed power is

$$\begin{aligned}
P_R(t) &= Ru^2 = R \left(\frac{\hat{u}}{2} e^{i\omega t} + \frac{\hat{u}^*}{2} e^{-i\omega t} \right)^2 \\
&= \frac{R}{4} (2\hat{u}\hat{u}^* + \hat{u}^2 e^{i2\omega t} + \hat{u}^{*2} e^{-i2\omega t}). \tag{2.80}
\end{aligned}$$

Here, the two last terms have a zero time-average sum. Thus

$$P_R = \overline{P_R(t)} = \frac{R}{2} |\hat{u}|^2 = \frac{R}{2} \hat{u}\hat{u}^* = \overline{P(t)}. \tag{2.81}$$

Hence, the consumed power and the delivered power are equal in time average. The instantaneous values are (if we choose $\varphi_F = 0$)

$$\begin{aligned}
P(t) &= \overline{P(t)} + \left(\tfrac{1}{4} \hat{F}\hat{u} e^{2i\omega t} + \text{c.c.} \right) \\
&= \overline{P(t)} + \tfrac{1}{2} |\hat{F}\hat{u}| \cos(2\omega t - \varphi) \tag{2.82}
\end{aligned}$$

for the delivered power, and

$$P_R(t) = \overline{P(t)} + (\tfrac{1}{4}R\hat{u}^2 e^{2i\omega t} + \text{c.c.})$$
$$= \overline{P(t)} + \tfrac{1}{2}R|\hat{u}|^2 \cos(2\omega t - 2\varphi) \tag{2.83}$$

for the consumed power. Thus, in general, $P_R(t) \neq P(t)$. The difference $P(t) - P_R(t) = P_k(t) + P_p(t)$ is instantaneous reactive power exchanged with stored energy in the system.

In the special case of resonance ($\omega = \omega_0, \varphi = 0, Z = R$), we have $P_R(t) = P(t) = (R/2)|\hat{u}|^2(1 + \cos 2\omega t) = R|\hat{u}|^2 \cos^2 \omega t$. Hence at resonance, there is no reactive power delivered. The stored energy is constant. It is alternating between kinetic energy and potential energy, which have equal maximum values.

In general, the stored instantaneous kinetic energy is

$$W_k(t) = \frac{m}{2}[u(t)]^2 = \frac{m}{2}|\hat{u}|^2 \cos^2(\omega t - \varphi)$$
$$= \frac{m}{4}|\hat{u}|^2 [1 + \cos(2\omega t - 2\varphi)], \tag{2.84}$$

where the average kinetic energy is

$$W_k \equiv \overline{W_k(t)} = \frac{m}{4}|\hat{u}|^2 = \frac{m}{4}\hat{u}\hat{u}^* \tag{2.85}$$

and the instantaneous potential energy is

$$W_p(t) = \frac{S}{2}[x(t)]^2 = \frac{S}{2}|\hat{x}|^2 \sin^2(\omega t - \varphi)$$
$$= \frac{S}{4}|\hat{x}|^2 [1 - \cos(2\omega t - 2\varphi)], \tag{2.86}$$

where the average potential energy is

$$W_p \equiv \overline{W_p(t)} = \frac{S}{4}|\hat{x}|^2 = \frac{S}{4\omega^2}|\hat{u}|^2. \tag{2.87}$$

The instantaneous total stored energy is

$$W(t) = W_k(t) + W_p(t)$$
$$= W_k + W_p + (W_k - W_p)\cos(2\omega t - 2\varphi), \tag{2.88}$$

which is of time average

$$W \equiv \overline{W(t)} = W_k + W_p = \frac{1}{4}(m|\hat{u}|^2 + S|\hat{x}|^2)$$
$$= \frac{m}{4}|\hat{u}|^2\left[1 + \left(\frac{\omega_0}{\omega}\right)^2\right]. \tag{2.89}$$

The amplitude of the oscillating part of the total stored energy is

$$W_k - W_p = \frac{1}{4\omega}\left(\omega m - \frac{S}{\omega}\right)|\hat{u}|^2 = \frac{X}{4\omega}|\hat{u}|^2 = \frac{X}{4\omega}\hat{u}\hat{u}^*. \tag{2.90}$$

The instantaneous reactive power is

$$P_k(t) + P_p(t) = \frac{d}{dt}W(t) = -2\omega(W_k - W_p)\sin(2\omega t - 2\varphi)$$

$$= -\frac{X}{2}|\hat{u}|^2 \sin(2\omega t - 2\varphi). \tag{2.91}$$

We observe that the reactance and the reactive power may be related to the difference between the kinetic energy and the potential energy stored in the system. At resonance $(\omega = \omega_0)$ the maximum kinetic energy $m|\hat{u}|^2/2$ equals the maximum potential energy $S|\hat{x}|^2/2 = S\omega_0^2|\hat{u}|^2/2$, and the reactance vanishes. If $\omega < \omega_0$ the maximum potential energy is larger than the maximum kinetic energy, and the mechanical reactance differs from zero (and is negative) because some of the stored energy has to be exchanged with the external energy, back and forth twice every oscillation period, as a result of the inbalance between the two types of energy store within the system. If $\omega > \omega_0$ the maximum kinetic energy is larger than the maximum potential energy, and the mechanical reactance is positive. It may be noted that the average values W_k and W_p are just half of the maximum values $m|\hat{u}|^2/2$ and $S|\hat{x}|^2/2$ of the kinetic and potential energies, respectively.

If we define the delivered "complex power" as

$$\mathcal{P} = \tfrac{1}{2}\hat{F}\hat{u}^* = \tfrac{1}{2}Z\hat{u}\hat{u}^* = \tfrac{1}{2}R|\hat{u}|^2 + \tfrac{i}{2}X|\hat{u}|^2, \tag{2.92}$$

we see from Eqs. (2.78) and (2.83) that $\text{Re}\{\mathcal{P}\} = P$ equals the average delivered power, the average consumed power and the amplitude of the oscillating part of the consumed power. Moreover, from Eq. (2.91) we see that $\text{Im}\{\mathcal{P}\}$ is the amplitude of the instantaneous reactive power. Finally, we see from Eq. (2.82) that $|\mathcal{P}|$ is the amplitude of the oscillating part of the delivered power. Note that the time-independent quantities $\text{Re}\{\mathcal{P}\}$, $\text{Im}\{\mathcal{P}\}$ and $|\mathcal{P}|$ are sometimes called *active power*, *reactive power* and *apparent power*, respectively.

2.4 State-Space Analysis

The simple oscillator shown in Figure 2.1 is represented by Eq. (2.2), which is a second-order linear differential equation with constant coefficients. It is convenient to reformulate it as the following set of two simultaneous differential equations of first order:

$$\dot{x}_1 = x_2, \quad \dot{x}_2 = -\frac{R}{m}x_2 - \frac{S}{m}x_1 + \frac{1}{m}u_1, \tag{2.93}$$

where we have introduced the state variables

$$x_1(t) = x(t), \quad x_2(t) = u(t) = \dot{x}(t) \tag{2.94}$$

and the input variable

$$u_1(t) = F(t). \tag{2.95}$$

Equations (2.93) may be written in matrix notation as

$$\begin{bmatrix} \dot{x}_1 \\ \dot{x}_2 \end{bmatrix} = \begin{bmatrix} 0 & 1 \\ -S/m & -R/m \end{bmatrix} \begin{bmatrix} x_1 \\ x_2 \end{bmatrix} + \begin{bmatrix} 0 \\ 1/m \end{bmatrix} u_1 \tag{2.96}$$

For a more general case of a linear system which is represented by linear differential equations with constant coefficients, these may be represented in the state-variable form[9]:

$$\dot{\mathbf{x}} = \mathbf{A}\mathbf{x} + \mathbf{B}\mathbf{u}, \tag{2.97}$$

$$\mathbf{y} = \mathbf{C}\mathbf{x} + \mathbf{D}\mathbf{u}. \tag{2.98}$$

If there are n state variables $x_1(t), \ldots, x_n(t)$, r input variables $u_1(t), \ldots, u_r(t)$ and m output variables $y_1(t), \ldots, y_m(t)$, then the system matrix (or state matrix) $\mathbf{A}$ is of dimension $n \times n$, the input matrix $\mathbf{B}$ is of dimension $n \times r$ and the output matrix $\mathbf{C}$ is of dimension $m \times n$. The $m \times r$ matrix $\mathbf{D}$, which is zero in many cases, represents direct coupling between the input and the output.

When the identity matrix is denoted by $\mathbf{I}$, the solution of the vectorial differential equation (2.97) may be written as (see Problem 2.9)

$$\mathbf{x}(t) = e^{\mathbf{A}(t-t_0)}\mathbf{x}(t_0) + \int_{t_0}^{t} e^{\mathbf{A}(t-\tau)}\mathbf{B}\mathbf{u}(\tau)\,d\tau, \tag{2.99}$$

where the matrix exponential is defined as the series

$$e^{\mathbf{A}t} = \mathbf{I} + \mathbf{A}t + \frac{1}{2!}\mathbf{A}^2 t^2 + \frac{1}{3!}\mathbf{A}^3 t^3 + \cdots, \tag{2.100}$$

which is a matrix of dimension $n \times n$, and which commutes with matrix $\mathbf{A}$. Hence we may perform algebraic manipulations (including integration and differentiation with respect to t) with $e^{\mathbf{A}t}$ in the same way as we do with e^{st} where s is a scalar constant. Inserting Eq. (2.99) into Eq. (2.98) gives the output vector

$$\mathbf{y}(t) = \mathbf{C}e^{\mathbf{A}(t-t_0)}\mathbf{x}(t_0) + \int_{t_0}^{t} \mathbf{C}e^{\mathbf{A}(t-\tau)}\mathbf{B}\mathbf{u}(\tau)\,d\tau + \mathbf{D}\mathbf{u}(t) \tag{2.101}$$

in terms of the input vector $\mathbf{u}(t)$ and the initial-value vector $\mathbf{x}(t_0)$. In many cases we analyse situations in which $\mathbf{u}(t) = 0$ for $t < t_0$ and $\mathbf{x}(t_0) = 0$. Then the first right-hand term in Eq. (2.101) vanishes. Note from the integral term in Eq. (2.101) that the output $\mathbf{y}$ at time t is influenced by the input $\mathbf{u}$ at an earlier time τ ($\tau < t$).

It may be difficult to make an accurate numerical computation of the matrix exponential, particularly if the matrix order n is large. Various methods have been proposed for computation and discussion of the matrix exponential.[10,11] Which method is best depends on the particular problem.

In some cases (but far from always), system matrix $\mathbf{A}$ has n linearly independent eigenvectors $\mathbf{m}_i$ ($i = 1, 2, \ldots, n$), where

$$\mathbf{A}\mathbf{m}_i = \lambda_i \mathbf{m}_i. \tag{2.102}$$

The eigenvalues λ_i, which may or may not be distinct, are solutions of (roots in) the nth degree equation:

$$\det(\mathbf{A} - \lambda \mathbf{I}) \equiv |\mathbf{A} - \lambda \mathbf{I}| = 0. \tag{2.103}$$

It may be mathematically convenient to transform system matrix $\mathbf{A}$ to a particular (usually simpler) matrix $\mathbf{A}'$ by means of a similarity transformation[9,12]:

$$\mathbf{A}' = \mathbf{M}^{-1} \mathbf{A} \mathbf{M}, \tag{2.104}$$

where $\mathbf{M}$ is a non-singular $n \times n$ matrix, which implies that its inverse $\mathbf{M}^{-1}$ exists. It is well known that $\mathbf{A}'$ has the same eigenvalues as $\mathbf{A}$. If the matrix $\mathbf{A}$ has n linearly independent eigenvectors $\mathbf{m}_i$, then $\mathbf{A}'$ is diagonal,

$$\mathbf{A}' = \mathrm{diag}(\lambda_1, \lambda_2, \ldots, \lambda_n), \tag{2.105}$$

if $\mathbf{M}$ is chosen as

$$\mathbf{M} = (\mathbf{m}_1, \mathbf{m}_2, \ldots, \mathbf{m}_n), \tag{2.106}$$

which is the $n \times n$ matrix whose ith column is the eigenvector $\mathbf{m}_i$. Because all the vectors $\mathbf{m}_i$ are linearly independent, this implies that the inverse matrix $\mathbf{M}^{-1}$ exists. In this case the matrix exponential is also simply a diagonal matrix,

$$e^{\mathbf{A}'t} = \mathrm{diag}(e^{\lambda_1 t}, e^{\lambda_2 t}, \ldots, e^{\lambda_n t}). \tag{2.107}$$

Also introducing the transformed state vector

$$\mathbf{x}' = \mathbf{M}^{-1}\mathbf{x} \quad (\text{i.e., } \mathbf{x} = \mathbf{M}\mathbf{x}') \tag{2.108}$$

and the transformed matrices

$$\mathbf{B}' = \mathbf{M}^{-1}\mathbf{B}, \quad \mathbf{C}' = \mathbf{C}\mathbf{M}, \tag{2.109}$$

we rewrite Eqs. (2.97) and (2.98) as

$$\dot{\mathbf{x}}' = \mathbf{A}'\mathbf{x}' + \mathbf{B}'\mathbf{u}, \tag{2.110}$$

$$\mathbf{y} = \mathbf{C}'\mathbf{x}' + \mathbf{D}\mathbf{u}. \tag{2.111}$$

In cases in which matrix $\mathbf{A}'$ and hence also its exponential are diagonal, as in Eqs. (2.105) and (2.107), then it is easy to compute the output $\mathbf{y}(t)$ from Eq. (2.101) with $\mathbf{A}$, $\mathbf{B}$, $\mathbf{C}$ and $\mathbf{x}_0$ replaced by $\mathbf{A}'$, $\mathbf{B}'$, $\mathbf{C}'$ and $\mathbf{x}_0'$, respectively.

The use of state-space analysis is a well-known mathematical tool, for instance in control engineering.[9] Except when the state-space dimension n is rather low, such as $n = 2$ in the example below, the state-space analysis is performed numerically.[10,11]

As a simple example we again use the oscillator shown in Figure 2.1, in which we shall consider the force $F(t)$ as input, $u_1(t) = F(t)$, and the displacement $x(t)$ as output, $y_1(t) = x(t) = x_1(t)$. The matrices to use with Eq. (2.98) are

$$\mathbf{C} = (1 \quad 0), \quad \mathbf{D} = (0 \quad 0), \tag{2.112}$$

and with Eq. (2.97),

$$\mathbf{A} = \begin{bmatrix} 0 & 1 \\ -S/m & -R/m \end{bmatrix}, \quad \mathbf{B} = \begin{bmatrix} 0 \\ 1/m \end{bmatrix}. \tag{2.113}$$

See Eq. (2.96). Thus, with $r = 1$ and $m = 1$, this is a SISO (single-input single-output) system. The state matrix is of dimension 2×2. With this example the eigenvalue equation (2.103) has the two solutions

$$\lambda_1 = -\delta + i\omega_d, \quad \lambda_2 = -\delta - i\omega_d, \tag{2.114}$$

where δ and ω_d are given by Eq. (2.5). The corresponding eigenvectors are (see Problem 2.10)

$$\mathbf{m}_1 = c_1 \begin{bmatrix} 1 \\ \lambda_1 \end{bmatrix}, \quad \mathbf{m}_2 = c_2 \begin{bmatrix} 1 \\ \lambda_2 \end{bmatrix}, \tag{2.115}$$

where c_1 and c_2 are arbitrary constants, which may by chosen equal to 1. These eigenvectors are linearly independent if $\lambda_2 \neq \lambda_1$ (i.e., $\omega_d \neq 0$). The application of Eq. (2.101) to Eq. (2.111) results in an output which agrees with Eqs. (2.4) and (2.6) for the case of free oscillations, that is, when $F(t) \equiv 0$ (see Problem 2.10). For the case of "critical damping" ($\omega_d = 0$ and $\lambda_1 = \lambda_2 = \lambda = -\delta$) there are not two linearly independent eigenvectors. Then neither $\mathbf{A}'$ nor $e^{\mathbf{A}'t}$ is diagonal (see Problem 2.11).

2.5 Linear Systems

We have discussed in detail the simple mechanical system in Figure 2.1 (or the analogous electric system shown in Figure 2.3). It obeys the constant-coefficient linear differential equation (2.2). Further, we have just touched upon another linear system, represented by a more general set (2.97) of linear differential equations with constant coefficients. Below let us discuss linear systems more generally.

A system may be defined as a collection of components, parts or units which influence each other mutually through relations between causes and resulting effects. Some of the interrelated physical variables denoted by $u_1(t), u_2(t), \ldots, u_r(t)$ may be considered as input to the system, whereas other variables, denoted by $y_1(t), y_2(t), \ldots, y_m(t)$, are considered as outputs. We may contract these variables into an r-dimensional input vector $\mathbf{u}(t)$ and an m-dimensional output vector $\mathbf{y}(t)$.

An example is the mechanical system shown in Figure 2.1, where the external force $F(t) = u_1(t)$ is the input variable, and output variables are the excursion $x(t) = y_1(t)$, the velocity $u(t) = y_2(t)$ and the acceleration $a(t) = y_3(t)$ of the oscillating mass. In this case the dimensions of the input and output vectors are $r = 1$ and $m = 3$, respectively. If the system parameters R, m and S are constant, or more precisely, independent of the oscillation amplitudes, then this particular system is linear. However, if the power $P_R(t) = R[u(t)]^2 = y_4(t)$ consumed by the resistance R had also been included as a fourth member of our chosen set of output variables, then our mathematically considered system would have been non-linear.

For instance, if the input $u_1(t) = F(t)$ had been doubled, then also $y_1(t)$, $y_2(t)$ and $y_3(t)$ would have been doubled, but $y_4(t)$ would have been increased by a factor of four.

The mathematical relationship between the variables may be expressed as

$$\mathbf{y}(t) = \mathbf{T}\{\mathbf{u}(t)\}, \tag{2.116}$$

where the operator symbol $\mathbf{T}$ designates the law for determining $\mathbf{y}(t)$ from $\mathbf{u}(t)$. If we in Eq. (2.101) set $\mathbf{x}(t_0) = 0$ and $\mathbf{D} = 0$, we have the following example of such a relationship:

$$\mathbf{y}(t) = \int_{t_0}^{t} \mathbf{C} e^{\mathbf{A}(t-\tau)} \mathbf{B} \mathbf{u}(\tau) \, d\tau. \tag{2.117}$$

With this example, the state variables obey a set of differential equations with constant coefficients and all the initial values at $t = t_0$ are zero. The output $\mathbf{y}$ at time t is a result of the input $\mathbf{u}(t)$ during the time interval (t_0, t) only.

A system is linear if, for any two input vectors $\mathbf{u}_a(t)$ and $\mathbf{u}_b(t)$ with corresponding output vectors $\mathbf{y}_a(t) = \mathbf{T}\{\mathbf{u}_a(t)\}$ and $\mathbf{y}_b(t) = \mathbf{T}\{\mathbf{u}_b(t)\}$, we have

$$\alpha_a \mathbf{y}_a(t) + \alpha_b \mathbf{y}_b(t) = \mathbf{T}\{\alpha_a \mathbf{u}_a(t) + \alpha_b \mathbf{u}_b(t)\} \tag{2.118}$$

for arbitrary constants α_a and α_b. The output and input in Eq. (2.118) are sums of two terms. It is straightforward to generalise this to a finite number of terms. Extension to infinite sums and integrals is an additional requirement which we shall include in our definition of a linear system. According to this definition the superposition principle applies for linear systems. Because $\mathbf{A}$, $\mathbf{B}$ and $\mathbf{C}$ are constant matrices, the system represented by Eq. (2.117) is linear.

In some simple cases there is an instantaneous relation between input and output, for instance between the voltage across an electric resistance and the resulting electric current, or between the acceleration of a mass and the net force applied to it. In most cases systems are dynamic. Then the instantaneous output variables depend both on previous and present values of the input variables. For instance, the velocity of a body depends on the force applied to the body at previous instants.

In some cases, it may be convenient to define a system in which both the input and the output are chosen to be variables caused by some other variable(s). Then it may happen that the present-time output variables may even be influenced by future values of the input variables. Such a system is said to be non-causal. Otherwise, for causal systems the output (the response) cannot exist before the input (the cause).

Let us, in the remaining part of this chapter, consider a linear system with a single input ($r = 1$) and single output ($m = 1$). The input function $u(t)$ and the output function $y(t)$ are then scalar functions. For this case we write

$$y(t) = L\{u(t)\} \tag{2.119}$$

instead of Eq. (2.116). (Because the system is linear, we have written L instead of

T for designating the law determining the system's output from its input.) Below, in Subsections 2.5.2 and 2.6.2, we introduce a mathematical function, the transfer function, which characterises the linear system.

2.5.1 The Delta Function and Related Distributions

As a preparation let us first give a few mathematical definitions. Let $\varphi(\tau)$ be an arbitrary function which is continuous and, infinitely many times, differentiable at $t = \tau$. Then the impulse function $\delta(t)$, also called the Dirac delta function, delta function, or, more properly, delta distribution, is defined[13] by the property

$$\varphi(0) = \int_{-\infty}^{\infty} \delta(t)\,\varphi(t)\,dt \qquad (2.120)$$

or more generally,

$$\varphi(t) = \int_{-\infty}^{\infty} \delta(\tau - t)\,\varphi(\tau)\,d\tau. \qquad (2.121)$$

In particular, if $\varphi(t) \equiv 1$ and $t = 0$, then this gives

$$\int_{-\infty}^{\infty} \delta(\tau)\,d\tau = 1. \qquad (2.122)$$

Although the delta distribution is not a mathematical function in the ordinary sense, it is a meaningful statement to say that for $t \neq 0$, $\delta(t) = 0$. The derivative $\dot{\delta}(t) = d\delta(t)/dt$ of the impulse function is defined by

$$\dot{\varphi}(0) = -\int_{-\infty}^{\infty} \dot{\delta}(t)\,\varphi(t)\,dt, \qquad (2.123)$$

and the nth derivative $\delta^{(n)}(t) = d^n\delta(t)/dt^n$ is defined by

$$\varphi^{(n)}(0) = (-1)^n \int_{-\infty}^{\infty} \delta^{(n)}(t)\,\varphi(t)\,dt. \qquad (2.124)$$

The impulse function $\delta(t)$ is even and the derivative $\delta^{(n)}(t)$ is even if n is even, and odd if n is odd.[13]
The function

$$\mathrm{sgn}(t) = 2U(t) - 1 = \begin{cases} 1 & \text{for } t > 0 \\ 0 & \text{for } t = 0 \\ -1 & \text{for } t < 0 \end{cases} \qquad (2.125)$$

is an odd function, whose derivative is $2\delta(t)$. In Eq. (2.125) we have introduced the (Heaviside) unit step function $U(t)$, which is 1 for $t > 0$ and 0 for $t < 0$. Note that $\dot{U}(t) = \delta(t)$.

2.5.2 Impulse Response: Time-Invariant System

Let us next assume that the input to a linear system is an impulse at time t_1. By this we mean that the input function is $u(t) = \delta(t - t_1)$. The system's output $y(t) = h(t, t_1)$, which corresponds to the impulse input, is called the *impulse response*. By means of Eq. (2.121) we may write an arbitrary input as

$$u(t) = \int_{-\infty}^{\infty} u(t_1)\, \delta(t - t_1)\, dt_1, \tag{2.126}$$

which may be interpreted as a superposition of impulse inputs $u(t_1)\,\delta(t - t_1)\,dt_1$. Because the system is linear, we may superpose the corresponding outputs $u(t_1)\, h(t, t_1)\, dt_1$. Thus, the resulting output is

$$y(t) = \int_{-\infty}^{\infty} u(t_1)\, h(t, t_1)\, dt_1 \equiv L\{u(t)\}. \tag{2.127}$$

If we know the impulse response $h(t, t_1)$, we can use Eq. (2.127) to find the output corresponding to an arbitrary given input.

A linear system in which the impulse response $h(t, t_1)$ depends on t and t_1 only through the time difference $t - t_1$ is called a time-invariant linear system. Then the integral in Eq. (2.127) becomes a convolution integral

$$y(t) = \int_{-\infty}^{\infty} u(t_1)\, h(t - t_1)\, dt_1 \equiv u(t) * h(t). \tag{2.128}$$

Note that convolution is commutative; that is, $h(t) * u(t) = u(t) * h(t)$.

In many cases, a system which is linear and time invariant may be represented by a set of simultaneous first-order differential equations with constant coefficients, as discussed in Section 2.4. Let us, as an example of such a case, consider a SISO system, for which Eq. (2.117) simplifies to

$$y(t) = \int_{0}^{t} \mathbf{C} e^{\mathbf{A}(t - \tau)} \mathbf{B}\, u(\tau)\, d\tau, \tag{2.129}$$

where input matrix $\mathbf{B}$ is of dimension $n \times 1$ and output matrix $\mathbf{C}$ is of dimension $1 \times n$, and where in Eq. (2.117) we have chosen $t_0 = 0$. Thus, for this special case, the impulse response is given by

$$h(t) = \begin{cases} 0 & \text{for } t < 0 \\ \mathbf{C} e^{\mathbf{A}t} \mathbf{B} & \text{for } t > 0 \end{cases} \tag{2.130}$$

as is easily seen if we compare Eq. (2.129) with Eq. (2.128). For this example, the system is causal.

We say that a linear, time-invariant system is causal if the impulse response vanishes for negative times, that is, if

$$h(t) = 0 \quad \text{for } t < 0. \tag{2.131}$$

Let us now return to the general case of a linear, time-invariant, SISO system.

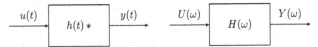

Figure 2.12: System with input signal $u(t)$ and output signal $y(t)$. The corresponding Fourier transforms are $U(\omega)$ and $Y(\omega)$, respectively. The linear system's impulse response $h(t)$ is the inverse Fourier transform of the transfer function $H(\omega)$.

If the input is $u_e(t) = e^{i\omega t}$ then, with $y_e(t) = L\{e^{i\omega t}\}$, we have

$$y_e(t_1 + t) = L\{e^{i\omega(t_1 + t)}\} = L\{e^{i\omega t_1}e^{i\omega t}\},$$
$$= e^{i\omega t_1} L\{e^{i\omega t}\} = u_e(t_1)\, y_e(t), \tag{2.132}$$

where t_1 is an arbitrary constant (of dimension time). Here we utilise the fact that the system is time invariant and linear. Inserting $t = 0$ and $t_1 = t$ into Eq. (2.132), we obtain

$$y_e(t) = y_e(0)\, e^{i\omega t} \equiv H(\omega)e^{i\omega t}, \tag{2.133}$$

which shows that if the input is an exponential function of time, then the output equals the input multiplied by a time-independent coefficient, which we have denoted $H(\omega)$. If the input is

$$u(t) = \tfrac{1}{2}(\hat{u}e^{i\omega t} + \hat{u}^*e^{-i\omega t}) \tag{2.134}$$

then the output is

$$y(t) = \tfrac{1}{2}(\hat{y}e^{i\omega t} + \hat{y}^*e^{-i\omega t}), \tag{2.135}$$

with

$$\hat{y} = H(\omega)\hat{u}, \qquad \hat{y}^* = H^*(\omega)\hat{u}^*. \tag{2.136}$$

Note that $H(\omega)$, which characterises the system, may be complex and depend on ω (see Figure 2.12). We shall see in the next section that $H(\omega)$, which we shall call the system's *transfer function*, is the Fourier transform of the impulse response $h(t)$.

Returning again to the example of forced oscillations in the system shown in Figure 2.1, where we consider the external force $F(t)$ as input and the velocity $u(t)$ as output, we see from Eq. (2.54) that the transfer function is $H(\omega) = 1/Z$, where $Z = Z(\omega)$ is the mechanical impedance, defined by Eq. (2.52).

2.6 Fourier Transform and Other Integral Transforms

We have previously studied forced sinusoidal oscillations (Section 2.1) and we have introduced complex amplitudes and phasors as convenient means to analyse sinusoidal oscillations (Section 2.2).

In cases in which linear theory is applicable, the obtained results may also be useful with quantities which do not necessarily vary sinusoidally with time. This follows from the following two facts. Firstly, functions of a rather general class may be decomposed into harmonic components according to Fourier analysis. Secondly, the superposition principle is applicable when linear theory is valid. A condition for the success of this method of approach is that the physical systems considered are time invariant. This implies that the inherent characteristics of the system remain the same at any time (see Section 2.5).

Instead of studying how the complex amplitude of a physical quantity varies with frequency, we shall now study the physical quantity's variation with time. The main content of this section includes a short review of Fourier analysis and of the connection between causality and the Kramers-Kronig relations.

2.6.1 Fourier Transformation in Brief

It is assumed that the reader is familiar with Fourier analysis. Let us, however, for convenience collect some of the main formulas related to the Fourier transformation. For a more rigorous treatment the reader may consult many textbooks, such as those by Papoulis[13] or Bracewell.[14]

If the function $f(t)$ belongs to a certain class of reasonably well-behaved functions, its Fourier transform $F(\omega)$ is defined by

$$F(\omega) = \int_{-\infty}^{\infty} f(t)\, e^{-i\omega t}\, dt \equiv \mathcal{F}\{f(t)\}, \tag{2.137}$$

and the inverse transform is

$$f(t) = \frac{1}{2\pi} \int_{-\infty}^{\infty} F(\omega)\, e^{i\omega t}\, d\omega \equiv \mathcal{F}^{-1}\{F(\omega)\}. \tag{2.138}$$

Note that these integrals, as well as other Fourier integrals in this book, must be interpreted as Cauchy principle value integrals when necessary (Papoulis,[13] p. 10).

Transformations (2.137) and (2.138) are linear. Other theorems concerning symmetry, time scaling, time shifting, frequency shifting, time differentiation and time integration are summarised in Table 2.2. The table also includes some transformation pairs relating to the delta function and its derivative (both of which belong to the class of generalised functions, also termed "distributions"), the unit step function (Heaviside function) and the signum function.

If $f(t)$ is a real function, then

$$F^*(-\omega) = F(\omega). \tag{2.139}$$

The real and imaginary parts of this Fourier transform are even and odd functions, respectively:

$$R(\omega) = \mathrm{Re}\{F(\omega)\} = \int_{-\infty}^{\infty} f(t)\cos(\omega t)\, dt = R(-\omega), \tag{2.140}$$

$$X(\omega) = \mathrm{Im}\{F(\omega)\} = -\int_{-\infty}^{\infty} f(t)\sin(\omega t)\, dt = -X(-\omega). \tag{2.141}$$

Moreover, if the real $f(t)$ is an even function, $f(t) = f_e(t)$, say, that is $f_e(-t) =$

Table 2.2. Some relations between functions of time and their corresponding Fourier transforms

Function of Time	Fourier Transform
$f(t)$	$F(\omega)$
$a_1 f_1(t) + a_2 f_2(t)$	$a_1 F_1(\omega) + a_2 F_2(\omega)$
$F(t)$	$2\pi f(-\omega)$
$f(at)$ (a real)	$(1/|a|) F(\omega/a)$
$f(t - t_0)$	$F(\omega) \exp(-it_0\omega)$
$f(t) \exp(i\omega_0 t)$	$F(\omega - \omega_0)$
$df(t)/dt = \dot{f}(t)$	$i\omega F(\omega)$
$\int_{-\infty}^{t} f(\tau)\, d\tau$	$F(\omega)/i\omega + \pi F(0)\, \delta(\omega)$
$\delta(t)$	1
1	$2\pi \delta(\omega)$
$\hat{u} \exp(i\omega_0 t) + \hat{u}^* \exp(-i\omega_0 t)$	$\hat{u} 2\pi \delta(\omega - \omega_0) + \hat{u}^* 2\pi \delta(\omega + \omega_0)$
$\mathrm{sgn}(t)$	$2/i\omega$
$U(t) = [\mathrm{sgn}(t) + 1]/2$	$1/i\omega + \pi \delta(\omega)$
$d\delta(t)/dt = \dot{\delta}(t)$	$i\omega$

$f_e(t)$, its Fourier transform is real and even,

$$F(\omega) = R(\omega) = 2 \int_0^\infty f_e(t) \cos(\omega t)\, dt, \tag{2.142}$$

and the inverse transform may be written as

$$f_e(t) = \frac{1}{\pi} \int_0^\infty R(\omega) \cos(\omega t)\, d\omega. \tag{2.143}$$

In contrast, if the real $f(t)$ is an odd function, $f(t) = f_o(t)$, say, that is $f_o(-t) = -f_o(t)$, its Fourier transform is purely imaginary and odd,

$$F(\omega) = i X(\omega) = -2i \int_0^\infty f_o(t) \sin(\omega t)\, dt, \tag{2.144}$$

and its inverse transform may be written as

$$f_o(t) = -\frac{1}{\pi} \int_0^\infty X(\omega) \sin(\omega t)\, d\omega. \tag{2.145}$$

Frequently used in connection with Fourier analysis is the convolution theorem. The function

$$f(t) = \int_{-\infty}^\infty f_1(t - \tau) f_2(\tau)\, d\tau \equiv f_1(t) * f_2(t) \tag{2.146}$$

is termed the "convolution" (convolution product) of the two functions $f_1(t)$ and $f_2(t)$. [See Eq. (2.128).] A very useful relation is given by the convolution theorem, which states that the Fourier transform of the convolution is

$$\mathcal{F}\{f_1(t) * f_2(t)\} = F_1(\omega) F_2(\omega) = \mathcal{F}\{f_1(t)\} \mathcal{F}\{f_2(t)\}. \tag{2.147}$$

Thus convolution in the time domain corresponds to ordinary multiplication in the frequency domain. It is implied that the convolution is commutative, that is, $f_1(t) * f_2(t) = f_2(t) * f_1(t)$. If we multiply Eq. (2.147) by $i\omega$, it is the Fourier transform of

$$f_1(t) * \frac{df_2(t)}{dt} = f_2(t) * \frac{df_1(t)}{dt}. \tag{2.148}$$

By using the symmetry theorem

$$\mathcal{F}\{F(t)\} = 2\pi f(-\omega) \tag{2.149}$$

(see Table 2.2), we can show that

$$\mathcal{F}\{f_1(t)\, f_2(t)\} = \frac{1}{2\pi} \int_{-\infty}^{\infty} F_1(\omega - y)\, F_2(y)\, dy = \frac{1}{2\pi} F_1(\omega) * F_2(\omega). \tag{2.150}$$

This is the frequency convolution theorem.

Let us next consider the relationship between Fourier integrals and Fourier series, as follows. Let $f_1(t) = 0$ for $|t| > T/2$. Then from Eq. (2.137) its Fourier transform is

$$F_1(\omega) = \int_{-T/2}^{T/2} f_1(t)\, e^{-i\omega t}\, dt. \tag{2.151}$$

Now we define a periodic function

$$f_T(t) = f_T(t + T) \tag{2.152}$$

for all t $(-\infty < t < \infty)$, where

$$f_T(t) = f_1(t) \quad -T/2 < t < T/2. \tag{2.153}$$

We may write the periodic function as a Fourier series

$$f_T(t) = \sum_{n=-\infty}^{\infty} c_n \exp(in\omega_0 t), \tag{2.154}$$

where

$$\omega_0 = 2\pi / T \tag{2.155}$$

and

$$c_n = \frac{1}{T} \int_{-T/2}^{T/2} f_T(t) \exp(-in\omega_0 t)\, dt = \frac{1}{T} F_1(n\omega_0). \tag{2.156}$$

If f_T is real, then

$$c_{-n} = c_n^*, \tag{2.157}$$

which corresponds to Eq. (2.139). The Fourier transform of $f_T(t)$ is

$$
\begin{aligned}
F_T(\omega) &= \int_{-\infty}^{\infty} \sum_{n=-\infty}^{\infty} c_n \exp\{i(n\omega_0 - \omega)t\} \, dt \\
&= 2\pi \sum_{n=-\infty}^{\infty} c_n \delta(\omega - n\omega_0) \\
&= \frac{2\pi}{T} \sum_{n=-\infty}^{\infty} F_1(n\omega_0) \, \delta(\omega - n\omega_0).
\end{aligned}
\tag{2.158}
$$

Thus, it follows that the Fourier transform of a periodic function is discontinuous, and it consists of impulses weighted by the Fourier transform of the function $f_1(t)$ for $\omega = n\omega_0$, $(n = 1, 2, \ldots)$. Note that $|F_1(in\omega_0)|/T$ may be interpreted as the amplitude of Fourier component number n of $f_T(t)$. See Eq. (2.156).

2.6.2 Time-Invariant Linear System

Fourier analysis may be applied to the study of linear systems. Let the input signal to a linear system be $u(t)$ and the corresponding output signal (response) be $y(t)$. We shall assume that the system is linear and time invariant; that is, it has the same characteristics now as in the past and in the future. We shall mostly consider systems which are causal, which means that there is no output response before a causing signal has been applied to the input. However, it is sometimes of practical interest to consider non-causal systems as well.

The Fourier transform of the impulse response $h(t)$ [see Eq. (2.128)] is

$$
H(\omega) = R(\omega) + iX(\omega) = \int_{-\infty}^{\infty} h(t) e^{-i\omega t} \, dt,
\tag{2.159}
$$

which we shall call the transfer function of the system. [If the system is not causal, i.e., if condition (2.131) is not satisfied, some authors use the term "frequency-response function", instead of "transfer function".] The real and imaginary parts of the transfer function are denoted by $R(\omega)$ and $X(\omega)$, respectively. If $h(t)$ is real, then

$$
H^*(-\omega) = H(\omega).
\tag{2.160}
$$

The Fourier transform of the output response $y(t)$ as given by the convolution product (2.128) is

$$
Y(\omega) = H(\omega) U(\omega)
\tag{2.161}
$$

according to the convolution theorem (2.147), where $U(\omega)$ is the Fourier transform of the input signal $u(t)$. A system (block diagram) interpretation of Eq. (2.161) is given in the right-hand part of Figure 2.12.

For the simple case in which the transfer function is independent of frequency,

$$
H(\omega) = H_0
\tag{2.162}
$$

where H_0 is constant, the impulse response is (see Table 2.2)

$$h(t) = H_0 \delta(t). \tag{2.163}$$

For an arbitrary input $u(t)$ the response then becomes

$$y(t) = \int_{-\infty}^{\infty} h(t - \tau) u(\tau) \, d\tau = \int_{-\infty}^{\infty} H_0 \delta(t - \tau) u(\tau) \, d\tau = H_0 u(t), \tag{2.164}$$

which, apart for the constant H_0, is a distortion-free reproduction of the input signal.

As another example, let us consider a harmonic input function

$$u(t) = \tfrac{1}{2} \hat{u} \exp(i\omega_0 t) + \tfrac{1}{2} \hat{u}^* \exp(-i\omega_0 t), \tag{2.165}$$

where $\hat{u}$ is the complex amplitude and $\hat{u}^*$ its conjugate. The Fourier transform is (see Table 2.2)

$$U(\omega) = \pi \hat{u} \delta(\omega - \omega_0) + \pi \hat{u}^* \delta(\omega + \omega_0). \tag{2.166}$$

The response is given by

$$\begin{aligned}
Y(\omega) &= H(\omega) U(\omega) \\
&= \pi \hat{u} \delta(\omega - \omega_0) H(\omega) + \pi \hat{u}^* \delta(\omega + \omega_0) H(\omega) \\
&= \pi \hat{u} \delta(\omega - \omega_0) H(\omega_0) + \pi \hat{u}^* \delta(\omega + \omega_0) H(-\omega_0).
\end{aligned} \tag{2.167}$$

Now, because $h(t)$ is real, we have $H(-\omega_0) = H^*(\omega_0)$ in accordance with Eq. (2.139). Hence we may write

$$y(t) = \tfrac{1}{2} \hat{y} \exp(i\omega t) + \tfrac{1}{2} \hat{y}^* \exp(-i\omega t), \tag{2.168}$$

where

$$\hat{y} = H(\omega_0)\hat{u}, \quad \hat{y}^* = H^*(\omega_0)\hat{u}^*, \tag{2.169}$$

in agreement with Eq. (2.136). It follows that the complex amplitude of the response equals the product of the transfer function and the complex amplitude of the input signal. According to Eq. (2.169) we may consider $\hat{u}$ as input to a linear system in the frequency domain, where H is the transfer function and $\hat{y}$ the output response.

Assuming that the impulse response $h(t)$ is a real function of time, we next wish to express the impulse response in terms of the real and imaginary parts of the transfer function:

$$\begin{aligned}
h(t) &= \frac{1}{2\pi} \int_{-\infty}^{\infty} H(\omega) e^{i\omega t} \, d\omega \\
&= \frac{1}{2\pi} \int_{-\infty}^{\infty} [R(\omega) \cos(\omega t) - X(\omega) \sin(\omega t)] \, d\omega \\
&\quad + \frac{i}{2\pi} \int_{-\infty}^{\infty} [R(\omega) \sin(\omega t) + X(\omega) \cos(\omega t)] \, d\omega.
\end{aligned} \tag{2.170}$$

If $h(t)$ is real, we have from Eqs. (2.139)–(2.141) that

$$R(-\omega) = R(\omega), \quad X(-\omega) = -X(\omega). \tag{2.171}$$

Thus, because $R(\omega)$ and $\cos(\omega t)$ are even functions of ω, whereas $X(\omega)$ and $\sin(\omega t)$ are odd functions of ω, the imaginary part in Eq. (2.170) vanishes. Moreover, we can rewrite the (real) impulse response as

$$h(t) = h_e(t) + h_o(t), \tag{2.172}$$

where

$$h_e(t) = \frac{1}{\pi} \int_0^\infty R(\omega) \cos(\omega t) \, d\omega, \tag{2.173}$$

$$h_o(t) = -\frac{1}{\pi} \int_0^\infty X(\omega) \sin(\omega t) \, d\omega. \tag{2.174}$$

Note that $h_e(t)$ and $h_o(t)$ are even and odd functions of t, respectively. Thus, according to Eq. (2.172), the real impulse response $h(t)$ is split into even and odd parts. The corresponding transforms are even and odd functions of ω, respectively:

$$\mathcal{F}\{h_e(t)\} = R(\omega), \quad \mathcal{F}\{h_o(t)\} = i X(\omega) \tag{2.175}$$

2.6.3 Kramers-Kronig Relations and Hilbert Transform

The function $h(t)$ is the response upon application of an input impulse at $t = 0$. Hence, for a causal system

$$h(t) = 0 \quad \text{for } t < 0, \tag{2.176}$$

because there can be no response before an input signal has been applied. Then Eq. (2.128) and the commutativity of the convolution product give

$$y(t) = h(t) * u(t) = \int_{-\infty}^t h(t-\tau) u(\tau) \, d\tau = \int_0^\infty h(\tau) u(t-\tau) \, d\tau. \tag{2.177}$$

Further, if also the input is causal (i.e., if $u(t) = 0$ for $t < 0$), then

$$y(t) = h(t) * u(t) = \int_0^t h(t-\tau) u(\tau) \, d\tau = \int_0^t h(\tau) u(t-\tau) \, d\tau. \tag{2.178}$$

This is the version of the convolution theorem used in the theory of the (one-sided) Laplace transform, which may be applied to causal systems assumed to be dead (quiescent) previous to an initial instant, $t = 0$. Using the Fourier transform when $h(t) = 0$ for $t < 0$, and inserting s for $i\omega$, we obtain the following relation between the Laplace transform H_L and the transfer function H:

$$H_L(s) \equiv \int_0^\infty h(t) e^{-st} \, dt = H(s/i) = H(-is). \tag{2.179}$$

Using the convolution theorem (2.147), we obtain the Laplace transform of

Eq. (2.178) as

$$Y_L(s) = H_L(s) U_L(s), \tag{2.180}$$

where $U_L(s)$ is the Laplace transform of $u(t)$, which is assumed to be vanishing for $t < 0$.

For the example of a constant transfer function $H(\omega) = H_0$, the impulse response is given by Eq. (2.163) as $H_0\delta(t)$. Note that this is an example of a causal impulse response. In this case the response to a causal input $u(t)$ is $y(t) = H_0 u(t)$, which means that $y(t) = 0$ for $t < 0$ if $u(t) = 0$ for $t < 0$. This example $H_0\delta(t)$ of a causal impulse response is an even function. Another example of a causal impulse response is $h(t) = H_{00}\dot{\delta}(t)$, for which the transfer function is $H(\omega) = i\omega H_{00}$ (cf. Table 2.2). This is an odd function.

However, in general, any causal impulse response $h(t)$ which "has a memory of the past" is neither an even nor an odd function, because $h(t)$ vanishes for all negative t, but not for all positive t. Below let us make some observations for this more general situation. However, for a while, let us exclude cases for which $H(\infty) \neq 0$, such as in the above example, where $H(\infty) = H_0$.

For a causal function $h(t)$ decomposed into even and odd parts according to Eq. (2.172)

$$h(t) = h_e(t) + h_o(t), \tag{2.181}$$

we have

$$h(t) = \begin{cases} 2h_e(t) = 2h_o(t) & \text{for } t > 0 \\ h_e(0) & \text{for } t = 0 \\ 0 & \text{for } t < 0 \end{cases} \tag{2.182}$$

and in general (for $t \neq 0$)

$$h_o(t) = h_e(t)\,\mathrm{sgn}(t), \quad h_e(t) = h_o(t)\,\mathrm{sgn}(t), \tag{2.183}$$

because the condition $h(t) = 0$ for $t < 0$ has to be satisfied. Using (2.173), (2.174) and (2.182) gives, for $t > 0$,

$$h(t) = 2h_e(t) = \frac{2}{\pi} \int_0^\infty R(\omega) \cos(\omega t)\, d\omega, \tag{2.184}$$

$$h(t) = 2h_o(t) = -\frac{2}{\pi} \int_0^\infty X(\omega) \sin(\omega t)\, d\omega. \tag{2.185}$$

It should be emphasised that these expressions apply for $t > 0$ only, whereas for $t < 0$ we have $h(t) = 0$. Note that the two alternative expressions (2.184) and (2.185) imply a relationship between $R(\omega)$ and $X(\omega)$, the real and imaginary parts of the Fourier transform of a causal, real function.

Using

$$\mathcal{F}\{\mathrm{sgn}(t)\} = \frac{2}{i\omega} \tag{2.186}$$

(see Table 2.2) and the frequency convolution theorem (2.150)

$$\mathcal{F}\{f_1(t)\, f_2(t)\} = \frac{1}{2\pi} F_1(\omega) * F_2(\omega) \tag{2.187}$$

to obtain the Fourier transforms of Eqs. (2.183), we have

$$\mathcal{F}\{h_e(t)\} = \mathcal{F}\{h_o(t)\,\mathrm{sgn}(t)\} = R(\omega)$$
$$= \frac{1}{2\pi} i\, X(\omega) * \frac{2}{i\omega} = \frac{1}{\pi\omega} * X(\omega) \tag{2.188}$$

and

$$\mathcal{F}\{h_o(t)\} = \mathcal{F}\{h_e(t)\,\mathrm{sgn}(t)\} = i\, X(\omega)$$
$$= \frac{1}{2\pi} R(\omega) * \frac{2}{i\omega} = -\frac{i}{\pi\omega} * R(\omega). \tag{2.189}$$

Writing convolutions explicitly in terms of integrals, we have

$$R(\omega) = \frac{1}{\pi} \int_{-\infty}^{\infty} \frac{X(y)}{\omega - y}\, dy, \tag{2.190}$$

$$X(\omega) = -\frac{1}{\pi} \int_{-\infty}^{\infty} \frac{R(y)}{\omega - y}\, dy. \tag{2.191}$$

Note that the integrand is singular for $y = \omega$ and that the integrals should be understood as principal value integrals (cf. footnote on p. 10 in Papoulis[13]). The relations (2.190) and (2.191) are called Kramers-Kronig relations.[15,16] If for a causal function the real/imaginary part of the Fourier transform is known for all frequencies, then the remaining imaginary/real part of the transform is given by a principal-value integral. For the integrals to exist it is necessary that

$$R(\omega) + i\, X(\omega) = H(\omega) \to 0 \quad \text{when } \omega \to \infty. \tag{2.192}$$

If this is not true, we have to subtract the singular part of $H(\omega)$, that is, $H(\infty)$. Let us consider the case in which $h(t)$ has an impulse singularity. Note (from Table 2.2) that

$$\mathcal{F}\{\delta(t)\} = 1, \quad \mathcal{F}\{\delta(t - t_0)\} = \exp(-i\omega t_0). \tag{2.193}$$

These Fourier transforms do not satisfy condition (2.192). If $H(\infty) \neq 0$ we define a modified transfer function,

$$H'(\omega) = H(\omega) - H(\infty), \tag{2.194}$$

and a new corresponding casual impulse response function,

$$h'(t) = h(t) - H(\infty)\,\delta(t). \tag{2.195}$$

For instance, if

$$H(\infty) = R(\infty) \neq 0, \quad X(\infty) = 0, \tag{2.196}$$

we have

$$X(\omega) = -\frac{1}{\pi} \int_{-\infty}^{\infty} \frac{R(y) - R(\infty)}{\omega - y} \, dy, \tag{2.197}$$

and

$$R(\omega) - R(\infty) = \frac{1}{\pi} \int_{-\infty}^{\infty} \frac{X(y)}{\omega - y} \, dy. \tag{2.198}$$

Multiplying the former of the two equations by i and summing yield the Hilbert transform

$$H'(\omega) = \frac{1}{i\pi} \int_{-\infty}^{\infty} \frac{H'(y)}{\omega - y} \, dy. \tag{2.199}$$

An alternative formulation of the Kramers-Kronig relations is obtained by noting that because $h(t)$ is real, we have

$$R(-\omega) = R(\omega), \quad X(-\omega) = -X(\omega) \tag{2.200}$$

[cf. Eqs. (2.140) and (2.141)]. Then

$$-\pi X(\omega) = \int_{-\infty}^{\infty} \frac{R(y)}{\omega - y} \, dy = \int_{-\infty}^{0} \frac{R(z)}{\omega - z} \, dz + \int_{0}^{\infty} \frac{R(y)}{\omega - y} \, dy. \tag{2.201}$$

Because

$$\int_{-\infty}^{0} \frac{R(z)}{\omega - z} \, dz = \int_{0}^{\infty} \frac{R(-y)}{\omega + y} \, dy = \int_{0}^{\infty} \frac{R(y)}{\omega + y} \, dy \tag{2.202}$$

and

$$\frac{1}{\omega + y} + \frac{1}{\omega - y} = \frac{2\omega}{\omega^2 - y^2}, \tag{2.203}$$

we obtain

$$X(\omega) = -\frac{2\omega}{\pi} \int_{0}^{\infty} \frac{R(y)}{\omega^2 - y^2} \, dy. \tag{2.204}$$

Similarly we find

$$\begin{aligned}
\pi R(\omega) &= \int_{-\infty}^{\infty} \frac{X(y)}{\omega - y} \, dy = \int_{0}^{\infty} \frac{X(-y)}{\omega + y} \, dy + \int_{0}^{\infty} \frac{X(y)}{\omega - y} \, dy \\
&= \int_{0}^{\infty} X(y) \left(\frac{1}{\omega - y} - \frac{1}{\omega + y} \right) dy,
\end{aligned} \tag{2.205}$$

which gives

$$R(\omega) = \frac{2}{\pi} \int_{0}^{\infty} \frac{y X(y)}{\omega^2 - y^2} \, dy. \tag{2.206}$$

2.6.4 An Energy Relation for Non-sinusoidal Oscillation

In Section 2.3 we derived formula (2.78) for the time-average power or rate of work associated with sinusoidal oscillations in a mechanical system, namely

$$P = \tfrac{1}{4} \hat{f} \hat{u}^* + \tfrac{1}{4} \hat{f}^* \hat{u} \tag{2.207}$$

(note that here we denote force by the variable f instead of with F as we did previously). For a more general oscillation, which is not periodic in time, it is more convenient to consider the total work done than the time average of the rate of work. The instantaneous rate of work is $P(t) = f(t) u(t)$, that is, the product of the instantaneous values of force f and velocity u. Hence, the total work done is

$$W = \int_{-\infty}^{\infty} f(t) u(t)\, dt. \tag{2.208}$$

By applying Parseval's theorem, or the frequency convolution theorem (2.150) with $\omega = 0$, we find that Eq. (2.208) gives

$$W = \frac{1}{2\pi} \int_{-\infty}^{\infty} F(\omega) U(-\omega)\, d\omega, \tag{2.209}$$

where $F(\omega)$ and $U(\omega)$ are the Fourier transforms of force $f(t)$ and velocity $u(t)$, respectively. We shall assume that $F(\omega)$ and $U(\omega)$ are related through the transfer function

$$Y(\omega) = 1/Z(\omega) = U(\omega)/F(\omega), \tag{2.210}$$

where

$$Z(\omega) = R(\omega) + iX(\omega) \tag{2.211}$$

is a mechanical impedance. Its real and imaginary parts are the mechanical resistance $R(\omega)$ and the mechanical reactance $X(\omega)$, respectively. It may, for instance, be given by Eq. (2.52).

Assuming that $f(t)$ and $u(t)$ are real functions, we find that Eq. (2.139) is applicable to $F(\omega)$ and $U(\omega)$. Further, also using Eqs. (2.210) and (2.211), we find that Eq. (2.209) becomes

$$
\begin{aligned}
W &= \frac{1}{2\pi} \int_{0}^{\infty} [F(\omega) U^*(\omega) + F^*(\omega) U(\omega)]\, d\omega \\
&= \frac{1}{\pi} \int_{0}^{\infty} \mathrm{Re}\{F(\omega) U^*(\omega)\}\, d\omega \\
&= \frac{1}{\pi} \int_{0}^{\infty} \mathrm{Re}\{Z(\omega) U(\omega) U^*(\omega)\}\, d\omega \\
&= \frac{1}{\pi} \int_{0}^{\infty} R(\omega) |U(\omega)|^2\, d\omega.
\end{aligned}
\tag{2.212}
$$

It may be interesting (see Problem 2.14) to compare this result with Eqs. (2.78) and (2.79), where the time-average mechanical power is expressed in terms of complex amplitudes.

Problems

Problem 2.1: Free Oscillation

Show that the equation

$$m\ddot{x} + R\dot{x} + Sx = 0$$

has the general solution

$$x = (C_1 \cos \omega_d t + C_2 \sin \omega_d t)e^{-\delta t}.$$

Further, show that the integration constants C_1 and C_2 as given by

$$C_1 = x_0, \quad C_2 = (u_0 + x_0\delta)/\omega_d$$

satisfy the initial conditions $x(0) = x_0$, $\dot{x}(0) = u_0$.

Problem 2.2: Quality Factor or Q Value for Resonator

Using definition (2.7), derive the equations

$$Q = \frac{\omega_0}{2\delta}\left(1 + \frac{\delta}{\omega_0} - \frac{1}{6}\frac{\delta^2}{\omega_0^2} + \mathcal{O}\left\{\frac{\delta^3}{\omega_0^3}\right\}\right)$$

$$\approx \frac{\omega_0}{2\delta} = \frac{\omega_0 m}{R} = \frac{S}{\omega_0 R} = \frac{(Sm)^{1/2}}{R}$$

and

$$\frac{\delta}{\omega_0} = \frac{1}{2Q}\left(1 + \frac{1}{2Q} + \frac{5}{24Q^2} + \mathcal{O}\{Q^{-3}\}\right) \approx \frac{1}{2Q}.$$

(Hint: use Taylor series for an exponential function and for a binomial and/or use the method of successive approximations.)

Problem 2.3: Forced Oscillation

Show that the equation

$$m\ddot{x} + R\dot{x} + Sx = F(t) = F_0 \cos(\omega t + \varphi_F)$$

has a particular solution

$$x(t) = x_0 \cos(\omega t + \varphi_x),$$

where the phase difference

$$\varphi = \varphi_F - \varphi_u = \varphi_F - \varphi_x - \pi/2$$

is an angle which is in quadrant no. 1 or no. 4, and which satisfies

$$\tan \varphi = (\omega m - S/\omega)/R.$$

Also show that

$$x_0 = u_0/\omega = \frac{F_0}{|Z|\omega}$$

where

$$|Z| = \{R^2 + (\omega m - S/\omega)^2\}^{1/2}.$$

Problem 2.4: Excursion Response Maximum

Using the results of Problem 2.3, discuss the non-dimensionalised excursion ratio $|\xi| \equiv Sx_0/F_0$ versus the frequency ratio $\gamma \equiv \omega/\omega_0 = \omega\sqrt{m/S}$. Find the frequency at which $|\xi|$ is maximum. Determine also the inequality which $\delta = R/2m$ has to satisfy in order that $1 < |\xi|_{max} < \sqrt{2}$. Observe that $|\xi|_{max} > |\xi(\gamma = 1)|$ and that $|\xi|_{max} \rightarrow |\xi(\gamma = 1)|$ as $(\delta/\omega_0) \rightarrow 0$.

Problem 2.5: Amplitude and Phase Constant

Determine the numerical values of amplitude A and phase constant φ for the oscillation

$$x = A\cos(62.8\, t + \varphi)$$

when the initial conditions at $t = 0$ are as follows: position $x(0) = 50$ mm, velocity $\dot{x}(0) = 0.8$ m/s. Determine the acceleration at $t = 0$. Draw a phasor diagram for the complex amplitudes of position (in scale 1:1), velocity (1 m/s $\hat{=}$ 20 mm) and acceleration (1 m/s^2 $\hat{=}$ 0.5 mm).

Problem 2.6: Complex Representations of Harmonic Oscillation

The following harmonic oscillation is given:

$$u(t) = 100 \sin(\omega t) - 50 \cos(\omega t).$$

Rewrite $u(t)$ as

(a) a cosine function with phase constant,
(b) a sine function with phase constant,
(c) the real part of a complex quantity,
(d) the imaginary part of a complex quantity,
(e) the sum of two complex conjugate quantities.

Problem 2.7: Superposed Oscillations of the Same Frequency

(a) Numerically determine the amplitude and the phase constant of the resultant oscillation obtained by superposition of the individual oscillations

$$x(t) = 3 \cos(\omega t + \pi/6) + 6 \sin(\omega t + 3\pi/2)$$
$$-3.5 \sin(\omega t + \pi/3) + 2.2 \sin(\omega t - \pi/9).$$

(b) Rewrite $x(t)$ in complex form.
(c) Draw phasors for the four individual oscillations in the same diagram, and construct the resultant phasor. Compare it with the above numerical computation.

Problem 2.8: Resonance Bandwidth

According to Eq. (2.54) the inverse of the mechanical impedance Z may be interpreted as the transfer function of a system in which applied force $\hat{F}$ is the input and velocity $\hat{u}$ is the output. The maximum modulus $|Y|_{\max}$ of this transfer function $Y = 1/Z$ is $1/R$, corresponding to the resonance frequency $\omega_0 = \sqrt{S/m}$. Derive an expression for the upper and lower frequencies, ω_u and ω_l, respectively, at which $|Y|/|Y|_{\max} = R|Y| = 1/\sqrt{2}$, in terms of ω_0 and $\delta = R/2m$. Also determine the resonance bandwidth $(\Delta\omega)_{\mathrm{res}} = \omega_u - \omega_l$. Compare the relative bandwidth $(\omega_u - \omega_l)/\omega_0$ with the inverse of the quality factor Q defined in Subsection 2.1.1.

Problem 2.9: Solving a System of Linear Differential Equations

Show that the solution

$$\mathbf{x}(t) = e^{\mathbf{A}(t-t_0)}\mathbf{x}(t_0) + \int_{t_0}^{t} e^{\mathbf{A}(t-\tau)}\mathbf{B}\mathbf{u}(\tau)\,d\tau$$

satisfies the vectorial differential equation

$$\dot{\mathbf{x}} = \mathbf{A}\mathbf{x} + \mathbf{B}\mathbf{u}.$$

Problem 2.10: State-Space Description of Oscillation

Let matrices $\mathbf{A}, \mathbf{B}, \mathbf{C}$ and $\mathbf{D}$ be given given by Eqs. (2.112) and (2.113). Determine the eigenvalues λ_1 and λ_2 in terms of the coefficients δ and ω_d given by Eq. (2.5). Assume that $\lambda_2 \neq \lambda_1$ and that $\lambda_1 - \lambda_2$ either has a positive imaginary part or otherwise that $\lambda_1 - \lambda_2$ is real and positive. Show that $\mathbf{A}$ has the linearly independent eigenvectors $\mathbf{m}_1$ and $\mathbf{m}_2$ with transposed vectors $\mathbf{m}_i^T = c_i(1\ \lambda_i)$ $(i = 1, 2)$, where c_1 and c_2 are arbitrary constants. Determine the inverse of matrix $\mathbf{M} = [\mathbf{m}_1\ \mathbf{m}_2]$ and show that $\mathbf{A}'$ defined by the similarity transformation (2.104) is a diagonal matrix. Next perform the transformations (2.108) and (2.109), find $\mathbf{x}'$, $\mathbf{B}'$ and $\mathbf{C}'$, and use Eq. (2.101) to obtain the solution of Eqs. (2.110) and (2.111). In the final answer, set $t_0 = 0$ and replace y_1, x_1, x_2 and u_1 by x, x, u and F, respectively. Finally, write down the solution for free oscillations, that is, for the case $F(t) \equiv 0$. Compare the result with Eqs. (2.4) and (2.6).

Problem 2.11: Critically Damped Oscillation

Discuss the solution of Problem 2.10 for the case in which $\lambda_1 = \lambda_2 = \lambda$. In this case **A** does not have two linearly independent eigenvectors, and the transformed matrix **A**′ is not diagonal. For the similarity transformation choose the matrix

$$\mathbf{M} = \begin{bmatrix} 1 & 0 \\ \lambda & 1 \end{bmatrix}$$

and perform the various tasks as in Problem 2.10. A special challenge is to determine matrix exponential $e^{\mathbf{A}'t}$ by using definition (2.100).

Problem 2.12: Mechanical Impedance and Power

A mass of $m = 6$ kg is suspended by a spring of stiffness $S = 100$ N/m. The oscillating system has a mechanical resistance of $R = 3$ Ns/m. The system is excited by an alternating force

$$F(t) = |\hat{F}|\cos(\omega t),$$

where the force amplitude is $|\hat{F}| = 1$ N, and the frequency $f = \omega/2\pi = 1$ Hz.

(a) Determine the mechanical impedance $Z = R + iX = |Z|e^{i\varphi}$. State numerical values for R, X, $|Z|$ and φ.
(b) Determine the complex amplitudes $\hat{F}$ for the force, $\hat{s}$ for the position, $\hat{u}$ for the velocity and $\hat{a}$ for the acceleration. Draw the complex amplitudes as vectors in a phasor diagram.
(c) Find the frequency $f_0 = \omega_0/2\pi$ for which the mechanical reactance vanishes.
(d) Find the time-averaged mechanical power P which the force $|\hat{F}| = 1$ N supplies to the system, at the two frequencies f and f_0.

Problem 2.13: Convolution with Sinusoidal Oscillation

Show that if $f(t)$ varies sinusoidally with an angular frequency ω_0, then the convolution product

$$g(t) = h(t) * f(t)$$

may be written as a linear combination of $f(t)$ and $\dot{f}(t)$. Assuming that the transfer function $H(\omega)$ is known, determine the coefficients of the linear combination.

Problem 2.14: Work in Terms of Complex Amplitude or Fourier Integral

Consider the particular case of harmonic oscillations. Discuss the mathematical relationship between Eq. (2.212) for the total work and the equation

$$P = \tfrac{1}{4}(\hat{F}\hat{u}^* + \hat{F}^*\hat{u}) = \tfrac{1}{2}\mathrm{Re}(\hat{F}\hat{u}^*)$$

for the power of a mechanical system performing harmonical oscillation [cf. Eq. (2.78)]. Consider also the dimensions (or SI units) of the various physical quantities.

Problem 2.15: Mechanical Impedance at Zero Frequency

When by Eq. (2.52) we defined the mechanical impedance

$$Z(\omega) = R + i(\omega m - S/\omega),$$

we tacitly assumed that $\omega \neq 0$. We define the transfer functions $Y(\omega) = 1/Z(\omega)$, $G(\omega) = i\omega Z(\omega)$ and $H(\omega) = 1/G(\omega) = Y(\omega)/i\omega$. Show that the corresponding impulse response functions $y(t)$, $g(t)$ and $h(t)$ are causal. (Hint: for transfer functions with poles, apply the method of contour integration when considering the inverse Fourier transform.) If the variables concerned are $s(t)$, $u(t)$ and $F(t)$, choose input and output variables for the four linear systems concerned. Further, show that in order to make $z(t) = \mathcal{F}^{-1}\{Z(\omega)\}$ a causal impulse response function, one must add a term to the impedance such that

$$Z(\omega) = R + i\omega m + S[1/i\omega + \pi\delta(\omega)].$$

Explain the physical significance of the last term.

CHAPTER THREE

Interaction Between Oscillations and Waves

There are many different types of waves in nature. Apart from the visible waves on the surface of oceans and lakes, there are, for instance, sound waves, light waves and other electromagnetic waves. This chapter gives a brief description of waves in general and compares surface waves on water with other types of waves. It also presents a simple generic discussion on the interaction between waves and oscillations. One phenomenon is generated waves radiated from an oscillator, and another phenomenon is oscillations excited by a wave incident upon the oscillating system. The radiation resistance is defined in terms of the power associated with the wave generated by an oscillator. For a mechanical oscillating system the "added mass" is related to added energy associated with the wave-generating process, not just to kinetic energy but to the difference between kinetic and potential energies.

3.1 Comparison of Waves on Water with Other Waves

Waves on water propagate along a surface. Acoustic waves in a fluid and electromagnetic waves in free space may propagate in any direction in a three-dimensional space. Waves on a stretched string propagate along a line (in a one-dimensional "space"). The same may be said about waves on water in a canal and about guided acoustic waves or guided electromagnetic waves along cylindrical structures, although in these cases the physical quantities (pressure, velocity, electric field, magnetic field, etc.) may vary in directions transverse to the direction of wave propagation.

As was mentioned in Chapter 2, there is an exchange of kinetic energy and potential energy in a mechanical oscillator (or magnetic energy and electric energy in the electric analogue). Also in a propagating wave there is interaction between different forms of energy, for instance magnetic and electric energy with electromagnetic waves, and kinetic and potential energy with mechanical waves, such as acoustic waves and water waves. With an acoustic wave the potential energy is associated with the elasticity of the medium in which the wave propagates. The potential energy with a water wave is due to gravity and surface tension. The contribution from the elasticity of water is negligible, because waves on water propagate rather slowly as compared with the velocity of sound in water. The gravity is responsible for the potential energy that is associated with the lifting of water from the wave troughs to the wave crests. As the waves increase the area of the interface between water and air, work done against the surface tension is converted to potential energy. For wavelengths in the range of 10^{-3} m to 10^{-1} m, both types of potential energy are important. For shorter waves, so-called capillary waves, the effect of gravity may be neglected (see Problems 3.2 and 4.1). For longer waves, so-called gravity waves, the surface tension may be neglected, as we shall do in the following, because we restrict our study to water waves of wavelengths exceeding 0.25 m.

Oscillations are represented by physical quantities which vary with time. For waves the quantities also vary with the spatial coordinates. Below let us consider waves which vary sinusoidally with time. Such waves are called "harmonic" or "monochromatic". When dealing with sea waves, they are also characterised as "regular" if their time variation is sinusoidal.

Let

$$p = p(x, y, z, t) = \text{Re}\{\hat{p}(x, y, z)e^{i\omega t}\} \tag{3.1}$$

represent a general harmonic wave, and let p denote the dynamic pressure in a fluid. (The total pressure is $p_{tot} = p_{stat} + p$, where the static pressure p_{stat} is independent of time.) The complex pressure amplitude $\hat{p}$ is a function of the spatial coordinates x, y and z.

For a plane acoustic wave propagating in a direction x, we have

$$p = p(x, t) = \text{Re}\{Ae^{i(\omega t - kx)} + Be^{i(\omega t + kx)}\}, \tag{3.2}$$

where $k = 2\pi/\lambda$ is the angular repetency (wave number) and λ is the wavelength. The first and second terms represent waves propagating in the positive and negative x direction, respectively. Assume that an observer moves with a velocity $v_p = \omega/k$ in the positive x direction. Then the observer will experience a constant phase $(\omega t - kx)$ of the first right-hand term in Eq. (3.2). If the observer moves with same speed in the opposite direction, he or she will experience a constant phase $(\omega t + kx)$ of the last term in Eq. (3.2). For this reason $v_p = \omega/k$ is called the phase velocity. At a certain instant the phase is constant on all planes perpendicular to the direction of wave propagation. For this reason the wave is called plane. In contrast, an acoustic wave

$$p(r, t) = \text{Re}\{(C/r)e^{i(\omega t - kr)}\} \tag{3.3}$$

radiated from a spherical loudspeaker in open air may be called a spherical wave, because the phase $(\omega t - kr)$ is the same everywhere on an envisaged sphere at distance r from the centre of the loudspeaker. Note that in this geometrical case the pressure amplitude $|C|/r$ decreases with the distance from the loudspeaker. The pressure amplitudes $|A|$ and $|B|$ are constant for the two oppositely propagating waves corresponding to the two right-hand terms in Eq. (3.2). An acoustic wave in a fluid is a longitudinal wave, because the fluid oscillates only in the direction of wave propagation. Acoustic waves are not discussed in detail here, but let us mention that if the sound pressure is given by Eq. (3.2), then the fluid velocity is

$$v = v_x = \frac{1}{\rho c} \, \mathrm{Re}\{Ae^{i(\omega t - kx)} - Be^{i(\omega t + kx)}\}, \tag{3.4}$$

where c is the sound velocity and ρ is the (static) mass density of the fluid. The readers who are interested in the derivation of Eqs. (3.2) and (3.4) and the wave equation for acoustic waves are referred to textbooks in acoustics.[17]

For a plane, harmonic, surface wave on water, propagating in a horizontal direction, an expression similar to Eq. (3.2) applies. However, then A and B cannot be constants; they depend on the vertical coordinate z (chosen to have positive direction upwards). On "deep water" A and B are then proportional to e^{kz}, as will be shown later in Chapter 4. Then $p = p(x, z, t)$, and

$$\hat{p} = \hat{p}(x, z) = A(z)e^{-ikx} + B(z)e^{ikx}. \tag{3.5}$$

In the case of a plane and linearly polarised electromagnetic wave the electric field is

$$E(x, t) = \mathrm{Re}\{Ae^{i(\omega t - kx)} + Be^{i(\omega t + kx)}\}, \tag{3.6}$$

where A and B are again complex constants.

3.2 Dispersion, Phase Velocity and Group Velocity

Both acoustic waves and electromagnetic waves are non-dispersive, which means that the phase velocity is independent of the frequency. The dispersion relation (the relationship between ω and k) is

$$\omega = ck, \tag{3.7}$$

where c is the constant speed of sound or light, respectively.

Gravity waves on water are, in general, dispersive. As will be shown later, in Chapter 4 (Section 4.3), the relationship for waves on deep water is

$$\omega^2 = gk, \tag{3.8}$$

where g is the acceleration of gravity. Using this dispersion relationship we find that the phase velocity is

$$v_p \equiv \omega/k = g/\omega = \sqrt{g/k}. \tag{3.9}$$

Note that in the study of the propagation of dispersive waves (for which the phase velocity depends on frequency), we have to distinguish between phase velocity and group velocity. Let us assume that the dispersion relationship may be written as $F(\omega, k) = 0$, where F is a differentiable function of two variables. Then the group velocity is defined as

$$v_g = \frac{d\omega}{dk} = -\frac{\partial F/\partial k}{\partial F/\partial \omega}. \tag{3.10}$$

As shown in Problem 3.1, if several propagating waves with slightly different frequencies are superimposed on each other, the result may be interpreted as a group of waves, each of them moving with the phase velocity, whereas the amplitude of the group of individual waves is modulated by an envelope moving with the group velocity. Here let us just consider the following simpler example of two superimposed harmonic waves of angular frequency $\omega \pm \Delta\omega$ and angular repetency $k \pm \Delta k$, namely

$$
\begin{aligned}
p(x, t) &= D\cos\{(\omega - \Delta\omega)t - (k - \Delta k)x\} + D\cos\{(\omega + \Delta\omega)t - (k + \Delta k)x\} \\
&= 2D\cos\{(\Delta\omega)t - (\Delta k)x\}\cos(\omega t - kx),
\end{aligned}
\tag{3.11}
$$

where an elementary trigonometric identity has been used in the last step. If $\Delta\omega \ll \omega$, this last expression for $p(x, t)$ is the product of a quickly varying function, representing a wave with propagation speed ω/k (the phase velocity), and a slowly varying function, representing an amplitude envelope, moving in the positive x direction with a propagation speed $\Delta\omega/\Delta k$, which tends to the group velocity if $\Delta\omega$ (and, correspondingly, Δk) tend to zero.

Using Eq. (3.8) we find that for a wave on deep water the group velocity is

$$v_g \equiv d\omega/dk = g/(2\omega) = v_p/2. \tag{3.12}$$

We shall see later, in Chapter 4, that this group velocity may be interpreted as the speed with which energy is transported by a deep-water wave. Although electromagnetic waves are non-dispersive in free space, such waves propagating along telephone lines or along optical fibres have, in general, some dispersion. In this case it is of interest to know that not only the energy but also the information carried by the wave are usually propagated with a speed equal to the group velocity.

3.3 Wave Power and Energy Transport

Next let us give some consideration to the energy, power and intensity associated with waves. Intensity I is the time-average energy transport per unit time and per unit area in the direction of wave propagation. Whereas the dimension in SI units is J (joule) for energy and W (watt) for power, it is J/s m^2 = W/m^2 for intensity. For surface waves on water and for acoustic waves the intensity is

$$I = \overline{p_{\text{tot}}v} = \overline{(p_{\text{stat}} + p)v}, \tag{3.13}$$

where the total pressure p_{tot} is the sum of the static pressure p_{stat} and the dynamic pressure p, and where $v = v_x$ is the fluid particle velocity component in the direction of wave propagation. (The overbar denotes time average.) Because the time-average particle velocity is zero, $\bar{v} = 0$, we have

$$I = p_{\text{stat}}\bar{v} + \overline{pv} = \overline{pv}. \tag{3.14}$$

Thus, for a harmonic wave we have

$$p = \text{Re}\{\hat{p}e^{i\omega t}\}, \quad v_x = \text{Re}\{\hat{v}_xe^{i\omega t}\}, \tag{3.15}$$

where $\hat{p} = \hat{p}(x, y, z)$ and $\hat{v}_x = \hat{v}_x(x, y, z)$ are complex amplitudes at (x, y, z) of the dynamic pressure p and of the x component of the fluid particle velocity, respectively. In analogy with the derivation of Eq. (2.78), we then have

$$I = I_x = I_x(x, y, z) = \overline{pv_x} = \tfrac{1}{2}\text{Re}\{\hat{p}\,\hat{v}_x^*\}. \tag{3.16}$$

Strictly speaking, v and hence I are vectors:

$$\vec{I} = \vec{I}(x, y, z) = \overline{p\vec{v}} = \tfrac{1}{2}\text{Re}\{\hat{p}\hat{\vec{v}}^*\}. \tag{3.17}$$

For an electromagnetic wave the intensity may be defined as the time average of the so-called Poynting vector, which is well known in electromagnetics (see, e.g., Panofsky and Phillips[18]).

For a plane acoustic wave propagating in the positive x direction, the sound pressure is as given by Eq. (3.2) and the oscillating fluid velocity is as given by Eq. (3.4), with $B = 0$ and the constant $|A|$ being equal to the pressure amplitude. Then the sound intensity as given by Eq. (3.16) is constant (independent of x, y and z).

For a plane gravity wave on water, propagating in the positive x direction, the dynamic pressure is as given by Eq. (3.5) with $B(z) \equiv 0$. Refer to Chapter 4 for a discussion of the oscillating fluid velocity $\vec{v}$ associated with this wave. It is just mentioned here that if the water is "deep", then both p and $\vec{v}$ are proportional to e^{kz}. Hence, the intensity is proportional to e^{2kz}, meaning that the intensity decreases exponentially with the distance downward from the water surface. Thus,

$$I_x = I_0e^{2kz}, \tag{3.18}$$

where I_0 is the intensity at the (average) water surface, $z = 0$. By integrating $I_x = I_x(z)$ from $z = -\infty$ to $z = 0$ we arrive at the wave-energy transport,

$$J = \int_{-\infty}^{0} I_x(z)\,dz = I_0 \int_{-\infty}^{0} e^{2kz}\,dz = I_0/2k, \tag{3.19}$$

which is the wave energy transported per unit time through an envisaged vertical strip of unit width, parallel to the wave front, that is parallel to the planes of constant phase of the propagating wave.

Let us now consider an arbitrary, envisaged, closed surface with a radiation source (or wave generator) inside, as indicated in Figure 3.1. The radiation source could be an oscillating body immersed in water, or another kind of wave generator. (In acoustics the source could be a loudspeaker, in radio engineering a transmitting antenna and in optics a light-emitting atom.)

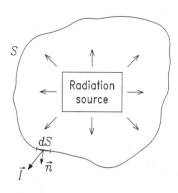

Figure 3.1: General source of radiation surrounded by an envisaged closed surface S.

The radiated power (energy per unit time) passing through the closed surface S may be expressed as an integral of the intensity over the surface,

$$P_r = \oiint \vec{I} \cdot \vec{n} \, dS = \oiint \vec{I} \cdot d\vec{S}, \tag{3.20}$$

where $\vec{n}$ is the unit normal, and $d\vec{S} \equiv \vec{n} \, dS$. When a spherical loudspeaker radiates isotropically in open air, the sound intensity $I(r)$ is independent of direction, and then the radiated power through an envisaged spherical surface of radius r is

$$P_r = I(r) \, 4\pi r^2. \tag{3.21}$$

Here we have neglected reflection and absorption of the acoustic wave from the ground and other obstacles. Assuming that the air does not absorb acoustic wave energy, we find that P_r must be independent of r. Hence, $I(r)$ is inversely proportional to the square of the distance r,

$$I(r) = I(a) \, (a/r)^2, \tag{3.22}$$

where $I(a)$ is the intensity for $r = a$. With this result in mind, it is easier to accept the statement, as implied in Eq. (3.3) for the spherical wave, that the amplitude of the sound pressure is inversely proportional to the distance r.

Assume now that an axisymmetric body, immersed in water of depth h, is performing vertical oscillations. Then an axisymmetric wave generation will take place, and trains of circular waves will radiate along the water surface outward from the oscillating body. The power radiated through an envisaged vertical cylinder of large radius r may, according to Eq. (3.20), be written as

$$P_r = \int_{-h}^{0} I(r, z) \, 2\pi r \, dz = J_r 2\pi r, \tag{3.23}$$

where

$$J_r = \int_{-h}^{0} I(r, z) \, dz \tag{3.24}$$

is the radiated wave-energy transport (per unit width of the wave front). Neglecting loss of wave energy in the water, P_r is independent of the distance r from the axis of the oscillating body. Consequently, J_r is inversely proportional to r,

$$J_r(r) = J_r(a)\,(a/r), \tag{3.25}$$

where $J_r(a)$ is the wave-energy transport at $r = a$. From this we would expect that the dynamic pressure and other physical quantities associated with the radiated wave have amplitudes that are inverse to the square root of r. As discussed in more detail later, in Chapters 4 and 5, this result is true, provided r is large enough. (There may be significant deviation from this result if the distance from the oscillating body is shorter than one wavelength.)

3.4 Radiation Resistance and Radiation Impedance

If the mass m indicated in Figure 2.1 is the membrane of a loudspeaker, an acoustic wave will be generated as a result of the oscillation of the system. Or, if the mass m_m is immersed in water, as indicated in Figure 3.2, a water wave will be generated. Let us now consider a wave-tank laboratory where such an immersed body of mass m_m is suspended through a spring S_m and a mechanical resistance R_m as indicated in Figure 2.1. Alternatively, the body could be a membrane suspended in a frame inside a loudspeaker cabinet. Assume that an external force

$$F(t) = \mathrm{Re}\{\hat{F}e^{i\omega t}\} \tag{3.26}$$

is applied to the body, resulting in a forced óscillatory motion with velocity

$$u(t) = \mathrm{Re}\{\hat{u}e^{i\omega t}\}. \tag{3.27}$$

The power consumed by the mechanical damper is (in time average)

$$P_m = \tfrac{1}{2}R_m|\hat{u}|^2, \tag{3.28}$$

which is in agreement with Eq. (2.81). The oscillating body generates a wave which carries away a radiated power P_r [cf. Figure 3.1 and Eq. (3.20)]. In analogy with Eq. (3.28) we write

$$P_r = \tfrac{1}{2}R_r|\hat{u}|^2, \tag{3.29}$$

which defines the so-called *radiation resistance* R_r.

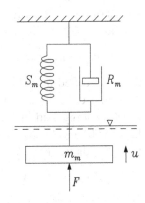

Figure 3.2: Body of mass m_m suspended in water through a spring S_m and a damper R_m.

As a result of the radiated wave, a reaction force F_r acts on the body in addition to the externally applied force F. Assuming that linear theory is valid, we find that F_r is also varying as a harmonic oscillation; that is, $F_r = \text{Re}\{\hat{F}_r e^{i\omega t}\}$. The dynamics of the system is then described by the following extension of Eq. (2.65) or Eq. (2.66):

$$i\omega m_m \hat{u} + R_m \hat{u} + (S_m/i\omega)\hat{u} = \hat{F} + \hat{F}_r, \tag{3.30}$$

or, in terms of the mechanical impedance Z_m,

$$Z_m \hat{u} = \hat{F} + \hat{F}_r. \tag{3.31}$$

Setting

$$\hat{F}_r = -Z_r \hat{u}, \tag{3.32}$$

we define an added impedance, or the so-called *radiation impedance* Z_r.

In general, Z_r is a complex function of ω:

$$Z_r = Z_r(\omega) = R_r(\omega) + i X_r(\omega), \tag{3.33}$$

which depends also on the geometry of the radiating system. Now we have from Eqs. (3.31) and (3.32) that

$$(Z_m + Z_r)\hat{u} = \hat{F}, \tag{3.34}$$

which gives the complex velocity amplitude

$$\hat{u} = \frac{\hat{F}}{Z_m + Z_r} = \frac{\hat{F}}{(R_m + R_r) + i(\omega m_m + X_r - S_m/\omega)}. \tag{3.35}$$

Comparing this result with Eq. (2.54), we observe that the oscillatory motion is modified because the oscillating mass has been immersed in water. The motion of the immersed body results in motion of the water surrounding the body. Some energy, represented by the radiated power (3.29), is carried away. Moreover, some energy is stored as kinetic energy, caused by the velocity of the water, and as potential energy, caused by gravity when the water surface is deformed and water is lifted from troughs to crests. The energy stored in the water is added to the energy stored in the mechanical system itself. Referring to Eq. (2.90), we may thus relate the *radiation reactance* $X_r(\omega)$ to the difference between the average values of the added kinetic energy and the added potential energy. The radiation reactance $X_r(\omega)$ is frequently written as ωm_r, where

$$m_r = m_r(\omega) = X_r(\omega)/\omega \tag{3.36}$$

is the so-called *added mass*, which is usually positive. There are, however, exceptional cases in which the added potential energy is larger than the added kinetic energy, and in such cases the added mass becomes negative.[19] Compare Eqs. (2.90) and (3.36). Combining Eqs. (3.33) and (3.36), we may write the radiation impedance as

$$Z_r(\omega) = R_r(\omega) + i\omega m_r(\omega). \tag{3.37}$$

Here, as well as in acoustics,[17] radiation impedance has the dimension of force divided by velocity, and hence the SI unit is $[Z_r] = [R_r] = $ Ns/m = kg/s. Analogously, in theory for radio antennae, the (electric) radiation impedance has dimension voltage divided by current, and the corresponding SI unit is Ω = V/A. The term "radiation impedance" is commonly used in connection with microphones and loudspeakers in acoustics[17] and also in connection with receiving and transmitting antennae in electromagnetics. A few authors[20,21] have also adopted the term "radiation impedance" in hydrodynamics, in connection with the generation and absorption of gravity waves. This term will also be used in the subsequent text.

3.5 Resonance Absorption

In the preceding section we assumed that an external force F was given [cf. Eq. (3.26) and Figure 3.2]. This external force could have been applied through a motor or some other mechanism, not shown in Figure 3.2. In addition to the external force, a reaction force F_r was taken into consideration in Eq. (3.30), and we assumed in Eq. (3.32) that this reaction force is linear in velocity u. If the motion had been prevented (by choosing at least one of the parameters S_m, R_m and m_m sufficiently large), then only the external force F would remain. Let us now assume that this force is applied through an incident wave. We shall adopt the term "excitation force" for the wave force F_e which acts on the immersed body when it is not moving, that is, when $u = 0$. For this case we replace $\hat{F}$ in Eqs. (3.30), (3.31), (3.34) and (3.35) by $\hat{F}_e$. According to Eq. (3.35) the body's velocity is then given by

$$\hat{u} = \frac{\hat{F}_e}{R_m + R_r + i[\omega(m_m + m_r) - S_m/\omega]}. \tag{3.38}$$

The power absorbed in the mechanical damper resistance R_m is [cf. Eq. (3.28)]

$$P_a = \frac{R_m}{2}|\hat{u}|^2 = \frac{(R_m/2)|\hat{F}_e|^2}{(R_m + R_r)^2 + (\omega m_m + \omega m_r - S_m/\omega)^2}. \tag{3.39}$$

Note that R_m could, in an ideal case, represent a load resistance and that P_a correspondingly represents useful power being consumed by the load resistance. We note that $P_a = 0$ for $R_m = 0$ and for $R_m = \infty$, and that $P_a > 0$ for $0 < R_m < \infty$. Thus there is a maximum of absorbed power when $\partial P_a / \partial R_m = 0$, which occurs if

$$R_m = \{R_r^2 + (\omega m_m + \omega m_r - S_m/\omega)^2\}^{1/2} \equiv R_{m,\mathrm{opt}}, \tag{3.40}$$

for which we have the maximum absorbed power:

$$P_{a,\mathrm{max}} = \frac{|\hat{F}_e|^2/4}{R_r + \{R_r^2 + (\omega m_m + \omega m_r - S_m/\omega)^2\}^{1/2}}. \tag{3.41}$$

See Problems 3.7 and 3.8.

Furthermore, we see by inspection of Eq. (3.39) that if we, for arbitrary R_m, can choose m_m and S_m such that

$$\omega m_m + \omega m_r - S_m/\omega = 0, \tag{3.42}$$

then the absorbed power has the maximum value

$$P_a = \frac{R_m |\hat{F}_e|^2 / 2}{(R_m + R_r)^2}. \tag{3.43}$$

If we now choose R_m in accordance with condition (3.40), which now becomes

$$R_m = R_r \equiv R_{m,\text{OPT}}, \tag{3.44}$$

the maximum absorbed power is

$$P_{a,\text{MAX}} = |\hat{F}_e|^2 / (8R_r) \tag{3.45}$$

and in this case Eq. (3.38) simplifies to

$$\hat{u} = \hat{F}_e / (2R_r) \equiv \hat{u}_{\text{OPT}}. \tag{3.46}$$

When condition (3.42) is satisfied we have resonance. We see from Eq. (3.38) that the oscillation velocity is in phase with the excitation force, because the ratio between the complex amplitudes $\hat{u}$ and $\hat{F}$ is then real. We may refer to Eq. (3.42) as the "resonance condition" or the "optimum phase condition". Note that this condition is independent of the chosen value of the mechanical damper resistance R_m, and the maximum absorbed power is as given by Eq. (3.43).

If the optimum phase condition cannot be satisfied, then the maximum absorbed power is as given by Eq. (3.41), provided that optimum amplitude condition (3.40) is satisfied.

If the optimum phase condition and the optimum amplitude condition can be satisfied simultaneously, then the maximum absorbed power is as given by Eq. (3.45), and the optimum oscillation is as given by Eq. (3.46). In later chapters (Chapters 6 and 7) we encounter situations analogous to this optimum case of power absorption from a water wave. Note, however, that the discussion in the present section is also applicable to absorption of energy from an acoustic wave by a microphone or from an electromagnetic wave by a receiving antenna. (In this latter case u is the electric current flowing in the electric terminal of the antenna, and F_e is the excitation voltage, that is, the voltage which is induced by the incident electromagnetic wave, at the terminal when $u = 0$.)

Let us now, for simplicity, neglect the frequency dependence of the radiation resistance R_r and of the added mass m_r. Introducing the natural angular frequency (eigenfrequency)

$$\omega_0 = \sqrt{S_m/(m_m + m_r)} \tag{3.47}$$

into Eq. (3.39), we rewrite the absorbed power as

$$P_a(\omega) = \frac{R_m |\hat{F}_e(\omega)|^2}{2(R_m + R_r)^2} \frac{1}{1 + (\omega_0/2\delta)^2 (\omega/\omega_0 - \omega_0/\omega)^2}, \tag{3.48}$$

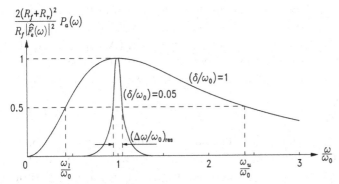

Figure 3.3: Frequency response of absorbed power for two different values of damping factor δ/ω_0.

where

$$\delta = \frac{R_m + R_r}{2(m_m + m_r)} \qquad (3.49)$$

is the so-called damping coefficient of the oscillator. Note that Eqs. (3.47) and (3.49) are extensions of definitions given in Eq. (2.5).

Referring to Eq. (2.19), we see that the relative absorbed-power response

$$\frac{P_a(\omega)/|\hat{F}_e(\omega)|^2}{P_a(\omega_0)/|\hat{F}_e(\omega_0)|^2} = \frac{1}{1 + (\omega_0/2\delta)^2(\omega/\omega_0 - \omega_0/\omega)^2}, \qquad (3.50)$$

which has its maximum value 1 at resonance ($\omega = \omega_0$), exceeds $\frac{1}{2}$ in a frequency interval $\omega_l < \omega < \omega_u$, where

$$\omega_u - \omega_l = (\Delta\omega)_{\text{res}} = 2\delta = (R_m + R_r)/(m_m + m_r). \qquad (3.51)$$

The relative absorbed-power response versus frequency is plotted in Figure 3.3 for two different values of the damping factor:

$$\frac{\delta}{\omega_0} = \frac{R_m + R_r}{2\omega_0(m_m + m_r)} = \frac{R_m + R_r}{2\sqrt{S_m(m_m + m_r)}}. \qquad (3.52)$$

Note that $|\hat{F}_e(\omega)|^2$ is a representation of the spectrum of the incident wave. An example is indicated in Figure 3.4, where $|\hat{F}_e(\omega)|^2$ is maximum at some angular

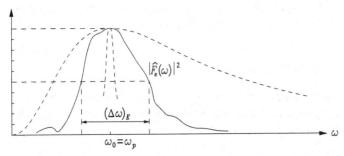

Figure 3.4: Representation of the wave spectrum (solid curve) compared with power absorption responses for the two cases given in Figure 3.3 (dashed curves).

frequency ω_p and exceeds half of its maximum in an interval of length $(\Delta\omega)_E$. Assume that an absorbing system has been chosen, for which $\omega_0 = \omega_p$. If a sufficiently large damping factor δ/ω_0 is chosen, we have that $(\Delta\omega)_{res} > (\Delta\omega)_E$. If we wish to absorb as much wave energy as possible, we should choose $R_m \geq R_r$ according to Eqs. (3.40) and (3.44). Then from Eq. (3.51) we have

$$(\Delta\omega)_{res} \geq \frac{2R_r}{m_m + m_r}. \tag{3.53}$$

The two dashed curves in Figure 3.4 represent two different wave-absorbing oscillators, one with a narrow bandwidth, the other with a wide one. Evidently, the narrow-bandwidth oscillator can absorb efficiently from only a small part of the indicated wave spectrum.

Problems

Problem 3.1: Group Velocity

(a) Show that the sum $s = s_1 + s_2$ of two waves of equal amplitudes and different frequencies, $s_j(x, t) = A \cos(\omega_j t - k_j x)$ (for $j = 1, 2$), may be written as

$$s(x, t) = 2A \cos\left(\frac{\omega_1 - \omega_2}{2}t - \frac{k_1 - k_2}{2}x\right) \cos\left(\frac{\omega_1 + \omega_2}{2}t - \frac{k_1 + k_2}{2}x\right)$$

either by using a trigonometric formula or by applying the method of complex representation of harmonic waves. Set $\omega_1 = \omega + \Delta\omega/2$ and $\omega_2 = \omega - \Delta\omega/2$ and discuss the case $\Delta\omega \to 0$.

(b) Further, consider the more general case

$$s(x, t) = \frac{1}{2}\sum_j A_j \exp\{i(\omega_j t - k_j x)\} + \text{c.c.}$$

Assume that $|A_j|$ has a maximum value $|A_m|$ for $j = m$ ($m \gg 1$) and that $|A_j|$ is negligible when ω_j deviates from ω_m by more than a relatively small amount $\Delta\omega$ ($\Delta\omega \ll \omega_m$, "narrow spectrum"). Show that the "signal" $s(x, t)$ may be interpreted as a "carrier wave", of angular frequency ω_m and angular repetency k_m, modulated by an "envelope wave" propagating with the group velocity $(d\omega/dk)_m$, provided the dispersion curve $\omega = \omega(k)$ or $k = k(\omega)$ may be approximated by its tangent at (k_m, ω_m) in the interval where $|A_j|$ is not negligible.

Problem 3.2: Capillary-Gravity Surface Wave

If the contribution to the wave's potential energy from capillary forces, in addition to the contribution from gravitational forces, is taken into consideration, then the phase velocity for waves on deep water is given by

$$v_p = \sqrt{g/k + \gamma k/\rho}$$

instead of by Eq. (3.9), where $\gamma = 0.07$ N/m is the surface tension on the air-water interface. Find the corresponding dispersion relationship which replaces Eq. (3.8). Also derive an expression for the group velocity. Find the numerical value of the phase velocity for a ripple of wavelength $2\pi/k = 10$ mm (assume $g = 9.8$ m/s^2 and $\rho = 1.0 \times 10^3$ kg/m^3). Finally, derive expressions and numerical values for the wavelength and the frequency of the ripple which has the minimum phase velocity.

Problem 3.3: Optical Dispersion in Gas

For a gas the refraction index $n = c/v_p$ (the ratio between the speed of light in vacuum c and the phase velocity v_p) is to a good approximation

$$n = 1 + p/(\omega_0^2 - \omega^2)$$

for frequencies where n is not significantly different from 1. In the above expression, p is a positive constant and ω_0 is a resonant angular frequency for the gas. Verify that the group velocity $v_g = d\omega/dk$ for an electromagnetic wave (light wave) in the gas (contrary to the phase velocity $v_p = \omega/k$) is smaller than c, for $\omega \ll \omega_0$ as well as for $\omega \gg \omega_0$.

Problem 3.4: Radiation Impedance for a Spherical Loudspeaker

An acoustic wave is radiated from (generated by) a pulsating sphere of radius a. The surface of the sphere oscillates with a radial velocity $u = \hat{u}e^{i\omega t}$. For $r > a$ the wave may be represented by the sound pressure

$$p = (A/r)e^{i(\omega t - kr)}$$

and the (radially directed) particle speed

$$v = v_r = [1 + 1/(ikr)](p/\rho c),$$

where the constants ρ and $c = \omega/k$ are the fluid density and the speed of sound, respectively (Kinsler and Frey,[17] pp. 114 and 163).

Use the boundary condition $u = [v]_{r=a}$ to determine the unknown coefficient A. Then derive expressions for radiation impedance Z_r, radiation resistance R_r and added mass m_r as functions of angular repetency k. (Base the derivation of Z_r on the reaction force from the wave on the surface of the sphere. Check the result for R_r by deriving an expression for the radiated power.)

Problem 3.5: Acoustic Point Absorber

Let us consider a pulsating sphere as a microphone. Let the radius a of the sphere be so small ($ka \ll 1$) that the microphone may be considered as a point absorber. When $ka \ll 1$ the radiation resistance is approximately

$$R_r \approx 4\pi a^4 k^2 \rho c.$$

The spherical shell of the microphone has a mass m and a stiffness S against radial displacements. Moreover, it has a mechanical resistance R representing conversion of absorbed acoustic power to electric power.

A plane harmonic wave

$$p_i = A_0 \, e^{i(\omega t - kx)}$$

is incident upon the sphere. For a point absorber we may disregard reflected or diffracted waves. At a certain frequency $(\omega = \omega_0)$ we have resonance. Find the value of R for which the absorbed power P has its maximum P_{max} and express P_{max} in terms of A_0, a, k and ρc. Also show that the maximum absorption cross section of the spherical microphone is $\lambda^2/4\pi$. (This might be compared with the absorption cross section $A_a = \lambda^2/2\pi$ for a microphone of the type of a plane vibrating piston in a plane stiff wall, as shown, for instance, in Meyer and Neumann.[22] The absorption cross section is defined as the absorbed power divided by the sound intensity $I_i = \frac{1}{2}|A_0|^2/\rho c$ of the incident wave.)

Problem 3.6: Short Dipole Antenna

A vertical grounded antenna of height $h = 4\,\text{m}$ has at frequency $\nu = \omega/2\pi = 3\,\text{MHz}$ (or wavelength $\lambda = 100\,\text{m}$) an effective height $h_{\text{eff}} = 2\,\text{m}$. The input port for the antenna is at the ground plane and the antenna is coupled to an electric circuit as indicated in Figure 3.5. Show that the radiation resistance is $R_r = 0.63\,\Omega$. The antenna is being used for reception of a plane electromagnetic wave with vertically polarised electric field of amplitude $|E_i| = 10^{-3}$ V/m.

The antenna circuit is resistance matched as well as tuned to resonance. Calculate the power which is absorbed by the antenna circuit. Also calculate the absorption cross section for the antenna.

[Hints: for the grounded antenna, the effective dipole length is $l_{\text{eff}} = 2h_{\text{eff}}$. For a Hertz dipole, of infinitesimal length l_{eff}, the radiation resistance is $R_r = (kl_{\text{eff}})^2 Z_0 / 6\pi$ where $Z_0 = (\mu_0/\epsilon_0)^{1/2} = 377\,\Omega$. The intensity (average power per unit area) of the incident electromagnetic wave is $\frac{1}{2}|E_i|^2/Z_0$.]

Problem 3.7: Optimum Load Resistance

Derive Eqs. (3.40) and (3.41) for the optimum load resistance R_m and the corresponding absorbed power $P_{a,\text{max}}$.

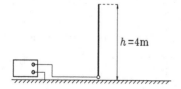

$h = 4\text{m}$

Figure 3.5: Vertical antenna connected to an electric circuit for transmission or reception of electromagnetic waves.

Problem 3.8: Maximum Absorbed Power

Show that Eq. (3.39) for the absorbed power may be reformulated as

$$\frac{8R_r}{|\hat{F}_e|^2} P_a = \frac{4R_m R_r}{(R_m + R_r)^2 + X^2} = 1 - \frac{(R_m - R_r)^2 + X^2}{(R_m + R_r)^2 + X^2},$$

where $X = \omega m_m + \omega m_r - S_m/\omega$. From this, Eqs. (3.42)–(3.45) may be obtained simply by inspection.

Problem 3.9: Power Radiated from Oscillating Submerged Body

Assume that the mass m of Problem 2.12 is submerged in water, where it generates a wave. Further, assume that the radiation resistance $R_r = 1$ Ns/m is included as one third of the total resistance $R = 3$ Ns/m. Similarly, assume that the added mass m_r is included in the total mass $m = 6$ kg. Here we neglect the fact that R_r and m_r vary with frequency. How large is the radiated power at frequencies f and f_0 defined in Problem 2.12?

CHAPTER FOUR

Gravity Waves on Water

The subject of this chapter is to study, mathematically, waves on an ideal fluid, namely a fluid which is incompressible and in which wave motion takes place without loss of mechanical energy. It is also assumed that the fluid motion is irrotational, and that the wave amplitude is so small that linear theory is applicable. Starting from basic hydrodynamics, the chapter derives the dispersion relationship for waves on water which is deep or otherwise has a constant depth. Plane and circular waves are discussed, and the transport of energy and momentum associated with wave propagation is considered. The final parts of the chapter introduce some concepts and derive some mathematical relations, which turn out to be very useful when, in subsequent chapters, interactions between waves and oscillating systems are discussed.

4.1 Basic Equations: Linearisation

Let us start with two basic hydrodynamic equations which express conservation of mass and momentum, namely the continuity equation

$$\frac{\partial \rho}{\partial t} + \nabla \cdot (\rho \vec{v}) = 0 \tag{4.1}$$

and the Navier-Stokes equation

$$\frac{D\vec{v}}{Dt} \equiv \frac{\partial \vec{v}}{\partial t} + (\vec{v} \cdot \nabla)\vec{v} = -\frac{1}{\rho}\nabla p_{\text{tot}} + \nu\nabla^2\vec{v} + \frac{1}{\rho}\vec{f}. \tag{4.2}$$

Here ρ is the mass density of the fluid, $\vec{v}$ is the velocity of the flowing fluid element, p_{tot} is the pressure of the fluid, and $\nu = \eta/\rho$ is the kinematic viscosity coefficient, which we shall neglect by assuming the fluid to be ideal. Hence, we set $\nu = 0$. Finally, $\vec{f}$ is external force per unit volume.

Here we consider only gravitational force, that is,

$$\vec{f} = \rho\vec{g}, \tag{4.3}$$

where $\vec{g}$ is the acceleration of gravity. For an incompressible fluid, ρ is constant and the continuity equation (4.1) gives

$$\nabla \cdot \vec{v} = 0. \tag{4.4}$$

With the introduced assumptions Eq. (4.2) becomes

$$\frac{\partial \vec{v}}{\partial t} + \vec{v} \cdot \nabla \vec{v} = -\frac{1}{\rho} \nabla p_{\text{tot}} + \vec{g}. \tag{4.5}$$

Using the vector identity

$$\vec{v} \times (\nabla \times \vec{v}) \equiv \tfrac{1}{2} \nabla v^2 - \vec{v} \cdot \nabla \vec{v} \tag{4.6}$$

and taking the curl of Eq. (4.5) gives

$$\frac{\partial}{\partial t}(\nabla \times \vec{v}) = \nabla \times \left(-\frac{1}{2} \nabla v^2 + \vec{v} \times (\nabla \times \vec{v}) - \frac{1}{\rho} \nabla p_{\text{tot}} + \vec{g} \right). \tag{4.7}$$

We use the vector identity $\nabla \times \nabla \varphi \equiv 0$ for any scalar function φ. Further, $\nabla \times \vec{g} = 0$, because $\vec{g}$ is the gradient of a gravitational potential, gz. Hence,

$$\frac{\partial}{\partial t}(\nabla \times \vec{v}) = \nabla \times [\vec{v} \times (\nabla \times \vec{v})]. \tag{4.8}$$

Assume that once the ideal fluid becomes irrotational, $\nabla \times \vec{v} = 0$ (e.g., if it were at rest), then it will continue to be irrotational forever, because $(\partial/\partial t)(\nabla \times \vec{v}) = 0$. Hence,

$$\nabla \times \vec{v} \equiv 0 \tag{4.9}$$

is assumed to be valid in the following. Because of the vector identity $\nabla \times \nabla \phi \equiv 0$, we can now write

$$\vec{v} = \nabla \phi, \tag{4.10}$$

where ϕ is the so-called *velocity potential*. Inserting into Eq. (4.5) gives

$$\nabla \left(\frac{\partial \phi}{\partial t} + \frac{v^2}{2} + \frac{p_{\text{tot}}}{\rho} + gz \right) = 0, \tag{4.11}$$

remembering the vector identity (4.6) and the relation $\vec{g} = -\nabla(gz)$. (The z axis is pointing upward.) Integration gives

$$\frac{\partial \phi}{\partial t} + \frac{v^2}{2} + \frac{p_{\text{tot}}}{\rho} + gz = C, \tag{4.12}$$

where C is an integration constant. This is (a non-stationary version of) the so-called Bernoulli equation.

Because $v^2 = \nabla \phi \cdot \nabla \phi$ according to Eq. (4.10), the scalar equation (4.12) for the scalar quantity ϕ replaces the vectorial equation (4.5) for the vectorial quantity $\vec{v}$. This is a mathematical convenience which is a benefit resulting from the assumption of irrotational flow – Eq. (4.9).

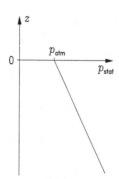

Figure 4.1: Hydrostatic pressure increases with depth below the water surface.

For the static case, when the fluid is not in motion, $\vec{v} = 0$ and $\phi = \text{constant}$, Eq. (4.12) gives

$$p_{tot} = p_{stat} = -\rho g z + \rho C. \tag{4.13}$$

At the free surface, $z = 0$, we have $p_{tot} = p_{atm}$, where p_{atm} is the atmospheric air pressure. Note that we here neglect surface tension on the air-fluid interface. This gives $C = p_{atm}/\rho$ and

$$p_{stat} = -\rho g z + p_{atm}. \tag{4.14}$$

The hydrostatic pressure increases linearly with the vertical displacement below the surface of the fluid, as shown in Figure 4.1.

The static air pressure in a submerged air chamber (Figure 4.2) is

$$p_{k0} = \rho g(-z_k) + p_{atm}, \tag{4.15}$$

where the water surface below the entrapped air is at depth $-z_k$.

The conditions of incompressibility (i.e., $\nabla \cdot \vec{v} = 0$) and of irrotational motion (i.e., $\vec{v} = \nabla \phi$) require that the Laplace equation

$$\nabla^2 \phi = 0 \tag{4.16}$$

must be satisfied throughout the fluid. Solutions of this partial differential equation must satisfy certain boundary conditions, which are considered below.

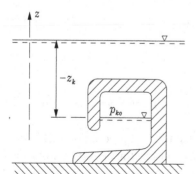

Figure 4.2: Submerged chamber containing entrapped air with elevated static pressure p_{k0}.

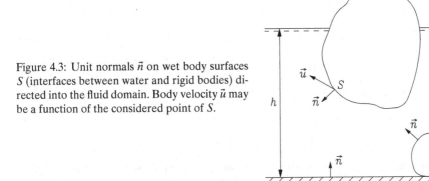

Figure 4.3: Unit normals $\vec{n}$ on wet body surfaces S (interfaces between water and rigid bodies) directed into the fluid domain. Body velocity $\vec{u}$ may be a function of the considered point of S.

At a solid-body boundary moving with velocity $\vec{u}$ (Figure 4.3), we have $u_n \equiv \vec{n} \cdot \vec{u} = v_n = \vec{n} \cdot \nabla\phi \equiv \partial\phi/\partial n$, as there is no fluid flow through the boundary. That is,

$$\partial\phi/\partial n = u_n. \tag{4.17}$$

In an ideal fluid which is inviscid ($\nu = 0$), no condition is required on the tangential component of $\vec{v} = \nabla\phi$. At a solid surface not in motion, we have

$$\partial\phi/\partial n = 0. \tag{4.18}$$

On a horizontal bottom of depth h, we have, in particular,

$$\partial\phi/\partial z = 0 \quad \text{at } z = -h. \tag{4.19}$$

We may note that the acceleration of gravity g, which is an important quantity for ocean waves, does not enter into the Laplace equation (4.16) or the body boundary condition (4.17). Moreover, these equations do not contain any derivative with respect to time. However, the fact that gravitational waves can exist on water is associated with the presence of the quantity g, as well as of a time derivative, in the following free-surface boundary condition.

On the free surface $z = \eta(x, y, t)$, which is the interface between water and open air (see Figure 4.4), the pressure in the fluid equals the air pressure (if we neglect capillary forces, which are, however, considered in Problem 4.1). Using $[p_{\text{tot}}]_{z=\eta} = p_{\text{atm}}$ in the Bernoulli equation (4.12) gives

$$\left[\frac{\partial\phi}{\partial t} + \frac{v^2}{2}\right]_{z=\eta} + g\eta = C - \frac{p_{\text{atm}}}{\rho}. \tag{4.20}$$

Figure 4.4: Free surface or the air-water interface when a wave with elevation η is present.

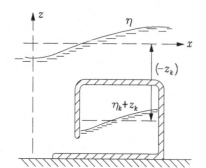

Figure 4.5: Submerged chamber for an OWC with an equilibrium water level below the mean sea level.

With constant air pressure, both sides of this equation must vanish, because the left-hand side depends on t whereas the right-hand side does not. Moreover, in the static case the left-hand side vanishes. Hence, $C = p_{atm}/\rho$, and with $\vec{v} = \nabla\phi$ [cf. Eq. (4.10)] we have the free-surface boundary condition

$$g\eta + \left[\frac{\partial\phi}{\partial t} + \frac{1}{2}\nabla\phi \cdot \nabla\phi\right]_{z=\eta} = 0. \tag{4.21}$$

In a wave-power converter of the oscillating-water-column (OWC) type with pneumatic power takeoff, the air pressure is not constant above the OWC. Let the dynamic part of the air pressure be denoted as p_k. The total air pressure is

$$p_{air} = p_{k0} + p_k = \rho g(-z_k) + p_{atm} + p_k, \tag{4.22}$$

where we have used Eq. (4.15). (Except for submerged OWCs, the equilibrium water level is at $z_k = 0$, which is also the case if air pressure fluctuation caused by wind were considered.) Let $\eta_k = \eta_k(x, y, t)$ denote the vertical deviation of the water surface from its equilibrium position below the entrapped air (see Figure 4.5). Thus, $\eta = \eta_k$ when $z_k = 0$. From the Bernoulli equation (4.12) we now have

$$\left[\frac{\partial\phi}{\partial t} + \frac{v^2}{2}\right]_{z=\eta_k+z_k} + g\eta_k + gz_k = C - \left[\frac{p_{tot}}{\rho}\right]_{z=\eta_k+z_k}$$
$$= C - \frac{p_{air}}{\rho} = C - (-gz_k) - \frac{p_{atm} + p_k}{\rho}. \tag{4.23}$$

Using $C - p_{atm}/\rho = 0$ and Eq. (4.10) gives

$$g\eta_k + \frac{p_k}{\rho} + \left[\frac{\partial\phi}{\partial t} + \frac{1}{2}\nabla\phi \cdot \nabla\phi\right]_{z=\eta_k+z_k} = 0. \tag{4.24}$$

This is the so-called *dynamic boundary condition*, which we linearise to

$$g\eta_k + \frac{p_k}{\rho} + \left[\frac{\partial\phi}{\partial t}\right]_{z=z_k} = 0 \tag{4.25}$$

when we assume that the dynamic variables such as ϕ, η, η_k and all their derivatives are small, and we neglect small terms of second or higher order, such as

$v^2 = \nabla\phi \cdot \nabla\phi$. Also the difference

$$\left[\frac{\partial\phi}{\partial t}\right]_{z=\eta_k+z_k} - \left[\frac{\partial\phi}{\partial t}\right]_{z=z_k} = \eta_k\left[\frac{\partial^2\phi}{\partial z\partial t}\right]_{z=z_k} + \cdots \tag{4.26}$$

may then be neglected. (Note that the right-hand side of this equation is a Taylor expansion where only the first non-vanishing term is explicitly written. It is a product of two small quantities, η_k and a derivative of ϕ.)

In addition to the dynamic boundary condition there is also a *kinematic bound-ary condition* on the interface between water and air. Physically it is the condition that a fluid particle on the interface stays on the interface as this is undulating as a result of wave motion. For the linearised case the kinetic boundary condition is simply

$$\left[\frac{\partial\phi}{\partial z}\right]_{z=z_k} = [v_z]_{z=z_k} = \frac{\partial\eta_k}{\partial t}, \tag{4.27}$$

where η_k may be replaced by η for the open-air case. (Readers interested in learning about the kinematic boundary condition for the more general non-linear case may consult, e.g., Mei,[1] Chap. 1.) Taking the time derivative of Eq. (4.25) and inserting into Eq. (4.27) gives

$$\left[\frac{\partial^2\phi}{\partial t^2} + g\frac{\partial\phi}{\partial z}\right]_{z=z_k} = -\frac{1}{\rho}\frac{dp_k}{dt}. \tag{4.28}$$

With zero dynamic air pressure (for $z = 0$) we get, in particular,

$$\left[\frac{\partial^2\phi}{\partial t^2} + g\frac{\partial\phi}{\partial z}\right]_{z=0} = 0. \tag{4.29}$$

Note that time t enters explicitly only into the free-surface boundary conditions. (It does not enter into the partial differential equation $\nabla^2\phi = 0$ and into the remaining boundary conditions.) Thus, without a free surface the solution $\phi = \phi(x, y, z, t)$ could not represent a wave.

We may also note that the boundary condition (4.17) has to be satisfied on the wet surface of the moving body. However, if the body is oscillating with a small amplitude, we may make the linearising approximation that the boundary condition (4.17) is to be applied at the time-average (or equilibrium) position of the wet surface of an oscillating body.

Any solution $\phi = \phi(x, y, z, t)$ for the velocity potential has to satisfy the Laplace equation (4.16) in the fluid domain and the inhomogeneous boundary conditions (4.17) and (4.28), which may sometimes or somewhere simplify to the homogeneous boundary conditions (4.18) and (4.29), respectively. For cases in which the fluid domain is of infinite extent, later sections (Sections 4.3 and 4.6) supplement the above boundary conditions with a "radiation condition" at infinite distance.

From the velocity potential $\phi(x, y, z, t)$, which is an auxiliary mathematical function, we can derive the following physical quantities. Everywhere in the fluid domain we can derive the fluid velocity from Eq. (4.10),

$$\vec{v} = \vec{v}(x, y, z, t) = \nabla\phi, \tag{4.30}$$

and the *hydrodynamic pressure* from the dynamic part of Eq. (4.12),

$$p = p(x, y, z, t) = -\rho\left(\frac{\partial\phi}{\partial t} + \frac{v^2}{2}\right) \approx -\rho\frac{\partial\phi}{\partial t}, \tag{4.31}$$

where in the last step we have neglected the small term of second order. Moreover, the elevation of the interface between water and entrapped air is given by the linearised dynamic boundary condition (4.25) as

$$\eta_k = \eta_k(x, y, t) = -\frac{1}{g}\left[\frac{\partial\phi}{\partial t}\right]_{z=z_k} - \frac{1}{\rho g}p_k. \tag{4.32}$$

The *wave elevation* (the elevation of the interface between the water and the open, constant-pressure air) is

$$\eta = \eta(x, y, t) = -\frac{1}{g}\left[\frac{\partial\phi}{\partial t}\right]_{z=0}. \tag{4.33}$$

4.2 Harmonic Waves on Water of Constant Depth

Except when otherwise stated, let us in the following consider the case of a plane horizontal sea bottom. If the water is sufficiently deep, the sea bottom does not influence the waves on the water surface. Then the shape of the bottom is of no concern for the waves. If the water is not sufficiently deep, and the sea bottom is not horizontal, an analysis differing from the following analysis is required. With sinusoidal time variation we write

$$\phi = \phi(x, y, z, t) = \text{Re}\{\hat{\phi}(x, y, z)e^{i\omega t}\}, \tag{4.34}$$

where $\hat{\phi}$ is the complex amplitude of the velocity potential at the space point (x, y, z). Similarly we define the complex amplitudes $\hat{\vec{v}} = \hat{\vec{v}}(x, y, z)$, $\hat{p} = \hat{p}(x, y, z)$, $\hat{\eta}_k = \hat{\eta}_k(x, y)$ and $\hat{\eta} = \hat{\eta}(x, y)$. The linearised basic equations, the Laplace equation (4.16) and the boundary conditions (4.17) and (4.28), now become

$$\nabla^2\hat{\phi} = 0 \tag{4.35}$$

everywhere in the water,

$$\left[\frac{\partial\hat{\phi}}{\partial n}\right]_S = \hat{u}_n \tag{4.36}$$

on the wet surface of solid bodies (Figure 4.3), and

$$\left[-\omega^2\hat{\phi} + g\frac{\partial\hat{\phi}}{\partial z}\right]_{z=z_k} = -\frac{i\omega}{\rho}\hat{p}_k \tag{4.37}$$

on the water-air surfaces (see Figure 4.5). Here, also $\hat{u}_n$ and $\hat{p}_k$ are complex amplitudes. Note that the normal component u_n of the motion of the wet body surface

(Figure 4.3) is a function of the point considered on that surface, whereas the dynamic air pressure p_k is assumed to have the same value everywhere inside the volume of entrapped air (Figure 4.5). Equations (4.30)–(4.33) for determining the physical variables become, in terms of complex amplitudes,

$$\hat{\vec{v}} = \nabla\hat{\phi}, \tag{4.38}$$

$$\hat{p} = -i\omega\rho\hat{\phi}, \tag{4.39}$$

$$\hat{\eta}_k = -\frac{i\omega}{g}[\hat{\phi}]_{z=z_k} - \frac{1}{\rho g}\hat{p}_k, \tag{4.40}$$

$$\hat{\eta} = -\frac{i\omega}{g}[\hat{\phi}]_{z=0}. \tag{4.41}$$

The remaining part of this section (and also Sections 4.3–4.6) discusses some particular solutions, which satisfy the Laplace equation and the homogeneous boundary conditions

$$\left[\frac{\partial\hat{\phi}}{\partial z}\right]_{z=-h} = 0, \tag{4.42}$$

$$\left[-\omega^2\hat{\phi} + g\frac{\partial\hat{\phi}}{\partial z}\right]_{z=0} = 0. \tag{4.43}$$

These solutions will thus satisfy the boundary conditions on a (non-moving) horizontal bottom of a sea of depth h and the boundary condition at the free water surface (the interface between water and air), where the air pressure is constant. Additional (inhomogeneous) boundary conditions will be imposed later (e.g. in Section 4.7).

Using the method of separation of variables, we seek a particular solution of the form

$$\hat{\phi}(x, y, z) = H(x, y)Z(z). \tag{4.44}$$

Inserting this into the Laplace equation (4.35) and dividing by $\hat{\phi}$, we get

$$0 = \frac{\nabla^2\hat{\phi}}{\hat{\phi}} = \frac{1}{H}\left[\frac{\partial^2 H}{\partial x^2} + \frac{\partial^2 H}{\partial y^2}\right] + \frac{1}{Z}\frac{d^2 Z}{dz^2} \tag{4.45}$$

or

$$-\frac{1}{Z}\frac{d^2 Z}{dz^2} = \frac{1}{H}\left[\frac{\partial^2 H}{\partial x^2} + \frac{\partial^2 H}{\partial y^2}\right] \equiv \frac{1}{H}\nabla_H^2 H. \tag{4.46}$$

The left-hand side of Eq. (4.46) is a function of z only. The right-hand side is a function of x and y. This is impossible unless it is a constant, $-k^2$, say. Thus from Eq. (4.46) we get the two equations

$$\frac{d^2 Z(z)}{dz^2} = k^2 Z(z), \tag{4.47}$$

$$\nabla_H^2 H(x, y) = -k^2 H(x, y). \tag{4.48}$$

We note here that the separation constant k^2 has the dimension of inverse length squared. Later, when discussing a solution of the two-dimensional Helmholtz equation (4.48), we shall see that k may be interpreted as the angular repetency of a propagating wave. We shall, however, first discuss the following solution of Eq. (4.47):

$$Z(z) = c_+ e^{kz} + c_- e^{-kz}. \tag{4.49}$$

Here c_+ and c_- are two integration constants. Because we have two new integration constants when solving Eq. (4.48) for H, we may choose $Z(0) = 1$, which means that $c_+ + c_- = 1$. Another relation to determine the integration constants is provided by the bottom boundary condition

$$\frac{\partial \hat{\phi}}{\partial z} = \frac{dZ(z)}{dz} H(x, y) = 0 \quad \text{for } z = -h. \tag{4.50}$$

It is then easy to show that (see Problem 4.2)

$$c_\pm = \frac{e^{\pm kh}}{e^{kh} + e^{-kh}}. \tag{4.51}$$

From Eq. (4.49) it now follows that

$$Z(z) = \frac{e^{k(z+h)} + e^{-k(z+h)}}{e^{kh} + e^{-kh}}. \tag{4.52}$$

Hence we have the following particular solution of the Laplace equation:

$$\hat{\phi} = H(x, y) e(kz), \tag{4.53}$$

where $H(x, y)$ has to satisfy the Helmholtz equation (4.48), and where

$$e(kz) = \frac{\cosh(kz + kh)}{\cosh(kh)} = \frac{e^{k(z+h)} + e^{-k(z+h)}}{e^{kh} + e^{-kh}} \tag{4.54}$$

$$= \frac{1 + e^{-2k(z+h)}}{1 + e^{-2kh}} e^{kz}. \tag{4.55}$$

To be strict, because this is a function of two variables, we should perhaps have denoted the function by $e(kz, kh)$. We prefer, however, to use the simpler notation $e(kz)$, in particular because, for the deep-water case, $kh \gg 1$, it tends to the exponential function: $e(kz) \approx e^{kz}$ (although it then approaches $2e^{kz}$ when z approaches $-h$, a z coordinate which is usually of little practical interest in the deep-water case). We may note that the solution (4.53) is applicable for the deep-water case even if the water depth is not constant, provided $kh_{\min} \gg 1$.

In order to satisfy the free-surface boundary condition (4.43) we require

$$\omega^2 = \omega^2 e(0) = g \left[\frac{de(kz)}{dz} \right]_{z=0} = gk \frac{\sinh(kh)}{\cosh(kh)} \tag{4.56}$$

or

$$\omega^2 = gk \tanh(kh), \tag{4.57}$$

which for deep water $(kh \gg 1)$ simplifies to

$$\omega^2 = gk. \tag{4.58}$$

This result has already been presented in Eq. (3.8).

Because we shall later interpret k as the angular repetency (wave number), we have now derived the dispersion equation (4.57), which is a relation between the angular frequency ω and the angular repetency k. We rewrite Eq. (4.57) as

$$\omega^2/(gk) = \tanh(kh) \tag{4.59}$$

and observe that, in the interval $0 < k < +\infty$, the right-hand side is monotonically increasing from 0 to 1, whereas the left-hand side, for given ω, is monotonically decreasing from $+\infty$ to 0. Hence, there is one, and only one, positive k which satisfies Eq. (4.59), and there is correspondingly one, and only one, negative solution, because both sides of the equation are odd functions of k. Thus there is only one possible positive value of the separation constant k^2 in Eqs. (4.47) and (4.48). We may raise the question whether negative or even complex values are possible.

Replacing k^2 by λ_n and $\hat{\phi}(x, y, z) = Z(z)H(x, y)$ by

$$\hat{\phi}_n(x, y, z) = Z_n(z)H_n(x, y), \tag{4.60}$$

we rewrite Eqs. (4.48) and (4.47) as

$$\nabla_H^2 H_n(x, y) = -\lambda_n H_n(x, y) \tag{4.61}$$
$$Z_n''(z) = \lambda_n Z_n(z), \tag{4.62}$$

where the integer subscript n is used to label the various possible solutions. Because $\hat{\phi}_n(x, y, z)$ has to satisfy boundary conditions (4.42) and (4.43), the functions $Z_n(z)$ are subject to the boundary conditions

$$Z_n'(-h) = 0, \tag{4.63}$$

$$Z_n'(0) = \frac{\omega^2}{g} Z_n(0). \tag{4.64}$$

In Eq. (4.62) the separation constant λ_n is an eigenvalue and $Z_n(z)$ is the corresponding eigenfunction. We shall now show that all eigenvalues have to be real. Let us consider two possible eigenvalues λ_n and λ_m with corresponding eigenfunctions $Z_n(z)$ and $Z_m(z)$. Replacing n by m in Eq. (4.62) and taking the complex conjugate give

$$Z_m^{*\prime\prime}(z) = \lambda_m^* Z_m^*(z). \tag{4.65}$$

Now, let us multiply Eq. (4.62) by $Z_m^*(z)$ and Eq. (4.65) by $Z_n(z)$. Subtracting and integrating then gives

$$I \equiv \int_{-h}^{0} [Z_m^*(z) Z_n''(z) - Z_n(z) Z_m^{*\prime\prime}(z)] \, dz$$

$$= (\lambda_n - \lambda_m^*) \int_{-h}^{0} Z_m^*(z) Z_n(z) \, dz. \tag{4.66}$$

Noting that the integrand in the first integral of Eq. (4.66) may be written as

$$Z_m^*(z) Z_n''(z) - Z_n(z) Z_m^{*''}(z)$$
$$= Z_m^*(z) Z_n''(z) - Z_m^{*'}(z) Z_n'(z) + Z_n'(z) Z_m^{*'}(z) - Z_n(z) Z_m^{*''}(z)$$
$$= \frac{d}{dz} [Z_m^*(z) Z_n'(z) - Z_n(z) Z_m^{*'}(z)] \tag{4.67}$$

and that the boundary conditions (4.63) and (4.64) apply to Z_m^* as well, we find that the integral I vanishes. This is true because

$$I = [Z_m^*(z) Z_n'(z) - Z_n(z) Z_m^{*'}(z)]_{-h}^0$$
$$= Z_m^*(0) \frac{\omega^2}{g} Z_n(0) - Z_n(0) \frac{\omega^2}{g} Z_m^*(0) - 0 - 0$$
$$= 0. \tag{4.68}$$

We have here, tacitly, assumed (as we shall do throughout) that ω is real; that is, $\omega^* = \omega$. Hence we have shown that for all m and n,

$$(\lambda_n - \lambda_m^*) \int_{-h}^0 Z_m^*(z) Z_n(z) \, dz = 0. \tag{4.69}$$

For $m = n$ we have, for non-trivial solutions, $|Z_n(z)| \neq 0$, that $(\lambda_n - \lambda_n^*) = 0$. This means that λ_n is real for all n.

For $\lambda_n \neq \lambda_m$, Eq. (4.69) gives the following orthogonality condition for the eigenfunctions $\{Z_n(z)\}$:

$$\int_{-h}^0 Z_m^*(z) Z_n(z) \, dz = 0. \tag{4.70}$$

We have shown that the eigenvalues are real and that there is only one positive eigenvalue k^2 satisfying dispersion equation (4.57). The other eigenvalues have to be negative. We shall label the eigenvalues in decreasing order as follows:

$$\lambda_0 = k^2 > \lambda_1 > \lambda_2 > \lambda_3 > \cdots > \lambda_n > \cdots. \tag{4.71}$$

Thus λ_n is negative if $n \geq 1$. A negative eigenvalue may be conveniently written as

$$\lambda_n = -m_n^2 \tag{4.72}$$

where m_n is real. Let us now in Eq. (4.49) replace k by $-im_n$ and $Z(z)$ by $Z_n(z)$. One of the two integration constants may then be eliminated by using boundary condition (4.63): $Z_n'(-h) = 0$. The resulting eigenfunction is (see Problem 4.3)

$$Z_n(z) = N_n^{-1/2} \cos(m_n z + m_n h), \tag{4.73}$$

where $N_n^{-1/2}$ is an arbitrary integration constant. This result also follows from Eqs. (4.52) and (4.54) if we observe that $\cosh[-im_n(z+h)] = \cos[-m_n(z+h)] = \cos[m_n(z+h)]$. The free-surface boundary condition (4.64) is satisfied if

(see Problem 4.3)

$$\omega^2/(gm_n) = -\tan(m_n h). \tag{4.74}$$

This equation also follows from Eq. (4.59) if k is replaced by $-im_n$. The left-hand side of Eq. (4.74) decreases monotonically from $+\infty$ to 0 in the interval $0 < m_n < +\infty$. The right-hand side decreases monotonically from $+\infty$ to 0 in the interval $(n - 1/2)\pi/h < m_n < n\pi/h$. Hence Eq. (4.74) has a solution in this latter interval. Noting that $n = 1, 2, 3, \ldots$, we find that there is an infinite, but numerable, number of solutions for m_n. For the solutions we have (see Problem 4.3)

$$m_n \to \frac{n\pi}{h} - \frac{\omega^2}{n\pi g} \to \frac{n\pi}{h} \quad \text{as } n \to \infty. \tag{4.75}$$

Further (as is also shown in Problem 4.3), if

$$\frac{1}{h} \int_{-h}^{0} |Z_n(z)|^2 dz = 1 \tag{4.76}$$

is chosen as a normalisation condition, then the integration constant in Eq. (4.73) is given by

$$N_n = \frac{1}{2}\left[1 + \frac{\sin(2m_n h)}{2m_n h}\right]. \tag{4.77}$$

For $n = 0$ we set $m_0 = ik$ and then we have, in particular,

$$N_0 = \frac{1}{2}\left[1 + \frac{\sinh(2kh)}{2kh}\right]. \tag{4.78}$$

We may note (see Problem 4.2) that

$$Z_0(z) = \sqrt{\frac{2kh}{D(kh)}}e(kz), \tag{4.79}$$

where $e(kz)$ is given by Eq. (4.54) and

$$D(kh) = \left[1 + \frac{2kh}{\sinh(2kh)}\right]\tanh(kh). \tag{4.80}$$

The orthogonal set of eigenfunctions $\{Z_n(z)\}$ is complete, if the function $Z_0(z)$ is included in the set, that is, if $n = 0, 1, 2, 3, \ldots$. The completeness follows[23] from the fact that the eigenvalue problem, Eqs. (4.62)–(4.64), is a Sturm-Liouville problem.

We have now discussed possible solutions of Eq. (4.47) or (4.62) with the homogeneous boundary conditions (4.42) and (4.43), or (4.63) and (4.64). It remains for us to discuss solutions of the Helmholtz equation (4.48) or (4.61). In the discussion below, let us start by considering the case with $n = 0$, that is, $m_n = m_0 = ik$. Then we take Eq. (4.48) as the starting point.

4.3 Plane Waves: Propagation Velocities

We consider a two-dimensional case with no variation in the y direction. Setting $\partial/\partial y = 0$ we have from Eqs. (4.46) and (4.48)

$$d^2 H(x)/dx^2 = -k^2 H(x), \tag{4.81}$$

which has the general solution

$$H(x) = ae^{-ikx} + be^{ikx}, \tag{4.82}$$

where a and b are arbitrary integration constants. Hence from Eq. (4.53)

$$\hat{\phi} = e(kz)(ae^{-ikx} + be^{ikx}) \tag{4.83}$$

and from Eq. (4.34)

$$\phi = \text{Re}\{\hat{\phi}e^{i\omega t}\} = \text{Re}\{[ae^{i(\omega t - kx)} + be^{i(\omega t + kx)}]\}e(kz), \tag{4.84}$$

which demonstrates that k is the angular repetency (wave number).

The first and second terms represent plane waves propagating in the positive and negative x directions, respectively, with a "phase velocity"

$$v_p \equiv \frac{\omega}{k} = \frac{g}{\omega}\tanh(kh) = \left\{\frac{g}{k}\tanh(kh)\right\}^{1/2}, \tag{4.85}$$

which is obtained from the dispersion relation (4.57). Note that v_p is the velocity by which the wave crest (or a line of constant phase) propagates (in a direction perpendicular to the wave crest). Later we prove (in Section 4.4) that the wave energy associated with a harmonic plane wave is transported with a velocity, the "group velocity" (3.10),

$$v_g = d\omega/dk, \tag{4.86}$$

which, in general, differs from the phase velocity. The quantity v_g bears its name because it is the velocity by which a wave group, composed of harmonic waves with slightly different frequencies, propagates (see Problem 3.1). For instance, a swell wave, which originates from a storm centre a distance L from a coast line, reaches this coast line a time L/v_g later (see Problem 4.4).

In Eq. (4.83) we might interpret the first term as an incident wave and the second term as a reflected wave. Introducing $\Gamma = b/a$, which is a complex reflection coefficient, and $A = -i\omega a/g$, which is the complex elevation amplitude at $x = 0$ in the case of no reflection, we may, in accordance with Eqs. (4.41) and (4.54), rewrite Eq. (4.83) as

$$\hat{\phi} = -\frac{g}{i\omega} Ae(kz)(e^{-ikx} + \Gamma e^{ikx}). \tag{4.87}$$

Correspondingly, the wave elevation $\eta = \eta(x, t)$ has a complex amplitude [cf. Eq. (4.41)]

$$\hat{\eta} = \hat{\eta}(x) = A[e^{-ikx} + \Gamma e^{ikx}]. \tag{4.88}$$

Setting

$$\hat{\eta}_f = Ae^{-ikx}, \tag{4.89}$$

$$\hat{\eta}_b = Be^{ikx} = A\Gamma e^{ikx}, \tag{4.90}$$

we have

$$\hat{\eta} = \hat{\eta}_f + \hat{\eta}_b, \tag{4.91}$$

$$\hat{\phi} = -\frac{g}{i\omega}\hat{\eta}e(kz), \tag{4.92}$$

$$\hat{p} = \rho g\hat{\eta}e(kz), \tag{4.93}$$

where we have also made use of Eq. (4.39). Note that the hydrodynamic pressure p decreases monotonically (exponentially for deep water) with the distance $(-z)$ below the mean free surface, $z = 0$. For the fluid velocity we have, from Eqs. (4.38) and (4.54),

$$\hat{v}_x = \frac{\partial\hat{\phi}}{\partial x} = g\frac{k}{\omega}e(kz)(\hat{\eta}_f - \hat{\eta}_b)$$

$$= \omega\frac{\cosh(kz + kh)}{\sinh(kh)}(\hat{\eta}_f - \hat{\eta}_b), \tag{4.94}$$

$$\hat{v}_z = \frac{\partial\hat{\phi}}{\partial z} = g\frac{ik}{\omega}e'(kz)(\hat{\eta}_f + \hat{\eta}_b)$$

$$= i\omega\frac{\sinh(kz + kh)}{\sinh(kh)}(\hat{\eta}_f + \hat{\eta}_b), \tag{4.95}$$

where

$$e'(kz) = \frac{de(kz)}{d(kz)} = \frac{\sinh(kz + kh)}{\cosh(kh)} = e(kz)\tanh(kz + kh). \tag{4.96}$$

We have used the dispersion relation (4.57) to obtain the last expressions for $\hat{v}_x$ and $\hat{v}_z$.

If we have a progressive wave in the forward (positive x) direction ($\hat{\eta}_b = 0$) for deep water ($kh \gg 1$), the motion of fluid particles is circularly polarised with a negative sense of rotation (in the clockwise direction in the xz plane), notably

$$\hat{v}_x = \omega\hat{\eta}_f e^{kz}, \tag{4.97}$$

$$\hat{v}_z = i\omega\hat{\eta}_f e^{kz}. \tag{4.98}$$

The displacement of the fluid particles (as indicated in Figure 4.6) is given by

$$\hat{\xi} = \hat{v}/i\omega, \tag{4.99}$$

$$\hat{\xi}_x = -i\hat{\eta}_f e^{kz}, \tag{4.100}$$

$$\hat{\xi}_z = \hat{\eta}_f e^{kz}. \tag{4.101}$$

For a progressive wave in the backward direction ($\hat{\eta}_f = 0$) on deep water, the motion is circularly polarised with a positive sense of rotation.

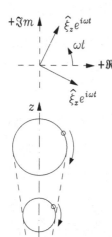

Figure 4.6: Phasor diagram of the x and z components of the fluid particle displacement (top) and fluid particle trajectories (bottom).

For a plane wave propagating in a direction making an angle β with the x axis, the wave elevation is given by

$$\hat{\eta} = A\exp\{-ik(x\cos\beta + y\sin\beta)\}. \tag{4.102}$$

The corresponding velocity potential $\hat{\phi}$, still given by Eq. (4.92) but now with the new $\hat{\eta}$, is a solution of the Laplace equation (4.35), satisfying the boundary conditions on the sea bed, $z = -h$, and on the free water surface, $z = 0$ (see Problem 4.5).

For deep water ($kh \gg 1$), the dispersion relation (4.57) is approximately $\omega^2 = gk$, in agreement with Eq. (4.58) and with the previous statement, Eq. (3.8). Note that

$$\tanh(kh) > 0.95 \quad \text{for } kh > 1.83 \quad \text{or } h > 0.3\lambda,$$
$$\tanh(kh) > 0.99 \quad \text{for } kh > 2.6 \quad \text{or } h > 0.42\lambda.$$

Hence, depending on the desired accuracy, we may assume deep water when the depth is at least one third or one half of the wavelength, respectively. For deep water the wavelength is

$$\lambda = 2\pi/k = 2\pi g/\omega^2 = (g/2\pi)T^2 = (1.56 \text{ m/s}^2)T^2. \tag{4.103}$$

As previously found [see Eqs. (3.9) and (3.12)] for deep water, we have

$$2v_g = v_p = g/\omega = \sqrt{g/k} \tag{4.104}$$

where v_p is the phase velocity and v_g the group velocity.

In contrast, for shallow water with a horizontal bottom, $kh \ll 1$, a power series expansion of Eqs. (4.57) gives

$$\omega^2 = gk(kh + \cdots) \approx ghk^2. \tag{4.105}$$

(Here the error is less than 1% or 5% if the water depth is less than 1/36 or 1/16 wavelength, respectively.) In this approximation a wave is not dispersive, because

$$v_g = v_p = \sqrt{gh} \tag{4.106}$$

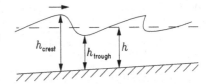

Figure 4.7: Increasing wave steepness as the wave
propagates on shallow water.

is independent of k and ω. Group velocity v_g differs, however, from phase velocity
v_p in the general dispersive case, as is considered below. The present theory is
linear, but let us nevertheless try to give a qualitative explanation of wave-breaking
on shallow water. Formula (4.106) indicates that the wave passes faster on the wave
crest than on the wave trough because $h_{crest} > h_{trough}$ (see Figure 4.7). Finally there
will be a vertical edge and then, of course, the linear theory does not apply because
$\partial \eta / \partial x \to \infty$.

For constant water depth h, the phase velocity is, in the general case,

$$v_p = \frac{\omega}{k} = \sqrt{\frac{g}{k}\tanh(kh)} = \frac{g}{\omega}\tanh(kh). \tag{4.107}$$

In order to obtain the group velocity v_g, we differentiate the dispersion equation
(4.57):

$$2\omega d\omega = gdk \tanh(kh) + \frac{gk}{\cosh^2(kh)}hdk$$

$$= \frac{dk}{k}gk \tanh(kh) + \frac{gk \tanh(kh)}{\cosh(kh)\sinh(kh)}h\,dk$$

$$= \frac{dk}{k}\omega^2 + \frac{2\omega^2 h\,dk}{\sinh(2kh)}. \tag{4.108}$$

Hence,

$$v_g = \frac{d\omega}{dk} = \frac{\omega}{2k}\left[1 + \frac{2kh}{\sinh(2kh)}\right]. \tag{4.109}$$

Because $v_p = \omega/k$, we then obtain

$$v_g = \frac{D(kh)}{2\tanh(kh)}v_p = \frac{g}{2\omega}D(kh), \tag{4.110}$$

where we have, for convenience, used the depth function $D(kh)$ defined by
Eq. (4.80). Because we shall frequently refer to this function later, here we present
some alternative expressions:

$$D(kh) = \left[1 + \frac{2kh}{\sinh(2kh)}\right]\tanh(kh)$$

$$= \tanh(kh) + \frac{kh}{\cosh^2(kh)}$$

$$= \tanh(kh) + kh - kh\tanh^2(kh)$$

$$= \left[1 - \left(\frac{\omega^2}{gk}\right)^2\right]kh + \frac{\omega^2}{gk} \tag{4.111}$$

We shall frequently also need the relationship (see Problem 4.12)

$$2k \int_{-h}^{0} e^2(kz)\, dz = D(kh),\tag{4.112}$$

where the vertical eigenfunction $e(kz)$ is given by Eq. (4.54) and the depth function $D(kh)$ is given by Eq. (4.111).

Note that for deep water, $kh \gg 1$, we have $D(kh) \approx 1$ and $\tanh(kh) \approx 1$. For small values of kh we have $\tanh(kh) = kh + \cdots$ and $D(kh) = 2kh + \cdots$. Although $\tanh(kh)$ is a monotonically increasing function, it can be shown that $D(kh)$ has a maximum $D_{\max} = x_0$ for $kh = x_0$ where $x_0 = 1.1996786$ is a solution of the transcendental equation $x_0\tanh(x_0) = 1$ (see Problem 4.12).

For arbitrary ω, we can obtain k from the transcendental dispersion equation (4.57), for instance by applying some numerical iteration procedure, and then we obtain v_p and v_g from Eqs. (4.107) and (4.110). The relationships are shown graphically in the curves of Figure 4.8.

We have now discussed a propagating wave, which corresponds to the case with $n = 0$ in Eqs. (4.73), (4.74) and (4.77). Now let n be an arbitrary non-negative integer. Setting again $\partial/\partial y = 0$, we find that Eq. (4.61), with Eq. (4.72) inserted, has a solution

$$H_n(x) = a_n \exp(-m_n x) + b_n \exp(m_n x),\tag{4.113}$$

where a_n and b_n are integration constants. The corresponding complex amplitude of the velocity potential is, from Eq. (4.60),

$$\hat{\phi}_n(x, z) = (a_n \exp\{-m_n x\} + b_n \exp\{m_n x\}) Z_n(z).\tag{4.114}$$

Note that this is a particular solution, which satisfies the homogeneous Laplace

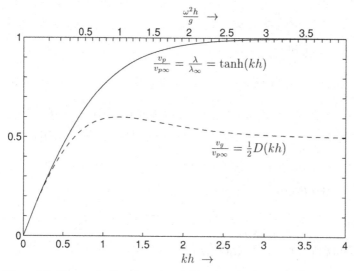

Figure 4.8: Phase velocity v_p, group velocity v_g and repetency $1/\lambda = k/2\pi$ as functions of depth h with given frequency, or as functions of frequency with given water depth. The subscript ∞ corresponds to (infinitely) deep water.

equation (4.35) and the homogeneous boundary conditions (4.42) and (4.43). Hence a superposition of such solutions as Eq. (4.114) is a solution

$$\hat{\phi}(x, z) = \sum_{n=0}^{\infty}(a_n \exp\{-m_n x\} + b_n \exp\{m_n x\})N_n^{-1/2} \cos(m_n z + m_n h),$$

$$(4.115)$$

where we have used Eq. (4.73).

Remember that for the term with $n = 0$ we set $m_n = ik$. Here we choose k and m_n (for $n \geq 1$) to be positive. Note that the term with a_n decays with increasing x, or is, for $n = 0$, a wave progressing in the positive x direction. The term with b_n decays with decreasing x, or is, for $n = 0$, a wave progressing in the negative x direction. If our region of interest is $0 < x < +\infty$, we thus have to set $b_n = 0$ for all n, except for the possibility that $b_0 \neq 0$ for a case in which a wave is incident from infinite distance, $x = +\infty$. For $n \geq 1$ this is a consequence of avoiding infinite values. The *radiation condition* was briefly mentioned previously (in Section 4.1). We have the opportunity to apply such a condition here. When $b_0 = 0$ the radiation condition of an outgoing wave at infinite distance is satisfied. This means that there is no contribution of the form e^{ikx} to the potential $\hat{\phi}$. Similarly, if our region of interest is $-\infty < x < 0$, then $a_n = 0$ for all n if the radiation condition is satisfied as $x \to -\infty$. However, if a wave is incident from $x = -\infty$, then $a_0 \neq 0$. If our region of interest is finite, $x_1 < x < x_2$, say, then we may have $a_n \neq 0$ and $b_n \neq 0$ for all n.

For the two-dimensional case, Eq. (4.115) represents a general plane-wave solution for the velocity potential in a uniform fluid of constant depth. The terms with $n = 0$ are propagating waves, whereas the terms with $n \geq 1$ are "evanescent waves". Propagation is along the x axis. If x in Eq. (4.115) is replaced by $(x \cos \beta + y \sin \beta)$, then the plane-wave propagation is along a direction which makes an angle β with the x axis. We can easily show this by considering transformation between two coordinate systems (x', y') and (x, y) which differ by a rotation angle β. See also Eq. (4.102).

4.4 Wave Transport of Energy and Momentum

4.4.1 Potential Energy

Let us consider the potential energy associated with the elevation of water from the wave troughs to the wave crests. Per unit (horizontal) area the potential energy relative to the sea bed equals the product of $\rho g(h + \eta)$, the water weight per unit area, and $(h + \eta)/2$, the height of the water mass centre above the sea bed:

$$(\rho g/2)(h + \eta)^2 = (\rho g/2)h^2 + \rho g h \eta + (\rho g/2)\eta^2. \tag{4.116}$$

The increase in relation to calm water is

$$\rho g h \eta + (\rho g/2)\eta^2, \tag{4.117}$$

where the first term has a vanishing average value.

Hence, the time-average potential energy per unit (horizontal) area is

$$E_P(x, y) = (\rho g/2)\overline{\eta^2(x, y, t)}. \tag{4.118}$$

In the case of a harmonic wave,

$$E_P(x, y) = (\rho g/4)|\hat{\eta}(x, y)|^2. \tag{4.119}$$

In particular, for the harmonic plane wave given by Eq. (4.91), we have

$$E_P(x) = \frac{\rho g}{4}|\hat{\eta}_f + \hat{\eta}_b|^2 = \frac{\rho g}{4}\left(|\hat{\eta}_f|^2 + |\hat{\eta}_f|^2 + \hat{\eta}_f\hat{\eta}_b^* + \hat{\eta}_f^*\hat{\eta}_b\right)$$
$$= \frac{\rho g}{4}\left(|A|^2 + |B|^2 + AB^*e^{-i2kx} + A^*Be^{i2kx}\right). \tag{4.120}$$

If $AB \neq 0$, this expression for $E_P(x)$ contains a term which varies sinusoidally with x with a "wavelength" $\lambda/2 = 2\pi/2k = \pi/k$. This sinusoidal variation does not contribute if we average over an x interval which is either very long or, alternatively, an integral number of $\lambda/2$. Denoting this average by $\langle E_p \rangle$, we have

$$\langle E_p \rangle = (\rho g/4)(|A|^2 + |B|^2), \tag{4.121}$$

and for a progressive plane wave $(B = 0)$,

$$\langle E_p \rangle = (\rho g/4)|A|^2. \tag{4.122}$$

4.4.2 Kinetic Energy

For simplicity we consider a progressive, plane, harmonic wave on deep water. The fluid velocity is given by Eqs. (4.97) and (4.98):

$$\hat{v}_x = -i\hat{v}_z = \omega\hat{\eta}_f e^{kz}. \tag{4.123}$$

The average kinetic energy per unit volume is

$$\frac{1}{2}\rho\frac{1}{2}\text{Re}\{|\hat{v}_x|^2 + |\hat{v}_z|^2\} = \frac{\rho}{4}\left(|\hat{v}_x|^2 + |\hat{v}_z|^2\right)$$
$$= \frac{\rho}{2}\omega^2|\hat{\eta}_f|^2 e^{2kz} = \frac{\rho}{2}\omega^2|A|^2 e^{2kz}. \tag{4.124}$$

By integrating from $z = -\infty$ to $z = 0$ we obtain the average kinetic energy per unit (horizontal) area:

$$E_k = \frac{\rho}{2}\omega^2|A|^2 \int_{-\infty}^{0} e^{2kz}dz = \frac{\rho}{2}\frac{\omega^2}{2k}|A|^2. \tag{4.125}$$

Using $\omega^2 = gk$ we obtain

$$E_k = (\rho g/4)|A|^2. \tag{4.126}$$

It can be shown (cf. Problem 4.8) that this expression for E_k is also valid for an arbitrary constant water depth h. Moreover, it can be shown (cf. Problem 4.8) that for a plane wave as given by Eq. (4.91) the kinetic energy per unit horizontal surface, averaged both over time and over the horizontal plane, is

$$\langle E_k \rangle = (\rho g/4)(|A|^2 + |B|^2).\tag{4.127}$$

4.4.3 Total Stored Energy

The total energy is the sum of potential energy and kinetic energy. For a progressive, plane, harmonic wave the time-average stored energy per unit (horizontal) area is

$$E = E_k + E_p = 2E_k = 2E_p = (\rho g/2)|A|^2.\tag{4.128}$$

When we also average over the horizontal plane, we have for a plane wave as given by Eq. (4.91) that

$$\langle E \rangle = 2\langle E_k \rangle = 2\langle E_p \rangle = (\rho g/2)(|A|^2 + |B|^2).\tag{4.129}$$

4.4.4 Wave-Energy Transport

Consider the energy transport of the plane harmonic wave propagating in the x direction. Per unit (vertical) area the time-average power propagating in the positive x direction equals the intensity [cf. Eq. (3.16)]:

$$I = \overline{p v_x} = \tfrac{1}{2}\mathrm{Re}\{\hat{p}\hat{v}_x^*\}\tag{4.130}$$

Note that because $\hat{p}\hat{v}_z^*$ is purely imaginary [cf. Eqs. (4.93) and (4.95)], the intensity has no z component.

Inserting for $\hat{p}$ and $\hat{v}_x$ from Eqs. (4.93) and (4.94), we need the product

$$(\hat{\eta}_f + \hat{\eta}_b)(\hat{\eta}_f - \hat{\eta}_b)^* = |\hat{\eta}_f|^2 - |\hat{\eta}_b|^2 + (\hat{\eta}_f^*\hat{\eta}_b - \hat{\eta}_f\hat{\eta}_b^*).\tag{4.131}$$

Because the last term here is purely imaginary, we get

$$I = (k\rho g^2/2\omega)(|\hat{\eta}_f|^2 - |\hat{\eta}_b|^2)e^2(kz).\tag{4.132}$$

Integrating from $z = -h$ to $z = 0$ gives the transported wave power per unit width of the wave front:

$$J = \int_{-h}^{0} I \, dz = \frac{\rho g^2}{4\omega}(|\hat{\eta}_f|^2 - |\hat{\eta}_b|^2)2k \int_{-h}^{0} e^2(kz)\, dz.\tag{4.133}$$

Using relation (4.112),

$$2k \int_{-h}^{0} e^2(kz)\, dz = D(kh),\tag{4.134}$$

we have

$$J = \frac{\rho g^2 D(kh)}{4\omega}(|\hat{\eta}_f|^2 - |\hat{\eta}_b|^2) = \frac{\rho g^2 D(kh)}{4\omega}(|A|^2 - |B|^2).\tag{4.135}$$

For a purely progressive wave ($\hat{\eta}_b = 0$, $\hat{\eta}_f = Ae^{-ikx}$), this gives

$$J = \frac{\rho g^2 D(kh)}{4\omega}|\hat{\eta}_f|^2 = \frac{\rho g^2 D(kh)}{4\omega}|A|^2. \tag{4.136}$$

Introducing the period $T = 2\pi/\omega$, and the wave height $H = 2|A|$, we have for deep water, when $kh \gg 1$, and hence, $D(kh) \approx 1$, that

$$J = \frac{\rho g^2}{32\pi}TH^2 = (976\text{W s}^{-1}\text{m}^3)TH^2, \tag{4.137}$$

with $\rho = 1020$ kg/m^3 for sea water. For $T = 10$ s and $H = 2$ m, this gives

$$J = 3.9 \times 10^4 \text{ W/m} \approx 40 \text{ kW/m}. \tag{4.138}$$

In the case of a reflecting fixed wall, for instance at a plane $x = 0$, we have $B = A$, which, according to Eq. (4.135), means that $J = 0$, because the incident power is cancelled by the reflected power. Imagine that a device at $x = 0$ extracts all the incident wave energy. Then it is necessary for the device to radiate a wave which cancels the otherwise reflected wave. In this situation a net energy transport as given by Eq. (4.136) would result, for $x < 0$.

We shall call the quantity J the *wave-energy transport*. (Note that some authors call this quantity "wave-energy flux" or "wave-power flux". This terminology will be avoided here, because of the confusion resulting from improper discrimination between "flux" and "flux density" in different branches of physics.) An alternative term for J could be *wave-power level*.[7]

4.4.5 Relation Between Energy Transport and Stored Energy

For a progressive, plane, harmonic wave, the energy transport J (energy per unit time and unit width of wave frontage) is given by Eq. (4.136), and the stored energy per unit horizontal area is E as given by Eq. (4.128). We may use this to define an energy transport velocity v_E by

$$J = v_E E. \tag{4.139}$$

Thus

$$v_E = J/E = gD(kh)/(2\omega) \tag{4.140}$$

Comparing with expression (4.110) for the group velocity v_g, we see that $v_E = v_g$. Hence,

$$J = v_g E. \tag{4.141}$$

Note that this simple relationship is valid for a purely progressive plane harmonic wave.

For a wave given by Eq. (4.91) we have

$$J = v_g \langle E \rangle \frac{|A|^2 - |B|^2}{|A|^2 + |B|^2}. \tag{4.142}$$

Thus, in this case measurement of

$$\langle E \rangle = \rho g \langle \eta^2(x, y, t) \rangle \tag{4.143}$$

does not, alone, determine wave-energy transport J. Indiscriminate use of Eq. (4.141) would then result in an overestimation of J.

4.4.6 Momentum Transport and Momentum Density of a Wave

As we have seen, waves are associated with energy transport J and with stored energy density E, which is equally divided between kinetic energy density E_k and potential energy density E_p. With a plane progressive wave $\hat{\eta} = Ae^{-ikx}$, these relations are stated mathematically by Eqs. (4.128), (4.136) and (4.139)–(4.141). Moreover, the wave is associated with a momentum, and hence a mean mass transport. A propagating wave induces a mean drift of water, the so-called Stokes drift (see Newman,[24] p. 251).

The x component of the momentum per unit volume is

$$\rho v_x = \rho \frac{\partial \phi}{\partial x}. \tag{4.144}$$

Per unit area of free water surface the momentum is given by

$$M_x = \int_{-h}^{\eta} \rho v_x \, dz. \tag{4.145}$$

Integration over $-h < z < 0$ (instead of $-h < z < \eta$) would yield a quantity with zero time average because v_x varies sinusoidally with time.

Hence, in lowest-order approximation, the time-average momentum density (per unit area of the free surface) is

$$\overline{M_x} = \overline{\int_0^{\eta} \rho v_x \, dz} = \overline{\eta[\rho v_x]_{z=0}} + \cdots \approx \overline{\eta[\rho v_x]_{z=0}}. \tag{4.146}$$

Thus in a linearised theory, J, E, E_k, E_p and $\overline{M_x}$ are quadratic in the wave amplitude. In analogy with Eq. (2.78) we have, with sinusoidal variation with time,

$$\overline{M_x} = \overline{\eta[\rho v_x]_{z=0}} = \frac{\rho}{2} \mathrm{Re}\{\hat{\eta}^*[\hat{v}_x]_{z=0}\}. \tag{4.147}$$

Let us now consider a plane wave with complex amplitudes of elevation, velocity potential and horizontal fluid velocity given by Eqs. (4.91), (4.92) and (4.94), respectively. Using these equations together with Eq. (4.146) gives

$$\overline{M_x} = \frac{\rho g k}{2\omega} \mathrm{Re}\{(Ae^{-ikx} - Be^{ikx})(A^*e^{ikx} + B^*e^{-ikx})\}$$

$$= \frac{\rho g k}{2\omega} \mathrm{Re}\{|A|^2 - |B|^2 + (AB^*e^{-i2kx} - A^*Be^{i2kx})\}. \tag{4.148}$$

Noting that the last term is purely imaginary, we obtain

$$\overline{M_x} = \frac{\rho g k}{2\omega}(|A|^2 - |B|^2). \tag{4.149}$$

If we compare this result with Eq. (4.135) and use Eqs. (4.107) and (4.110), we find the following relationship between the average momentum density and the wave-energy transport:

$$J = v_p v_g \overline{M}_x. \tag{4.150}$$

For a progressive wave, Eq. (4.141) is applicable and then

$$E = J/v_g = v_p \overline{M}_x. \tag{4.151}$$

Let us next consider the momentum transport through a vertical plane normal to the direction of wave propagation. Per unit width of the wave front we have

$$\mathcal{I}_x = \int_h^0 \overline{(\rho v_x) v_x} \, dz = \int_h^0 \rho \overline{(v_x^2)} \, dz, \tag{4.152}$$

where $\mathcal{I}_x$ denotes momentum transport in the x direction. Note that we have a non-vanishing time average even if we integrate over the interval $-h < z < 0$ only. This expression then represents, to lowest-order approximation, the momentum transport per unit wave frontage. With sinusoidal time variation we have, in analogy with Eq. (2.78),

$$\mathcal{I}_x = \int_h^0 \frac{\rho}{2} \text{Re}\{\hat{v}_x \hat{v}_x^*\} \, dz = \frac{\rho}{2} \int_{-h}^0 |\hat{v}_x|^2 \, dz. \tag{4.153}$$

With the plane wave as given by Eqs. (4.91), (4.92) and (4.94), we have

$$\mathcal{I}_x = \frac{\rho}{2}\left(\frac{gk}{\omega}\right)^2 \int_{-h}^0 e^2(kz) \, dz \, [|A|^2 + |B|^2 - (AB^* e^{-i2kx} + A^* B e^{i2kx})]. \tag{4.154}$$

Using Eq. (4.112) gives

$$\mathcal{I}_x = \frac{\rho g^2 k D(kh)}{4\omega^2} [|A|^2 + |B|^2 - (AB^* e^{-i2kx} + A^* B e^{i2kx})]. \tag{4.155}$$

For a progressive wave ($B = 0$) this simplifies to

$$\mathcal{I}_x = \frac{\rho g^2 k D(kh)}{4\omega^2} |A|^2. \tag{4.156}$$

Comparing this with Eqs. (4.149)–(4.151) gives

$$\mathcal{I}_x = \frac{g D(kh)}{2\omega} \overline{M}_x = v_g \overline{M}_x = \frac{J}{v_p} = \frac{E v_g}{v_p}. \tag{4.157}$$

Thus, for a purely progressive wave, the momentum transport equals the momentum density multiplied by the group velocity. This may be interpreted as follows: The momentum associated with the wave is propagated with a speed which equals the group velocity.

If $AB \neq 0$, the expression for $\mathcal{I}_x$ contains a term which varies sinusoidally with x with wavelength $\lambda/2 = 2\pi/2k = \pi/k$. With a purely standing wave ($|B| = |A|$)

we have $v_x = 0$ and hence $\mathcal{I}_x = 0$ in the antinodes, whereas $\mathcal{I}_x$ and $|\hat{v}_x|$ have their maxima in the nodes.

We shall neglect this sinusoidal variation by considering the average over an x interval which is either very long or alternatively an integral number of $\lambda/2$. Denoting this average by $\langle \mathcal{I}_x \rangle$ we have

$$\langle \mathcal{I}_x \rangle = \frac{\rho g^2 k D(kh)}{4\omega^2} (|A|^2 + |B|^2) = v_g \overline{M}_{x,+} + v_g \overline{M}_{x,-}, \tag{4.158}$$

where we have defined

$$\overline{M}_{x,+} = \frac{\rho g k}{2\omega} |A|^2, \quad \overline{M}_{x,-} = \frac{\rho g k}{2\omega} |B|^2. \tag{4.159}$$

With this definition we may rewrite Eq. (4.149) as

$$\overline{M}_x = \overline{M}_{x,+} - \overline{M}_{x,-}. \tag{4.160}$$

Note the minus sign associated with the wave propagating in the negative x direction in the expression for $\overline{M}_x$. In the expression for $\langle \mathcal{I}_x \rangle$ this minus sign is cancelled by the negative group velocity for this wave.

4.4.7 Drift Forces Caused by the Absorption and Reflection of Wave Energy

If an incident wave $\hat{\eta} = Ae^{-ikx}$ is completely absorbed at the plane $x = 0$, then $\mathcal{I}_x = 0$ for $x > 0$ whereas $\mathcal{I}_x > 0$ for $x < 0$. Thus the momentum transport is stopped at $x = 0$. In a time Δt a momentum $\overline{M}_x \Delta x = F_d' \Delta t$ disappears. Here $\Delta x = v_g \Delta t$ is the distance of momentum transport during the time Δt. A force F_d' per unit width of wave front is required to stop the momentum transport. This (time-average) drift force is given by

$$F_d' = \frac{\Delta x}{\Delta t} \overline{M}_x = v_g \overline{M}_x = \mathcal{I}_x = \frac{\rho g^2 k D(kh)}{4\omega^2} |A|^2. \tag{4.161}$$

Using the dispersion relationship (4.57), we find that this gives

$$F_d' = \frac{\rho g}{4} \frac{D(kh)}{\tanh(kh)} |A|^2. \tag{4.162}$$

Note that for deep water [see Eqs. (4.104), (4.107) and (4.110)], with

$$F_d' = (\rho g/4)|A|^2, \tag{4.163}$$

the ratio between F_d' and $|A|^2$ is independent of frequency.

The drift force must be taken up by the anchor or mooring system of the wave absorber. Note that the drift force is a time-average force usually much smaller than the amplitude of the oscillatory force. With complete absorption of a wave with amplitude $|A| = 1$ m on deep water, the drift force is

$$F_d' = [(1020 \times 9.81)/4]1^2 = 2500 \text{ N/m}. \tag{4.164}$$

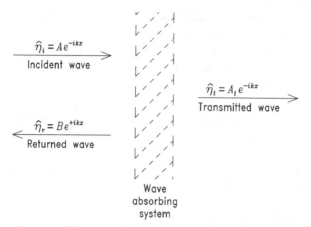

Figure 4.9: Wave which is incident upon a wave-absorbing system is partly absorbed, partly transmitted downstream and partly reflected and/or radiated upstream.

The drift force is twice as large if the wave is completely reflected instead of completely absorbed, because of the equally large but oppositely directed momentum transport that is due to the reflected wave.

A two-dimensional wave-energy absorber (or an infinitely long array of equally interspaced three-dimensional wave absorbers) may partly reflect and partly transmit the incident wave, as indicated in Figure 4.9. In this case the drift force is given by

$$F_d' = \langle \mathcal{I}_x \rangle_{\text{lhs}} - \langle \mathcal{I}_x \rangle_{\text{rhs}}, \tag{4.165}$$

where the two terms represent the average momentum transport on the left-hand side (lhs) and right-hand side (rhs) of the absorbing system indicated in Figure 4.9.

Assuming, for mathematical simplicity, that we have deep water ($kh \gg 1$), the application of Eq. (4.163) gives for this case

$$F_d' = \frac{\rho g}{4} \left(|A|^2 + |B|^2 - |A_t|^2 \right), \tag{4.166}$$

where A_t denotes the complex amplitude of the transmitted wave. An expression[25] also exists for finite water depth, corresponding to Eq. (4.166) for infinite depth. Applying Eqs. (4.135) and (4.136) for the wave-energy transport, remembering that $D(kh) = 1$ for deep water and using the principle of conservation of energy, we obtain from Eq. (4.166)

$$\begin{aligned} F_d' &= \frac{\omega}{g} \frac{\rho g^2}{4\omega} (|A|^2 + |B|^2 - |A_t|^2) = \frac{\omega}{g}(J_i + J_r - J_t) \\ &= \frac{\omega}{g}(J_i - J_r - J_t + 2J_r) = \frac{\omega}{g}(P' + 2J_r), \end{aligned} \tag{4.167}$$

where P' is the absorbed power per unit width. Further, J_i, J_r and J_t are the wave-energy transport – see Eq. (4.136) – for the incident, reflected and transmitted waves, respectively. Assuming that J_r is positive, we have $F_d' > (\omega/g)P' = F_{d,\text{min}}'$.

As an example, let us consider absorption of 0.5 MW from a wave of period $T = 10$ s. This is associated with a drift force

$$F_d > F_{d,\min} = \frac{2\pi}{10}\frac{1}{9.81} 5 \times 10^5 = 32 \times (10^3 \text{ N} \cong 3.2 \text{ tons}). \tag{4.168}$$

4.5 Real Ocean Waves

Harmonic waves were the main subject of the previous section. Such waves are also called "regular waves", as opposed to the real "irregular" waves of the ocean. The swells, that is, travelling waves which have left their regions of generation by winds, are closer to harmonic waves than the more irregular, locally wind-generated sea waves.

Irregular waves are of a stochastic nature, and they may be considered, at least approximately, as a superposition of many different frequencies. Usually only statistical information is, at best, available for the amplitude, the phase and the direction of propagation for each individual harmonic wave. Most of the energy content of ocean waves is associated with waves of periods in the interval from 5 s to 15 s. According to Eq. (4.103) the corresponding interval of deep-water wavelengths is from 40 m to 350 m.

The present section relates wave spectra to superposition of plane waves on deep water, or otherwise water of finite, but constant, depth. For more thorough studies, readers may consult other literature.[26,27]

For a progressive harmonic plane wave the stored energy per unit surface is

$$E = E_k + E_p = 2E_k = 2E_p = \frac{\rho g}{2}|\hat{\eta}_f|^2 = \rho g \overline{\eta^2(x, y, t)}, \tag{4.169}$$

according to linear wave theory. See Eqs. (4.89), (4.119) and (4.128). Note that for a harmonic plane wave, $\overline{\eta^2}$ is independent of x and y. Still assuming linear theory, the superposition principle is applicable, which means that a real sea state may be described in terms of components of harmonic waves. Correspondingly we may generalise and rewrite Eq. (4.118) as

$$E = 2E_p = \rho g \overline{\eta^2(x, y, t)} = \rho g \int_0^\infty S(f)\, df, \tag{4.170}$$

where

$$\overline{\eta^2(x, y, t)} = \int_0^\infty S(f)\, df \equiv \frac{H_s^2}{16}. \tag{4.171}$$

Here $S(f)$ is called the energy spectrum, or simply the spectrum. The so-called *significant wave height* H_s is defined as four times the square root of the integral of the spectrum. Sometimes the wave spectrum is defined in terms of the angular frequency $\omega = 2\pi f$. The corresponding spectrum may be written $S_\omega(\omega)$, where

$$\int_0^\infty S_\omega(\omega)\, d\omega = \int_0^\infty S(f)\, df. \tag{4.172}$$

Thus

$$S(f) = S_f(f) = 2\pi S_\omega(2\pi f) = 2\pi S_\omega(\omega), \tag{4.173}$$

where the units of S and S_ω are m^2/Hz and m^{2}s/rad, respectively.

In a more precise spectral description the direction of propagation and the phase of each harmonic component should also be taken into account. A harmonic plane wave with direction of incidence given by β (the angle between the propagation direction and the x axis) has an elevation given by

$$\begin{aligned}
\eta(x, y, t) &= \mathrm{Re}\{\eta_i \exp\{i(\omega t - kx\cos\beta - ky\sin\beta)\}\} \\
&= \tfrac{1}{2}\eta_i \exp\{i(\omega t - kx\cos\beta - ky\sin\beta)\} + \text{c.c.}
\end{aligned} \tag{4.174}$$

where c.c. means a term which is the complex conjugate of the preceding term [cf. Eqs. (4.102), (2.34) and (2.35)]. Note that the complex parameter $\eta_i = \eta_i(\omega)$ contains information on the phase as well as of the amplitude $|\eta_i|$. Consider a general sea state $\eta(x, y, t)$ decomposed into harmonic components

$$\begin{aligned}
\eta(x, y, t) &= \sum_{\omega_m > 0}\sum_{\beta_n} \frac{\eta_i(\omega_m, \beta_n)}{2} \exp\{i(\omega_m t - k_m x\cos\beta_n - k_m y\sin\beta_n)\} + \text{c.c.} \\
&= \mathrm{Re}\left\{\sum_{\omega_m > 0}\sum_{\beta_n} \eta_i(\omega_m, \beta_n)\exp\{i(\omega_m t - k_m x\cos\beta_n - k_m y\sin\beta_n)\}\right\},
\end{aligned} \tag{4.175}$$

where

$$\omega_m^2 = gk_m\tanh(k_m h) \tag{4.176}$$

is the general dispersion relationship, which simplifies to

$$\omega_m^2 = gk_m \tag{4.177}$$

in the case of deep water. Although any frequency ω_m ($\omega_m > 0$) and any angle of incidence β_n ($-\pi < \beta_n \leq \pi$) are, in principle, possible in this summation, in most cases only certain finite intervals for ω_m and β_n contribute significantly to the complete irregular wave. For convenience, we shall assume that in the sums over ω_m and β_n we have $\omega_{m+1} - \omega_m = \Delta\omega$ independent of m, and $\beta_{n+1} - \beta_n = \Delta\beta$ independent of n.

Alternatively, we may reformulate the sums as integrals. For convenience, we admit both negative and positive values for the integration variable ω. Thus we have

$$\eta(x, y, t) = \int_{-\infty}^{\infty} d\omega \int_{-\pi}^{\pi} \frac{A(\omega, \beta)}{2} \exp\{i(\omega t - kx\cos\beta - ky\sin\beta)\}\, d\beta, \tag{4.178}$$

where

$$\eta_i(\omega_m, \beta_n) = A(\omega_m, \beta_n)\Delta\omega\Delta\beta,$$
$$\eta_i^*(\omega_m, \beta_n) = A(-\omega_m, \beta_n)\Delta\omega\Delta\beta. \tag{4.179}$$

If we sum over negative as well as positive frequencies, we have for the elevation

$$\eta(x, y, t) = \sum_{\omega_m}\sum_{\beta_n}\frac{1}{2}A(\omega_m, \beta_n)\exp\{i(\omega_m t - k_m r_n)\}\Delta\omega\Delta\beta, \tag{4.180}$$

with

$$r_n = x\cos\beta_n + y\sin\beta_n. \tag{4.181}$$

The square of the elevation is

$$\eta^2(x, y, t) = \frac{1}{4}\sum_{\omega_m}\sum_{\omega_{m'}}\sum_{\beta_n}\sum_{\beta_{n'}}(\Delta\omega)^2(\Delta\beta)^2 A(\omega_m, \beta_n) A(\omega_{m'}, \beta_{n'})$$
$$\times \exp\{i[(\omega_m + \omega_{m'})t - k_m r_n - k_{m'} r_{n'}]\}. \tag{4.182}$$

Summing over $\omega_{m'}$, we find that most of the terms have a vanishing time average. The only exceptions are the terms for which $\omega_{m'} = -\omega_m$. Thus the time average of the square of the elevation is

$$\overline{\eta^2(x, y, t)} = \frac{1}{4}\sum_{\omega_m}\sum_{\beta_n}\sum_{\beta_{n'}}(\Delta\omega)^2(\Delta\beta)^2 A(\omega_m, \beta_n) A(-\omega_m, \beta_{n'})$$
$$\times \exp\{-ik_m(r_n - r_{n'})\}, \tag{4.183}$$

because

$$k_{m'} = k(\omega_{m'}) = k(-\omega_m) = -k(\omega_m) = -k_m. \tag{4.184}$$

Observe that we require k_m to have the same sign as ω_m in order to ensure that the wave component propagates in the direction given by β_m and not in the opposite direction.

Next we also average with respect to the horizontal coordinates x and y. Then the non-vanishing contributions to the variance $\overline{\eta^2}$ of the elevation result from those terms for which $r_{n'} = r_n$; that is, $\beta_{n'} = \beta_n$. Moreover, from the theory of Fourier transform of real functions of the time, we know that

$$A(-\omega_m, \beta_n) = A^*(\omega_m, \beta_n), \tag{4.185}$$

where the asterisk denotes the complex conjugate. Hence

$$\langle\eta^2\rangle = \sum_{\omega_m}\sum_{\beta_n}\left|\frac{1}{2}A(\omega_m, \beta_n)\right|^2 (\Delta\omega)(\Delta\beta)(\Delta\omega)(\Delta\beta)$$

$$= \sum_{\omega_m>0}\sum_{\beta_n}\frac{1}{2}|A(\omega_m, \beta_n)|^2 (\Delta\omega)(\Delta\beta)(\Delta\omega)(\Delta\beta). \tag{4.186}$$

In the last step here it was summed over only positive frequencies.

Introducing the direction-resolved energy spectrum

$$s(f, \beta) = 2\pi s_\omega(\omega, \beta) \tag{4.187}$$

where

$$s_\omega(\omega, \beta) = \tfrac{1}{2}|A(\omega_m, \beta_n)|^2(\Delta\omega)(\Delta\beta), \tag{4.188}$$

and replacing the sums by integrals, we rewrite the wave-elevation variance as

$$
\begin{aligned}
\langle \eta^2 \rangle &= \int_{-\infty}^{\infty} d\omega \int_{-\pi}^{\pi} d\beta \frac{1}{2} s_\omega(\omega, \beta) \\
&= \int_{0}^{\infty} d\omega \int_{-\pi}^{\pi} d\beta s_\omega(\omega, \beta) = \int_{0}^{\infty} \int_{-\pi}^{\pi} s(f, \beta)\, df\, d\beta.
\end{aligned} \tag{4.189}
$$

Note that $s_\omega(\omega, \beta)$ is an even function of ω; that is, $s_\omega(-\omega, \beta) = s_\omega(\omega, \beta)$.

The (direction-integrated) energy spectrum is

$$S(f) = \int_{-\pi}^{\pi} s(f, \beta)\, d\beta. \tag{4.190}$$

In analogy with Eq. (4.170) the average stored energy per unit horizontal surface is

$$\langle E \rangle = \rho g \langle \eta^2 \rangle = \rho g \int_{0}^{\infty} df \int_{-\pi}^{\pi} d\beta\, s(f, \beta). \tag{4.191}$$

From wave measurements, much statistical information has been obtained for the spectral functions $S(f)$ and $s(f, \beta)$ from various ocean regions.

It has been found that wave spectra have some general characteristics which may be approximately described by semiempirical mathematical relations. The most well-known functional relation is the Pierson-Moskowitz (PM) spectrum,

$$S(f) = (A/f^5)\exp\{-B/f^4\} \tag{4.192}$$

(for $f > 0$). Various parametrisations have been proposed for A and B. One possible variant is

$$A = BH_s^2/4, \quad B = (5/4)f_p^4. \tag{4.193}$$

Here H_s is the "significant wave height" and f_p is the "peak frequency" (the frequency for which S has its maximum). The corresponding wave period $T_p = 1/f_p$ is called the "peak period".

For the direction-resolved spectrum we may write

$$s(f, \beta) = D(\beta, f)S(f) \tag{4.194}$$

where

$$\int_{-\pi}^{\pi} D(\beta, f)\, d\beta = 1. \tag{4.195}$$

One proposal for the directional distribution (neglecting its possible frequency dependence) is

$$D(\beta) = \begin{cases} (2/\pi)\cos^2(\beta - \beta_0) & \text{for } |\beta - \beta_0| < \pi/2 \\ 0 & \text{otherwise} \end{cases} \qquad (4.196)$$

where β_0 is the predominant angle of incidence.

The real-valued, non-negative function $s(f, \beta)$ provides information on the modulus of the complex-valued function $A(\omega, \beta)$. From Eq. (4.188) we have

$$|A(\omega, \beta)| = \left[\frac{2s_\omega(\omega, \beta)}{\Delta\omega\Delta\beta}\right]^{1/2} = \frac{1}{2\pi}\left[\frac{2s(f, \beta)}{\Delta f\Delta\beta}\right]^{1/2}. \qquad (4.197)$$

That is, only the amplitudes of the harmonic waves are given, while the spectrum provides no information on the phase of the individual harmonic waves. We may then write

$$A(\omega, \beta) = |A(\omega, \beta)|e^{i\psi(\omega,\beta)}, \qquad (4.198)$$

where the unknown phase ψ is assumed to be uniformly statistically distributed in the interval $(-\pi, \pi)$. That is, its distribution function is

$$f(\psi) = \begin{cases} (1/2\pi) & \text{when } -\pi < \psi < \pi \\ 0 & \text{otherwise} \end{cases}. \qquad (4.199)$$

When a real irregular wave is simulated, the amplitudes $|A(\omega, \beta)|$ are selected according to the appropriate spectrum $s(f, \beta)$, whereas the phases $\psi(f, \beta)$ are chosen as random numbers in the interval $-\pi < \psi < \pi$.

4.6 Circular Waves

Using the method of separation of variables in a Cartesian (x, y, z) coordinate system, we have studied plane-wave solutions of the Laplace equation (4.35) with the free-surface and sea-bed homogeneous boundary conditions (4.43) and (4.42), respectively. Explicit expressions for plane-wave solutions are given, for instance, by Eq. (4.83) or by Eq. (4.92) with Eq. (4.102). These are examples of plane waves propagating along the x axis, or in a direction which differs by an angle β from the x axis, respectively. Another plane-wave solution which includes evanescent plane waves is given by Eq. (4.115).

In the following paragraphs let us discuss solutions of the problem by using the method of separation of variables in a cylindrical coordinate system (r, θ, z). The form of the Helmholtz equation (4.48) or (4.61) will be affected. The differential equation (4.47) or (4.62) with boundary conditions (4.63) and (4.64) remains, however, unchanged. Hence we may still utilise the functions $e(kz)$ defined by Eq. (4.54) and the orthogonal set of functions $\{Z_n(z)\}$ which is defined by Eq. (4.73), and which is a complete set if $n = 0, 1, 2, \ldots$. In the following we shall concentrate on the propagating wave (that is, $n = 0$) and leave the evanescent waves aside.

The coordinates in the horizontal plane are related by $(x, y) = (r \cos\theta, r \sin\theta)$, and we replace $H(x, y)$ by $H(r, \theta)$. The Helmholtz equation (4.48) now becomes

$$-k^2 H = \nabla_H^2 H = \frac{\partial^2 H}{\partial x^2} + \frac{\partial^2 H}{\partial y^2} = \frac{\partial^2 H}{\partial r^2} + \frac{1}{r}\frac{\partial H}{\partial r} + \frac{1}{r^2}\frac{\partial^2 H}{\partial \theta^2}. \tag{4.200}$$

Once more we use the method of separation of variables to obtain a particular solution

$$H(r, \theta) = R(r)\Theta(\theta). \tag{4.201}$$

Inserting into Eq. (4.200) and dividing by H/r^2 gives

$$\frac{r^2}{R}\left(\frac{d^2 R}{dr^2} + \frac{1}{r}\frac{dR}{dr} + k^2 R\right) = -\frac{1}{\Theta}\frac{d^2\Theta}{d\theta^2} = m^2, \tag{4.202}$$

where m is a constant because the left-hand side is a function of r only, whereas the right-hand side is independent of r. Hence,

$$\frac{d^2\Theta}{d\theta^2} = \Theta''(\theta) = -m^2\Theta, \tag{4.203}$$

$$\frac{d^2 R}{dr^2} + \frac{1}{r}\frac{dR}{dr} + \left(k^2 - \frac{m^2}{r^2}\right)R = 0. \tag{4.204}$$

The general solution of Eq. (4.203) may be written as

$$\Theta = \Theta_m = c_c \cos(m\theta) + c_s \sin(m\theta), \tag{4.205}$$

where c_c and c_s are integration constants. We require an unambiguous solution; that is,

$$\Theta_m(\theta + 2\pi) = \Theta_m(\theta). \tag{4.206}$$

This requires m to be an integer:

$$m = 0, 1, 2, 3, \ldots. \tag{4.207}$$

Because there is also an integration constant in the solution for factor $R(r)$, we can set

$$\Theta_{\max} = 1 \tag{4.208}$$

such that

$$c_c^2 + c_s^2 = 1. \tag{4.209}$$

Then we may write

$$\Theta_m(\theta) = \cos(m\theta + \psi_m)$$
$$= \cos(\psi_m)\cos(m\theta) - \sin(\psi_m)\sin(m\theta), \tag{4.210}$$

where ψ_m is an integration constant which is not currently specified.

Before considering the general solution of the Bessel differential equation (4.204), we note that in an ideal fluid which is non-viscous, and hence loss free, the radiated power P_r which passes an envisaged cylinder with large radius r has to be independent of r; that is,

$$P_r = J2\pi r \tag{4.211}$$

is independent of r. Hence, the energy transport J is inversely proportional to the distance, r. For large r the circular wave is approximately plane (the curvature of the wave front is of little importance). Then Eq. (4.136) for J applies. This shows that the wave-elevation amplitude $|\hat{\eta}|$ is proportional to $r^{-1/2}$. Hence $\hat{\phi}$ is proportional to $r^{-1/2}$ [because $\hat{\phi} = (-g/i\omega)e(kz)\hat{\eta}$ according to Eq. (4.92)] and thus $|R|$ is proportional to $r^{-1/2}$. However, $R(r)$ also has, for an outgoing wave, a phase that varies with r as $-kr$. For large r we then expect that

$$\hat{\phi} \propto r^{-1/2} e^{-ikr} e(kz) \tag{4.212}$$

where the coefficient of proportionality may be complex.

Bessel's differential equation (4.204) has two linearly independent solutions $J_m(kr)$ and $N_m(kr)$, which are Bessel functions of order m. $J_m(kr)$ is of the first kind and $N_m(kr)$ is of the second kind. $J_m(kr)$ is finite for $r = 0$, whereas $N_m(kr) \to \infty$ when $r \to 0$. Other singularities do not exist for $r \neq \infty$. For large r, $J_m(kr)$ and $N_m(kr)$ approximate cosine and sine functions multiplied by $r^{-1/2}$. In analogy with Euler's formulas,

$$e^{ikr} = \cos(kr) + i\sin(kr), \tag{4.213}$$
$$e^{-ikr} = \cos(kr) - i\sin(kr), \tag{4.214}$$

the Hankel functions of order m and of first kind and second kind are defined by

$$H_m^{(1)}(kr) = J_m(kr) + i N_m(kr), \tag{4.215}$$
$$H_m^{(2)}(kr) = J_m(kr) - i N_m(kr), \tag{4.216}$$

respectively. Note that

$$H_m^{(1)}(kr) = \left[H_m^{(2)}(kr) \right]^*. \tag{4.217}$$

Asymptotically we have [28]

$$H_m^{(2)}(kr) \approx \sqrt{\frac{2}{\pi kr}} \exp\left\{ -i\left(kr - \frac{m\pi}{2} - \frac{\pi}{4} \right) \right\} \left(1 + \mathcal{O}\left\{ \frac{1}{kr} \right\} \right) \tag{4.218}$$

as $kr \to \infty$. Hence the solution of the differential equation (4.204) can be written as

$$R(r) = a_m H_m^{(2)}(kr) + b_m H_m^{(1)}(kr), \tag{4.219}$$

where a_m and b_m are integration constants. From the asymptotic expression (4.218) we see that $H_m^{(2)}(kr)$ represents an outgoing (divergent) wave and from Eq. (4.217) that $H_m^{(1)}(kr)$ represents an incoming (convergent) wave. (It might here be compared with the two terms in each of Eqs. (4.83), (4.84) and (4.87) for the case of propagating plane waves.)

In Section 4.1 it was mentioned that the boundary conditions might have to be supplemented by a *radiation condition*. Let us now introduce the radiation condition of outgoing wave at infinity, which means we have to set $b_m = 0$. Then we have from Eqs. (4.53), (4.201) and (4.219) the solution

$$\hat{\phi} = a_m H_m^{(2)}(kr)\Theta_m(\theta)e(kz) \tag{4.220}$$

for the complex amplitude of the velocity potential. A sum of similar solutions

$$\hat{\phi} = \sum_{m=0}^{\infty} a_m H_m^{(2)}(kr)\Theta_m(\theta)e(kz) \tag{4.221}$$

is also a solution which satisfies the radiation condition, the dispersion equation (4.57), and the Laplace equation (4.35). This solution satisfies the homogeneous boundary conditions (4.42) and (4.43) at the sea bed ($z = -h$) and at the free water surface ($z = 0$).

According to the asymptotic expression (4.218), the particular solution (4.221) [of the partial differential equation (4.200)] is, in the asymptotic limit for $kr \gg 1$,

$$\hat{\phi} \approx A(\theta)e(kz)(kr)^{-1/2}e^{-ikr} \quad \text{for } kr \gg 1, \tag{4.222}$$

where

$$A(\theta) = \sqrt{\frac{2}{\pi}} \sum_{m=0}^{\infty} a_m \Theta_m(\theta)\exp\left\{i\left(\frac{m\pi}{2} + \frac{\pi}{4}\right)\right\} \tag{4.223}$$

is the so-called far-field coefficient of the outgoing wave. If we insert for $\Theta_m(\theta)$ from Eq. (4.210) we may note that Eq. (4.223) is a Fourier-series representation of the far-field coefficient.

The outgoing wave could originate from a radiation source in or close to the origin $r = 0$. The z variation according to factor $e(kz)$ as in the particular solution (4.221) can satisfy the inhomogeneous boundary conditions, such as Eq. (4.17) or (4.28), at the radiation source only in simple particular cases. A more general solution which satisfies the Laplace equation (4.35), the boundary conditions and the radiation condition as $r \to \infty$ is the following:

$$\hat{\phi} = \hat{\phi}_l(r, \theta, z) + A(\theta)e(kz)(kr)^{-1/2}e^{-ikr}. \tag{4.224}$$

Here the last term represents the so-called *far field* while the first term $\hat{\phi}_l$ represents the *near field*, a local velocity potential which connects the far field to the radiation source in such a way that the inhomogeneous boundary conditions at the radiation source are satisfied. The term $\hat{\phi}_l(r, \theta, z)$ may contain terms corresponding to evanescent waves. For large r such terms decay exponentially as $\exp(-m_1 r)$ (or faster) with increasing r, where m_1 is the smallest positive solution of Eq. (4.74), that is, $\pi/(2h) < m_1 < \pi/h$. (See Problems 4.3 and 5.6.) It can be shown that, in general, the local potential $\hat{\phi}_l$ decreases with distance r at least as fast as $1/r$ when $r \to \infty$ (cf. Wehausen and Laitone,[29] pp. 475–478).

The wave elevation corresponding to Eq. (4.224) is given by

$$\hat{\eta} = \hat{\eta}(r,\theta) = -\frac{i\omega}{g}\hat{\phi}_l(r,\theta,0) - \frac{i\omega}{g}A(\theta)(kr)^{-1/2}e^{-ikr},$$ (4.225)

which is obtained by using Eq. (4.41). For large values of kr, where $\hat{\phi}_l$ is negligible, $(kr)^{-1/2}$ varies relatively little over one wavelength, which means that the curvature of the wave front is of minor importance. Then we may use the energy-transport formula (4.136) for plane waves and there replace the wave-elevation amplitude $|A|$ by $|(\omega/g)A(\theta)(kr)^{-1/2}|$. The radiated energy transport per unit wave frontage then becomes

$$J(r,\theta) = \frac{1}{kr}\frac{D(kh)\rho g^2}{4\omega}\frac{\omega^2}{g^2}|A(\theta)|^2 = \frac{\omega\rho\, D(kh)}{4kr}|A(\theta)|^2.$$ (4.226)

The radiated power is

$$P_r = \int_0^{2\pi} J(r,\theta)r\, d\theta = \frac{\omega\rho\, D(kh)}{4k}\int_0^{2\pi}|A(\theta)|^2\, d\theta.$$ (4.227)

For a circularly symmetric radiation source, the far-field coefficient $A(\theta)$ must be independent of θ. Hence according to Eq. (4.223), we have

$$A(\theta) = \sqrt{2/\pi}\, a_0 e^{i\pi/4}.$$ (4.228)

Hence, the power associated with the outgoing wave is

$$P_r = \frac{\omega\rho\, D(kh)}{k}|a_0|^2 = \frac{\pi\omega\rho\, D(kh)}{2k}|A|^2,$$ (4.229)

where A is now the θ independent far-field coefficient. This is a relation we may use in combination with Eq. (3.29) to obtain an expression for the radiation resistance for an oscillating body which generates an axisymmetric (circularly symmetric) wave. See also Subsection 5.5.3 and in particular Eq. (5.181).

4.7 A Useful Integral Based on Green's Theorem

This section is a general mathematical preparation for further studies in the following sections and chapters. Let us start this section by stating that Green's theorem

$$\oiint \left(\varphi_i \frac{\partial \varphi_j}{\partial n} - \varphi_j \frac{\partial \varphi_i}{\partial n} \right) dS = 0$$ (4.230)

applies to two arbitrary differentiable functions φ_i and φ_j, both of which satisfy the (three-dimensional) Helmholtz equation

$$\nabla^2 \varphi = \lambda\varphi$$ (4.231)

within the volume V contained inside the closed surface of integration. In the integrand, $\partial/\partial n = \vec{n}\cdot\nabla$ is the normal component of the gradient on the surface of integration. The "eigenvalue" λ is a constant, whereas φ_i and φ_j depend on the spatial coordinates. In particular, Green's theorem is applicable to two functions ϕ_i and ϕ_j which satisfy the Laplace equation (corresponding to $\lambda = 0$).

Green's theorem follows from Gauss' divergence theorem

$$\iiint_V \nabla \cdot \vec{A} \, dV = \oiint A_n \, dS = \oiint \vec{n} \cdot \vec{A} \, dS \tag{4.232}$$

if we consider

$$\vec{A} = \varphi_i \nabla \varphi_j - \varphi_j \nabla \varphi_i. \tag{4.233}$$

Then

$$A_n = \varphi_i \frac{\partial \varphi_j}{\partial n} - \varphi_j \frac{\partial \varphi_i}{\partial n} \tag{4.234}$$

and

$$\nabla \cdot \vec{A} = \varphi_i \nabla^2 \varphi_j + \nabla \varphi_i \cdot \nabla \varphi_j - \varphi_j \nabla^2 \varphi_i - \nabla \varphi_j \cdot \nabla \varphi_i$$
$$= \varphi_i \nabla^2 \varphi_j - \varphi_j \nabla^2 \varphi_i. \tag{4.235}$$

In view of the Helmholtz equation (4.231) we have

$$\nabla \cdot \vec{A} = \varphi_i \lambda \varphi_j - \varphi_j \lambda \varphi_i = 0, \tag{4.236}$$

from which Green's theorem follows as a corollary to Gauss' theorem.

Let us consider a finite region of the sea containing one or more structures, fixed or oscillating, as indicated in Figure 4.10. It is assumed that outside this region, the sea is unbounded in all horizontal directions. Moreover, the water is deep there, or otherwise the water has a constant depth h, outside the mentioned finite region. The structures, some of which may contain air chambers above OWCs, will diffract incoming waves. The phenomenon of diffraction occurs when the incident plane wave alone, such as given by Eq. (4.92) for example, violates the boundary conditions (4.36) and (4.37) on the structures. Furthermore, waves may be generated by oscillating bodies or by oscillating air pressures

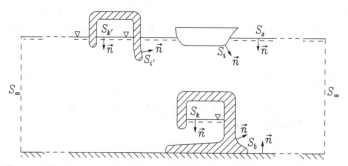

Figure 4.10: System of bodies and chambers for air-pressure distributions (OWCs) contained within an imaginary cylindrical control surface S_∞. Wetted surfaces of oscillating bodies are indicated by $S_{i'}$ and S_i, whereas S_k and $S_{k'}$ denote internal water surfaces. Fixed surfaces, including the sea bed, are given as S_b and S_0 denotes the external free water surface. The arrows indicate unit normals pointing into the fluid region.

above the water, such as in the entrapped air within the chambers indicated in Figure 4.10.

Let us next apply Green's theorem to the fluid region shown in Figure 4.10. The fluid region is contained inside a closed surface composed of the following surfaces.

$S = \sum_{k=1}^{N_k} S_k + \sum_{i=1}^{N_i} S_i$, which is the sum (union) of all N_k internal water surfaces S_k above the OWCs and of the wet surface S_i of all N_i bodies,

S_0, which is the free water surface (at $z = 0$) external to bodies and chamber structures for OWCs,

S_b, which is the sum (union) of all wet surfaces of fixed rigid structures, including fixed OWC chamber structures and the sea bed (which is the plane $z = -h$ in case the sea bed is horizontal), and

S_∞, which is a "control" surface, an envisaged vertical cylinder. If the cylinder base is circular we may denote its radius by r. In many cases we consider the limit $r \to \infty$.

If the extension of S_0 and of the fluid region is not infinite, that is, if the fluid is contained in a finite basin, the surfaces S_0 and S_b intersect along a closed curve, and S_∞ does not come into play. We may also consider a case in which an infinite coast line intersects the chosen control surface S_∞.

Let ϕ_i and ϕ_j be two arbitrary functions which satisfy the Laplace equation

$$\nabla^2 \phi_{i,j} = 0 \tag{4.237}$$

and the homogeneous boundary conditions

$$\frac{\partial}{\partial n}\phi_{i,j} = 0 \quad \text{on } S_b \tag{4.238}$$

and

$$\left(\omega^2 - g\frac{\partial}{\partial z}\right)\phi_{i,j} = \left(\omega^2 + g\frac{\partial}{\partial n}\right)\phi_{i,j} = 0 \quad \text{on } S_0. \tag{4.239}$$

We define the very useful integral

$$I(\phi_i, \phi_j) \equiv \iint_S \left(\phi_i \frac{\partial \phi_j}{\partial n} - \frac{\partial \phi_i}{\partial n}\phi_j\right) dS. \tag{4.240}$$

Next, we shall show that, instead of integrating over S, the totality of wave-generating surfaces, we may integrate over S_∞, a "control" surface in the far-field region of waves radiated from S. Note that the integrand in Eq. (4.240) vanishes on S_b and on S_0 because, firstly $(\partial \phi_{i,j}/\partial n) = 0$ on S_b and secondly

$$\phi_i \frac{\partial \phi_j}{\partial n} - \frac{\partial \phi_i}{\partial n}\phi_j = -\phi_i \frac{\omega^2}{g}\phi_j + \frac{\omega^2}{g}\phi_i\phi_j = 0 \quad \text{on } S_0. \tag{4.241}$$

Hence, it follows from Green's theorem (4.230) that if the fluid is contained within a finite basin, then the integral (4.240) vanishes. Otherwise, we have

$$I(\phi_i, \phi_j) = -\iint_{S_\infty} \left(\phi_i \frac{\partial \phi_j}{\partial n} - \frac{\partial \phi_i}{\partial n}\phi_j \right) dS \tag{4.242}$$

or

$$I(\phi_i, \phi_j) = \iint_{S_\infty} \left(\phi_i \frac{\partial \phi_j}{\partial r} - \frac{\partial \phi_i}{\partial r}\phi_j \right) dS. \tag{4.243}$$

Note that if control surface S_∞ is a cylinder which is not circular, this formula still applies provided $\partial/\partial r$ is interpreted as the normal component of the gradient on S_∞ pointing in the outward direction. Note that Eq. (4.240) is a definition, whereas Eq. (4.243) is a theorem.

Let us make the following comments.

(a) ϕ_i and ϕ_j may represent velocity potentials of waves generated by the oscillating rigid-body surfaces S_i or oscillating internal water surfaces S_k. Alternatively, they may represent other kinds of waves. The constraints on ϕ_i and ϕ_j are that they have to satisfy the homogeneous boundary conditions (4.238) and (4.239) on S_b and S_0.

(b) Further ϕ_i and/or ϕ_j may be matrices/matrix provided they/it (i.e., all matrix elements) satisfy the mentioned homogeneous conditions. If both ϕ_i and ϕ_j are matrices which do not commute, their order of appearance in products is important.

(c) We may (for real ω) replace ϕ_i and/or ϕ_j by their complex conjugate, because ϕ_i^* and ϕ_j^* too satisfy the Laplace equation and the homogeneous boundary conditions on S_b and S_0.

(d) If at least one of ϕ_i and ϕ_j is a scalar function, or if they otherwise commute, we have

$$I(\phi_j, \phi_i) = -I(\phi_i, \phi_j). \tag{4.244}$$

(e) Observe that we could include some finite part of surface S_b to belong to surface S. For instance, the surface of a fixed piece of rock on the sea bed (or of the fixed sea-bed-mounted OWC structure shown in Figure 4.10) could be included in the surface S as a wet surface oscillating with zero amplitude.

(f) The definition (4.240) and the theorem (4.243) are also valid if the boundary condition (4.238) on S_b is replaced by

$$\frac{\partial}{\partial n}\phi_{i,j} + C_n\phi_{i,j} = 0, \tag{4.245}$$

where C_n is a complex constant or a complex function given along the surface S_b. For this to be true when C_n is not real, the two functions ϕ_i and ϕ_j have, however, to satisfy the same radiation condition, that means it would not be allowable to replace just one of the two functions by its complex conjugate.

We next consider the integral (4.243) when ϕ_i and ϕ_j in addition to satisfying the homogeneous boundary conditions (4.238) and (4.239) also satisfy a radiation condition as $r \to \infty$.

Waves satisfying the radiation condition of outgoing waves at infinite distance from the wave source, such as, for example, diffracted waves or radiated waves, have – in the three-dimensional case – an asymptotic expression of the type

$$\phi = \psi \sim A(\theta)e(kz)(kr)^{-1/2}e^{-ikr} \tag{4.246}$$

as $kr \to \infty$ [see Eqs. (4.222) and (4.224)]. For a while let us use the symbol ψ to denote the velocity potential of a general wave which satisfies the radiation condition. We have

$$\frac{\partial \psi}{\partial r} \sim -ik\psi \tag{4.247}$$

when we include only the dominating term of the asymptotic expansion in the far-field region. Thus, for the two waves ψ_i and ψ_j we have

$$\lim_{r \to \infty} \iint_{S_\infty} \left(\psi_i \frac{\partial \psi_j}{\partial r} - \frac{\partial \psi_i}{\partial r} \psi_j \right) dS = 0. \tag{4.248}$$

Hence, according to Eq. (4.243),

$$I(\psi_i, \psi_j) = 0. \tag{4.249}$$

Note that also

$$I(\psi_i^*, \psi_j^*) = 0 \tag{4.250}$$

because

$$\frac{\partial \psi^*}{\partial r} \sim ik\psi^*. \tag{4.251}$$

However, because ψ_i and ψ_j^* satisfy opposite radiation conditions,

$$I(\psi_i, \psi_j^*) \neq 0. \tag{4.252}$$

We have

$$I(\psi_i, \psi_j^*) = \lim_{r \to \infty} \iint_{S_\infty} A_i(\theta)A_j^*(\theta)e^2(kz)\frac{2ik}{kr} dS$$

$$= 2i \int_{-h}^{0} e^2(kz)\, dz \int_{0}^{2\pi} A_i(\theta)A_j^*(\theta)\, d\theta. \tag{4.253}$$

Using Eq. (4.112) gives

$$I(\psi_i, \psi_j^*) = i \frac{D(kh)}{k} \int_0^{2\pi} A_i(\theta) A_j^*(\theta) \, d\theta. \tag{4.254}$$

Note that because the two terms of the integrand $\psi_i(\partial \psi_j^*/\partial r) - (\partial \psi_i/\partial r) \psi_j^*$ have equal contributions in the far-field region, we also have

$$I(\psi_i, \psi_j^*) = \lim_{r \to \infty} 2 \iint_{S_\infty} \psi_i \frac{\partial \psi_j^*}{\partial r} \, dS. \tag{4.255}$$

For the two-dimensional case, possible evanescent waves set up as a result of inhomogeneous boundary conditions in the finite part of the fluid region (see Figure 4.10) are negligible in the far-field region. Thus all terms corresponding to $n \geq 1$ in Eq. (4.115) may be omitted there. The remaining terms for $n = 0$ correspond to Eq. (4.87). Consequently, for waves satisfying the radiation condition, we have the far-field asymptotic expression

$$\phi = \psi \sim -\frac{g}{i\omega} A^{\pm} e(kz) \, e^{\mp ikx} = -\frac{g}{i\omega} A^{\pm} e(kz) \, e^{-ik|x|}. \tag{4.256}$$

The upper signs apply for $x \to +\infty$ ($\theta = 0$) and the lower signs apply for $x \to -\infty$ ($\theta = \pi$). Comparison with Eq. (4.92) shows that the constants A^+ and A^- represent wave elevation.

Now consider a vertical cylinder S_∞ with rectangular cross section of width d in the y direction and of an arbitrarily large length in the x direction. Then

$$\begin{aligned} I(\psi_i, \psi_j^*) &= \iint_{S_\infty} \left(\psi_i \frac{\partial \psi_j^*}{\partial r} - \frac{\partial \psi_i}{\partial r} \psi_j^* \right) dS \\ &= \left(\frac{-g}{i\omega} \right) \left(\frac{-g}{-i\omega} \right) \left(A_i^+ A_j^{+*} + A_i^- A_j^{-*} \right) 2ik \int_{-h}^0 e^2(kz) \, dz \int_{-d/2}^{d/2} dy. \end{aligned} \tag{4.257}$$

Per unit width in the y direction we have

$$I'(\psi_i, \psi_j^*) \equiv \frac{I(\psi_i, \psi_j^*)}{d} = i \left(\frac{g}{\omega} \right)^2 D(kh) \left(A_i^+ A_j^{+*} + A_i^- A_j^{-*} \right). \tag{4.258}$$

Also in the two-dimensional case it is easily seen that, for two waves satisfying the same radiation condition, we have

$$I'(\psi_i, \psi_j) = 0, \quad I'(\psi_i^*, \psi_j^*) = 0. \tag{4.259}$$

4.8 Far-Field Coefficients and Kochin Functions

In a previous section we discussed circular waves and derived the asymptotic expression (4.222) for an outgoing wave, where the angular frequency is implicitly

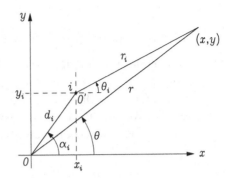

Figure 4.11: Horizontal coordinates, Cartesian (x, y) and polar (r, θ). Global coordinates are with respect to the common origin O. Local coordinates are with respect to the local origin O' of the (radiating and/or diffracting) structure number i.

contained in the far-field coefficient $A(\theta)$. At very large distance $(r \to \infty)$ all the waves diffracted and radiated from the structures shown in Figure 4.10 seem to originate from a region near the origin $(r = 0)$. The resulting outgoing wave is, in linear theory, given by an asymptotic expression such as Eq. (4.222) where the far-field coefficient is a superposition of outgoing waves from the individual diffracting and radiating structures; that is,

$$A(\theta) = \sum_i A_i(\theta), \tag{4.260}$$

where it is summed over all the sources for outgoing circular waves. Let

$$(x_i, y_i) = (d_i \cos \alpha_i, d_i \sin \alpha_i) \tag{4.261}$$

represent the local origin (or the vertical line through a reference point, e.g., centre of gravity) of source number i for the outgoing circular wave. See Figure 4.11 for definition of the lengths d_i and r_i and of the angles α_i, θ_i, and θ. The horizontal coordinates of an arbitrary point may be written in various ways as

$$(x, y) = (r \cos \theta, r \sin \theta) = (x_i + r_i \cos \theta_i, y_i + r_i \sin \theta_i). \tag{4.262}$$

We shall assume that all sources for outgoing waves are located in a bounded region in the neighbourhood of the origin, that

$$r \gg \max(d_i) \quad \text{for all } i \tag{4.263}$$

and that

$$kr \gg 1. \tag{4.264}$$

Then for very large distances, each term of the velocity potential $\sum \phi_i = \sum \psi_i$ of the resulting outgoing wave is given by an asymptotic expression such as [cf. Eq. (4.222)]:

$$\phi_i = \psi_i \sim B_i(\theta_i) \, e(kz) \, (kr_i)^{-1/2} \exp\{-ikr_i\} \tag{4.265}$$

or

$$\phi_i = \psi_i \sim A_i(\theta)\, e(kz)\, (kr)^{-1/2} \exp\{-ikr\}. \tag{4.266}$$

Here, the complex function $B_i(\theta)$ is the far-field coefficient referred to the source's local origin, and $A_i(\theta)$ for the same wave source (number i) is the far-field coefficient referred to the common origin. For very distant field points (x, y), that is, $r \gg d_i$, we have the following asymptotic approximations for the relation between local and global coordinates (see Figure 4.11):

$$\theta_i \approx \theta, \tag{4.267}$$

$$r_i \approx r - d_i \cos(\alpha_i - \theta), \tag{4.268}$$

$$(kr_i)^{-1/2} \approx (kr)^{-1/2}. \tag{4.269}$$

For the outgoing wave from source number i we have

$$\begin{aligned}
\phi_i &\sim B_i(\theta_i)\, e(kz)\, (kr_i)^{-1/2} \exp\{-ikr_i\} \\
&\sim B_i(\theta)\, e(kz)\, (kr)^{-1/2} \exp\{ikd_i \cos(\alpha_i - \theta)\} \exp\{-ikr\}.
\end{aligned} \tag{4.270}$$

Note that we use a higher-order approximation for r_i in the phase than in the amplitude (modulus), as is also usual in interference and diffraction theories, for instance in optics.

Hence we have the following relations between the far-field coefficients:

$$\begin{aligned}
A_i(\theta) &= B_i(\theta) \exp\{ikd_i \cos(\alpha_i - \theta)\}, \\
&= B_i(\theta) \exp\{ik(x_i \cos\theta + y_i \sin\theta)\}.
\end{aligned} \tag{4.271}$$

For a single axisymmetric body oscillating in heave only, the far-field coefficient B_0, say, is independent of θ. If the vertical symmetry axis coincides with the z axis, also the corresponding A_0 is independent of θ. However, if $d_i \neq 0$, A_0 varies with θ, as a result of the exponential factor in Eq. (4.271).

Radiated waves and diffracted waves satisfy the radiation condition of outgoing waves at infinite distance. They are represented asymptotically by expressions such as (4.265) or (4.266), where the amplitude, the phase and the direction dependence are determined by the far-field coefficient.

An incident wave Φ, for instance [see Eqs. (4.92) and (4.102)]

$$\Phi = \hat{\phi}_0 = -\frac{g}{i\omega}\, Ae(kz) \exp\{-ik(x \cos\beta + y \sin\beta)\}, \tag{4.272}$$

does not satisfy the radiation condition. Let ϕ_i and ϕ_j be two arbitrary waves, where

$$\phi_{i,j} = \Phi_{i,j} + \psi_{i,j}. \tag{4.273}$$

Here ψ_i and ψ_j are two waves which satisfy the radiation condition (of outgoing waves at infinity), and Φ_i and Φ_j are two arbitrary incident waves. From the definition of the integral (4.240) we see that

$$I(\phi_i, \phi_j) = I(\Phi_i, \Phi_j) + I(\Phi_i, \psi_j) + I(\psi_i, \Phi_j) + I(\psi_i, \psi_j). \tag{4.274}$$

Because ψ_i and ψ_j satisfy the same radiation condition, we have from Eq. (4.249) that $I(\psi_i, \psi_j) = 0$. Further, because Φ_i and Φ_j satisfy the homogeneous boundary conditions on the planes $z = 0$ and $z = -h$, and because they satisfy the Laplace equation everywhere in the volume region between these planes, *including* the volume region occupied by the oscillator structures, it follows from Green's theorem and Eq. (4.243) that also $I(\Phi_i, \Phi_j) = 0$. Hence

$$I(\phi_i, \phi_j) = I(\Phi_i, \psi_j) + I(\psi_i, \Phi_j)$$
$$= I(\Phi_i, \psi_j) - I(\Phi_j, \psi_i), \tag{4.275}$$

where we have, in the last step, also used Eq. (4.244), for which it is necessary to assume that Φ_j commutes with ψ_i.

Let us now consider the two incident waves

$$\Phi_{i,j} = -\frac{g}{i\omega} A_{i,j} e(kz) \exp\{-ikr(\beta_{i,j})\}, \tag{4.276}$$

where

$$r(\beta) \equiv x \cos\beta + y \sin\beta. \tag{4.277}$$

Note that the two waves have different complex elevation amplitudes A_i and A_j, and different angles of incidence β_i and β_j. They have, however, equal period $T = 2\pi/\omega$ and hence equal wavelengths $\lambda = 2\pi/k$.

Following Newman,[30] we define the so-called Kochin function

$$H_j(\beta) \equiv -\frac{k}{D(kh)} I\{e(kz) e^{ikr(\beta)}, \psi_j\}. \tag{4.278}$$

Note that $H_j(\beta)$ depends on a chosen angle β. Otherwise $H_j(\beta)$ is a property of the wave ψ_j which satisfies the radiation condition. We shall show that $H_j(\theta)$ is proportional to the far-field coefficient $A_j(\theta)$ of the wave ψ_j. Because, from Eq. (4.277),

$$r(\beta \pm \pi) = -r(\beta), \tag{4.279}$$

we have

$$H_j(\beta \pm \pi) = -\frac{k}{D(kh)} I\{e(kz) e^{-ikr(\beta)}, \psi_j\}. \tag{4.280}$$

Using this in combination with the two chosen incident waves Φ_i and Φ_j, given by Eq. (4.276), we have from Eq. (4.275)

$$I(\phi_i, \phi_j) = \frac{gD(kh)}{i\omega k} \{A_i H_j(\beta_i \pm \pi) - A_j H_i(\beta_j \pm \pi)\}. \tag{4.281}$$

Note that this result is based on Eq. (4.275) and hence on Eq. (4.244). Thus observe that A_j has to commute with H_i when Eq. (4.281) is being applied. Also note that the given complex wave-elevation amplitude A_j here, and in Eq. (4.276), should not be confused with the function $A_j(\theta)$, which is the far-field coefficient.

Next we shall show that

$$H_j(\theta) = \sqrt{2\pi} A_j(\theta) e^{i\pi/4}. \tag{4.282}$$

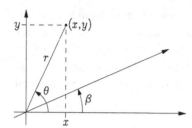

Figure 4.12: Cartesian and polar coordinates in the horizontal plane, with β as the angle of wave incidence.

Using Eqs. (4.243) and (4.266) in Eq. (4.278) gives

$$H_j(\beta) = -\frac{k}{D(kh)} I\{e(kz)\,e^{ikr(\beta)},\ A_j(\theta)\,e(kz)\,(kr)^{-1/2}e^{-ikr}\}, \tag{4.283}$$

where the integration surface is now S_∞ (cf. Figure 4.10). It should be observed here that r is the (horizontal) radial coordinate (see Figure 4.11), whereas $r(\beta)$ is as given by Eq. (4.277).

We shall assume that the control surface S_∞ is a circular cylinder of radius r. The integration variables are then θ and z (cf. Figure 4.12), whereas r is a constant which tends to infinity. Noting that $r(\beta) \equiv x\cos\beta + y\sin\beta = r\cos\theta\cos\beta + r\sin\theta\sin\beta = r\cos(\theta - \beta)$, we have [cf. Eqs. (4.243) and (4.283)]

$$H_j(\beta) = -\frac{1}{2D(kh)}\,2k\int_{-h}^{0} e^2(kz)\,dz \lim_{kr\to\infty}\int_{0}^{2\pi}(G_1 - G_2)r\,d\theta, \tag{4.284}$$

where

$$G_1 = -ik\,A_j(\theta)\,(kr)^{-1/2}\exp(-ikr)\exp[ikr\cos(\theta - \beta)] \tag{4.285}$$

and

$$G_2 = -G_1\cos(\theta - \beta). \tag{4.286}$$

Using Eq. (4.112), we find that Eq. (4.284) becomes

$$H_j(\beta) = \lim_{kr\to\infty}\frac{i}{2}\int_{0}^{2\pi}\sqrt{kr}A_j(\theta)\,[1 + \cos(\theta - \beta)]$$
$$\times \exp\{-ikr\,[1 - \cos(\theta - \beta)]\}\,d\theta. \tag{4.287}$$

We now substitute $\varphi = \theta - \beta$ as the new integration variable. Further, using the asymptotic expression (for $u \to \infty$)

$$\lim_{u\to\infty}\int_{-\beta}^{2\pi-\beta} f(\varphi)\exp\{-iu\,(1 - \cos\varphi)\}d\varphi = \lim_{u\to\infty}\sqrt{\pi/u}\,[(1 - i)f(0)$$
$$+ (1 + i)e^{-i2u}f(\pi)] \tag{4.288}$$

which is derived below, we obtain

$$H_j(\beta) = \tfrac{i}{2}(1 - i)\sqrt{\pi}\,A_j(\beta)\,(1 + \cos 0)$$
$$= (1 + i)\sqrt{\pi}\,A_j(\beta) = \sqrt{2\pi}\,A_j(\beta)\,e^{i\pi/4}, \tag{4.289}$$

in accordance with the above statement (4.282) which was to be proved.

The above-mentioned mathematical relation (4.288) is derived as follows, by the "method of stationary phase". To be slightly more general, we shall first consider the asymptotic limit (as $u \to \infty$) of the integral

$$I_{ab} \equiv \int_a^b f(\varphi)\, e^{-iu\psi(\varphi)}\, d\varphi \qquad (4.290)$$

where $f(\varphi)$ and $\psi(\varphi)$ are analytic functions in the interval $a < \varphi < b$. Moreover, $\psi(\varphi)$ is real and has one, and only one, extremum $\psi(c)$ in the interval, such that $\psi'(c) = 0$ and $\psi''(c) \neq 0$, where $a < c < b$. If we take a constant $e^{-iu\psi(c)}$ outside the integral (as with the next integral below), and if we then let $u \to \infty$, the integrand oscillates infinitely fast with φ, except near the "stationary" point $\varphi = c$, where the imaginary exponent varies slowly with φ. Hence, the contribution to the integral is negligible outside the interval $c - \epsilon < \varphi < c + \epsilon$, where ϵ is a small positive number. Using a Taylor expansion for $\psi(\varphi)$ around $\varphi = c$ we have

$$\lim_{u \to \infty} I_{ab} = \lim_{u \to \infty} f(c)\, e^{-iu\psi(c)} \int_{c-\epsilon}^{c+\epsilon} \exp[-i\, u\psi''(c)(\varphi - c)^2/2]\, d\varphi. \qquad (4.291)$$

We now take

$$\alpha = (\varphi - c)\sqrt{u\psi''(c)\operatorname{sgn}\psi''(c)/2} = (\varphi - c)\sqrt{u|\psi''(c)|/2} \qquad (4.292)$$

as the new integration variable. Note that the new integration limits are given by

$$\alpha = \pm\epsilon\sqrt{u|\psi''(c)|/2} \to \pm\infty \quad \text{as } u \to \infty. \qquad (4.293)$$

This gives

$$\lim_{u \to \infty} I_{ab} = \lim_{u \to \infty} f(c)\sqrt{\frac{2}{u|\psi''(c)|}}\, e^{-iu\psi(c)} \int_{-\infty}^{\infty} \exp[-i\alpha^2 \operatorname{sgn}\psi''(c)]\, d\alpha$$

$$= \lim_{u \to \infty} f(c)\sqrt{\frac{2}{u|\psi''(c)|}}\, e^{-iu\psi(c)} \left[\int_{-\infty}^{\infty} \cos\alpha^2\, d\alpha \right.$$

$$\left. - i\operatorname{sgn}\psi''(c) \int_{-\infty}^{\infty} \sin\alpha^2\, d\alpha \right]$$

$$= \lim_{u \to \infty} f(c)\sqrt{\frac{\pi}{u|\psi''(c)|}}\, e^{-iu\psi(c)}\, [1 - i\operatorname{sgn}\psi''(c)], \qquad (4.294)$$

because both of the two last integrals equal $\sqrt{\pi/2}$. Note that for $\psi(\varphi) = 1 - \cos\varphi$ we have $\psi'(\varphi) = 0$ for $\varphi = 0$ and for $\varphi = \pi$. Moreover, observing that $\psi(0) = 0$, $\psi''(0) = 1$, $\psi(\pi) = 2$ and $\psi''(\pi) = -1$, we can easily see that Eq. (4.288) follows from Eqs. (4.290) and (4.294).

In the two-dimensional case ($\partial/\partial y \equiv 0$) the bodies and air chambers shown as cross sections in Figure 4.10 are assumed to have infinite extension in the y direction. To make the integral (4.240) finite we integrate over that part of the surface S which is contained within the interval $-d/2 < y < d/2$, where we shall later let the width d tend to infinity. The corresponding integral per unit width is $I' = I/d$, as in Eq. (4.258). The surface S_∞ may be taken as the two planes $x = r^\pm$, where the constants $r^\pm$ are sufficiently large ($r^\pm \to \pm\infty$). The two-dimensional Kochin function $H'_j(\beta)$ is defined as in Eq. (4.278) with I replaced by I'. In accordance with Eq. (4.243) we take the integral over S_∞ instead of S, and we use the far-field approximation (4.256) to obtain

$$H_j(\beta) = -\frac{k}{D(kh)} \int_{-h}^{0} e^2(kz)\, dz \int_{-d/2}^{d/2} \exp\{iky\sin\beta\}\, dy\, (G^+ + G^-),$$

(4.295)

where

$$G^\pm = -(ik \pm ik\cos\beta)\left(\frac{-g}{i\omega}\right) A_j^\pm \exp\{ikr^\pm \cos\beta\} \exp\{\mp ikr^\pm\}.$$
(4.296)

The integral over z is $D(kh)/2k$ in accordance with Eq. (4.112), and moreover,

$$\lim_{d\to\infty} \frac{1}{d} \int_{-d/2}^{d/2} \exp(iky\sin\beta)\, dy = \begin{cases} 1 & \text{if } \sin\beta = 0 \\ 0 & \text{if } \sin\beta \neq 0. \end{cases}$$
(4.297)

Note that for a two-dimensional case the angle of incidence is either $\beta = 0$ or $\beta = \pi$. In both cases $\sin\beta = 0$. Thus the two-dimensional Kochin function per unit width, $H'_j(\beta) = H_j(\beta)/d$, is given by

$$H'_j(0) = -\frac{1}{2}\frac{-g}{i\omega} A_j^+(-ik - ik\cos 0) = -\frac{gk}{\omega} A_j^+,$$
(4.298)

$$H'_j(\pi) = -\frac{1}{2}\frac{-g}{i\omega} A_j^-(-ik + ik\cos\pi) = -\frac{gk}{\omega} A_j^-.$$
(4.299)

Otherwise, if $\sin\beta \neq 0$, we have $H'_j(\beta) = 0$. Here we have derived the two-dimensional version of relation (4.282) between the Kochin function and the far-field coefficient.

Let us finally consider some relations among the Kochin functions, the far-field coefficients and the useful integral (4.240). For source number j for outgoing waves we have $H_j = H_j(\theta)$ as given by Eqs. (4.278) and (4.282). For the outgoing wave we have the asymptotic approximation

$$\psi_j \sim A_j(\theta)\, e(kz)\, (kr)^{-1/2} e^{-ikr}$$
$$= H_j(\theta)\, e(kz)\, (2\pi kr)^{-1/2} \exp\{-ikr - i\pi/4\}.$$
(4.300)

For two waves ψ_i and ψ_j satisfying the same radiation condition we have Eq. (4.254):

$$I(\psi_i, \psi_j^*) = i\,\frac{D(kh)}{k} \int_0^{2\pi} A_i(\theta)\, A_j^*(\theta)\, d\theta = \frac{i D(kh)}{2\pi k} \int_0^{2\pi} H_i(\theta)\, H_j^*(\theta)\, d\theta.$$
(4.301)

Using relation (4.271) between the far-field coefficients referred to the global origin and the local origin for wave source number j, we have

$$I(\psi_i, \psi_j^*) = i\frac{D(kh)}{k} \int_0^{2\pi} A_i(\theta) A_j^*(\theta)\, d\theta$$

$$= i\frac{D(kh)}{k} \int_0^{2\pi} B_i(\theta)\, B_j^*(\theta) \exp\{ik(r_j - r_i)\}\, d\theta. \tag{4.302}$$

The exponent in the latter integrand may be rewritten by using the geometrical relations

$$r_j - r_i = (x_i - x_j)\cos\theta + (y_i - y_j)\sin\theta$$
$$= d_i \cos(\alpha_i - \theta) - d_j \cos(\alpha_j - \theta)$$
$$= d_{ij} \cos(\alpha_{ij} - \theta). \tag{4.303}$$

where distances and angles are defined in Figure 4.13 ($d_{ij} = d_{ji}$, $\alpha_{ij} = \alpha_{ji} + \pi$).
The two-dimensional version of Eq. (4.301) is

$$I'(\psi_i, \psi_j^*) = \frac{i D(kh)}{k^2}\{H_i'(0)\, H_j'^*(0) + H_i'(\pi)\, H_j'^*(\pi)\}, \tag{4.304}$$

which is obtained by combination of Eqs. (4.258), (4.298) and (4.299).

For waves not satisfying the radiation condition, such as ϕ_i and ϕ_j in Eq. (4.274), we shall now consider $I(\phi_i, \phi_j^*)$. Assuming that Φ_j^* commutes with ψ_i, we obtain

$$I(\phi_i, \phi_j^*) = I(\Phi_i, \psi_j^*) - I(\Phi_j^*, \psi_i) + I(\psi_i, \psi_j^*) \tag{4.305}$$

in a similar way as we derived Eq. (4.275). Combining Eq. (4.305) with Eqs. (4.276) and (4.278), from which also follows

$$\Phi_j^* = \frac{g}{i\omega}\, A_j^* e(kz)\exp\{ikr(\beta_j)\}, \tag{4.306}$$

$$H_j^*(\beta) = -\frac{k}{D(kh)}\, I\{e(kz)\, e^{-ikr(\beta)}, \psi_j^*\}, \tag{4.307}$$

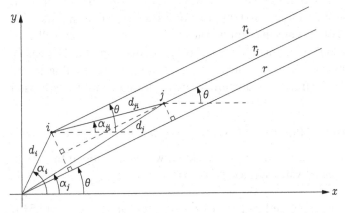

Figure 4.13: Definition sketch for horizontal distances and angles determining the position of the common origin and the local origins for sources number i and number j, for outgoing waves.

we find

$$I(\phi_i, \phi_j^*) = \frac{g\,D(kh)}{i\omega k}\{A_i\,H_j{}^*(\beta_i) + A_j^*\,H_i(\beta_j)\} + I(\psi_i, \psi_j^*). \qquad (4.308)$$

Here we may use Eq. (4.301) to express the last term in terms of Kochin functions. For the two-dimensional case we instead use Eq. (4.304), and then we also replace I and $H_{i,j}$ by I' and $H'_{i,j}$, respectively.

If we compare Eqs. (4.281) and (4.308), we notice two differences for the conjugation of ϕ_j. Firstly, as a result of Eq. (4.249), Eq. (4.281) contains no term corresponding to the last term in Eq. (4.308). Secondly, in contrast to Eq. (4.281), there is not a shift of an angle of π in the argument of the Kochin functions in Eq. (4.308).

Equations (4.281) and (4.308) were used by Newman[30] as a base for deriving a set of reciprocity relations between parameters associated with various waves. The next chapter derives and discusses some reciprocity relations.

4.9 Waves in the Time Domain

(The text in this section is partly taken from *Applied Ocean Research*, Vol. 17, J. Falnes, "On non-causal impulse response functions related to propagating water waves", pp. 379–389, 1995, with permission from Elsevier Science.)[31] In analysing waves we have the opportunity to choose between a frequency-domain or a time-domain approach. Traditionally the frequency domain has received greater attention, as it also does here in this chapter from Section 4.2 and onward. More recently, however, time-domain investigations have also been carried out.[32,33] Such investigations may be based on a numerical solution of the time-domain Laplace equation (4.16) with boundary conditions (4.17), (4.18), (4.28) and (4.29). It is outside the scope of this chapter to discuss such investigations in detail.

In this section, let us, however, make a few comments based on applying an inverse Fourier transform to some of the frequency-domain results. To be specific, let us discuss two linear systems in which the wave elevation at a particular point is the input. We shall see that these systems are non-causal. The reason is that this input is not the cause of the chosen outputs. The real cause of output, as well as of input, may be a wavemaker (in the laboratory) or a distant storm (on the ocean).

Let us start by assuming that an incident plane wave is propagating in the positive x direction. In the frequency domain the wave elevation may be expressed as

$$\eta(x, y; \omega) = A(\omega)\exp\{-ikx\} \equiv \eta(x, \omega), \qquad (4.309)$$

where $A(\omega)$ is the Fourier transform of the incident wave elevation at the origin $(x, y) = (0, 0)$ of the mean water surface,

$$A(\omega) = \eta(0, 0; \omega) = \eta(0, \omega) = \int_{-\infty}^{\infty} \eta(0, t)\exp\{-i\omega t\}\,dt. \qquad (4.310)$$

To ensure wave propagation in the positive x direction, we choose a solution of the dispersion equation, $k = k(\omega)$, for which the angular repetency k has the same sign as the angular frequency ω.

4.9.1 Relation Between Wave Elevations at Two Locations

Because real sea waves have a finite coherence time, it is possible to predict, with a certain probability, the wave elevation at a given point on the basis of wave measurement at the same point. However, a more deterministic type of prediction of the incident wave elevation is possible if the wave is measured at some distance l from the body, in the "upstream" direction (i.e., opposite to the direction of wave propagation). Let the measurement take place at $x = x_A$, where

$$x_B - x_A = l > 0. \tag{4.311}$$

Considering $\eta(x_A, \omega)$ and $\eta(x_B, \omega)$ as input and output, respectively, of a linear system, it follows from Eq. (4.309) that the transfer function of the system is

$$H_l(\omega) = \exp\{-ik(\omega)l\}. \tag{4.312}$$

The corresponding impulse response function $h_l(t)$ is strictly speaking not causal,[34] although it is approximately causal, and particularly so if l is large. With $l = 400$ m, remarkably good prediction of computer-simulated sea swells may be predicted for a time at least half a minute into the future.[35] Previous experimental work[36] on real sea waves gave a less accurate prediction, but a more satisfactory prediction of the hydrodynamic pressure on the sea bottom at the location x_B.

The transfer function $H_l(\omega)$ as given by Eq. (4.312), or the corresponding impulse response function

$$h_l(t) = \frac{1}{2\pi} \int_{-\infty}^{\infty} H_l(\omega) \exp\{i\omega t\} \, d\omega = \frac{1}{2\pi} \int_{-\infty}^{\infty} \exp\{-ik(\omega)l + i\omega t\} \, d\omega, \tag{4.313}$$

relates the values at two positions (x_B and $x_A = x_B - l$) of any physical quantity of a general propagating wave as represented by Eq. (4.309). The physical quantity could, for instance, be an electric-field component associated with an electromagnetic wave, propagating in free space with speed c. Then the dispersion equation is

$$\omega^2 = c^2 k^2 \tag{4.314}$$

and, combining this with Eq. (4.313), we find

$$h_l(t) = \frac{1}{2\pi} \int_{-\infty}^{\infty} \exp\left\{i\omega\left(t - \frac{l}{c}\right)\right\} d\omega = \delta\left(t - \frac{l}{c}\right), \tag{4.315}$$

which is an impulse that is time shifted by l/c. This was, of course, to be expected, because with this dispersion-free wave all frequency components propagate with the same constant speed c. It follows that $h_l(t) = 0$ for $t < l/c$, and consequently for $t < 0$, because $l > 0$. Hence, the function $h_l(t)$ is causal for the dispersion-free wave. This is so, because with a dispersion-free wave every frequency component is propagated with a constant amplitude and with a phase change which is proportional to the frequency and to the distance of propagation. Any multifrequency wave is then propagated without distortion. As interpreted through Fourier analysis, an impulse at $x = x_A$ at time $t = 0$ results from a continuum of sinusoidal waves which are of equal amplitude and which are all in phase there at $t = 0$ and out of phase in a unique way for all other values of t. The unique summation of (inverse Fourier integration over) the wave components results in zero when t is different from zero. At a later time $t = l/c$, the impulse has moved without distortion to $x = x_B = x_A + l$.

Thus, the impulse response function $h_l(t)$ cannot be non-causal unless the wave is dispersive. In such a case, in which a multifrequency wave may be distorted, the above-mentioned continuum of sinusoidal waves (each of them extending to infinity in space and time) do not, after propagation of the original impulse at $(x, t) = (0, 0)$, necessarily satisfy the phase conditions required for an impulse to result. Hence, it is not guaranteed that the function $h_l(t)$ vanishes for $t < 0$. It seems plausible that the impulse response function $h_l(t)$ is non-causal when the wave is dispersive. In the following paragraphs let us show explicitly that this is true for water waves.

For gravity waves on (inviscid, incompressible) water of depth h, the dispersion equation is

$$\omega^2 = gk \tanh(kh) \tag{4.316}$$

as given by Eq. (4.57). For deep water ($h \to \infty$) this simplifies to

$$\omega^2 = g|k|. \tag{4.317}$$

The requirement that $h_l(t)$ must be real means [cf. Eq. (4.313)] that

$$H_l^*(\omega) = H_l(-\omega) \tag{4.318}$$

and, hence [cf. Eqs. (4.312) and (2.160)] that

$$\text{sgn}(k) = \text{sgn}(\omega). \tag{4.319}$$

This latter condition also ensures that the wave (4.309) propagates in the positive x direction.

For waves satisfying the dispersion relationship (4.316), neither the phase velocity $v_p = \omega/k$ nor the group velocity $v_g = d\omega/dk$ can be negative for any ω. The maximum propagation velocity is

$$v_{g,\text{max}} = v_{p,\text{max}} = (gh)^{1/2} \tag{4.320}$$

for $k = 0$; that is, $\omega = 0$. For the "shallow-water approximation" the group velocity and phase velocity are equal, to $(gh)^{1/2}$, and independent of frequency. In the integral (4.313) all frequencies (even $\omega \to \infty$) are involved. Hence it must be borne in mind that the gravity wave on water is dispersive, even when h is small (but $h > 0$). For the deep-water case, $\omega^2 h/g \gg 1$, we have [cf. Eq. (4.104)]

$$v_g \approx \tfrac{1}{2} v_p \approx \tfrac{1}{2} g/|\omega| \approx \tfrac{1}{2} (g/|k|)^{1/2}. \tag{4.321}$$

If the water depth is infinite, this holds for all ω. Note the singularity in this case, for $\omega = 0$. Let us, for a while, consider the case of waves on deep water ($h \to \infty$); that is, we assume that the dispersion relation (4.317)

$$k = |\omega|\omega/g \tag{4.322}$$

is valid for all ω and all k. Using Eq. (4.313) we may write the impulse response as

$$h_l(t) = \frac{1}{2\pi} \int_{-\infty}^{\infty} \{\cos(\omega t - kl) + i \sin(\omega t - kl)\} \, d\omega. \tag{4.323}$$

Because k has the same sign as ω – compare Eq.(4.319) – and because the sine function is odd, the imaginary part of the integral vanishes. Moreover, the cosine function is even, and for $\omega > 0$ we have $k = \omega^2/g$. Hence

$$h_l(t) = \frac{1}{\pi} \int_0^{\infty} \cos\left(\frac{\omega t - \omega^2 l}{g}\right) d\omega = h_e(t) + h_o(t). \tag{4.324}$$

The even part and the odd part of the impulse response function are

$$h_e(t) = \frac{1}{\pi} \int_0^{\infty} \cos\frac{\omega^2 l}{g} \cos(\omega t) \, d\omega, \tag{4.325}$$

$$h_o(t) = \frac{1}{\pi} \int_0^{\infty} \sin\frac{\omega^2 l}{g} \sin(\omega t) \, d\omega, \tag{4.326}$$

respectively. Using a table of Fourier transforms,[37] we find, for $t > 0$,

$$h_e(t) = \frac{1}{4}\left(\frac{2g}{\pi l}\right)^{1/2} \left(\cos\frac{gt^2}{4l} + \sin\frac{gt^2}{4l}\right), \tag{4.327}$$

$$h_o(t) = \frac{1}{2}\left(\frac{2g}{\pi l}\right)^{1/2} \left(C_2\left\{t\left(\frac{g}{2\pi l}\right)^{1/2}\right\}\cos\frac{gt^2}{4l} + S_2\left\{t\left(\frac{g}{2\pi l}\right)^{1/2}\right\}\sin\frac{gt^2}{4l}\right), \tag{4.328}$$

where C_2 and S_2 are Fresnel integrals, defined as

$$C_2(x) = (2\pi)^{-1/2} \int_0^x t^{-1/2} \cos(t) \, dt, \tag{4.329}$$

$$S_2(x) = (2\pi)^{-1/2} \int_0^x t^{-1/2} \sin(t) \, dt \tag{4.330}$$

for $x > 0$.

Because $|h_e(t)| \neq |h_o(t)|$, the impulse response $h(t) = h_e(t) + h_o(t)$ is not causal. From the asymptotic behaviour of the Fresnel integrals, we have that $C_2(x) \to 1/2$ and $S_2(x) \to 1/2$ as $x \to +\infty$. Thus

$$h_e(t) - h_o(t) \to 0 \quad \text{as } t \to +\infty. \tag{4.331}$$

Hence,

$$h_l(t) \to 0 \quad \text{as } t \to -\infty \tag{4.332}$$

and

$$h_l(t) \to \frac{1}{2}\left(\frac{2g}{\pi l}\right)^{1/2}\left(\cos\frac{gt^2}{4l} + \sin\frac{gt^2}{4l}\right) \quad \text{as } t \to +\infty. \tag{4.333}$$

In this limit, the impulse response is an oscillation, with a decreasing time interval between consecutive zero crossings. Qualitatively, this is physically reasonable because the phase velocity and the group velocity are monotonically decreasing functions of the frequency. It has become apparent that the impulse response has a non-vanishing, albeit small, value for $t < 0$.

It is less easy to find a simple expression for the impulse response function $h_l(t)$ for the case of a finite water depth. However, if we can show that $h_l(0) \neq 0$ and that $h_l(t)$ is a continuous function, it must follow that $h_l(t)$ is non-causal. Because $v_p = \omega/k$ and $v_g = d\omega/dk$ are never negative, we take k as a new integration variable in Eq. (4.313), which becomes

$$\begin{aligned} h_l(t) &= \frac{1}{2\pi}\int_{-\infty}^{\infty} \exp\{-ik(\omega)l\}\exp\{i\omega t\}\,d\omega \\ &= \frac{1}{2\pi}\int_{-\infty}^{\infty} \exp\{-ikl\}\exp\{i\omega(k)t\}v_g\,dk \\ &= \frac{1}{2\pi}\int_{-\infty}^{\infty} v_g(k)\exp\{-ik[l - v_p(k)t]\}\,dk. \end{aligned} \tag{4.334}$$

At time $t = 0$ the impulse response is

$$h_l(0) = \frac{1}{2\pi}\int_{-\infty}^{\infty} v_g(k)\exp\{-ikl\}\,dk = \frac{1}{2\pi}V_g(l), \tag{4.335}$$

where

$$V_g(l) = \int_{-\infty}^{\infty} v_g(k)\exp\{-ikl\}\,dk \tag{4.336}$$

is a (spatial) Fourier transform of $v_g(k)$. If v_g is constant, $V_g(l)$ is a delta function, and hence $h_l(0) = 0$ if $l > 0$. However, in the case of dispersion $v_g(k)$ varies with k, and then we have that $h_l(0) \neq 0$, apart for exceptional cases in which $V_g(l) = 0$. Because $v_g(k)$ is finite, see Eq. (4.320), a continuous function of t results if the integral (4.334) is taken over a finite interval $-k_\infty < k < k_\infty$. For waves on finite water depth the group velocity behaves as that with infinite water depth when $k \to \infty$. Moreover, we have already derived a continuous function $h_l(t)$ for the case of infinite water depth: compare Eqs. (4.324)–(4.328). Hence, we know that the integral (4.334) exists, and that, consequently, the impulse response function $h_l(t)$ is continuous. Thus, for gravity waves on water, the impulse response function $h_l(t)$ is obviously non-causal whenever $h_l(0) \neq 0$.

4.9.2 Relation Between Hydrodynamic Pressure and Wave Elevation

Let us now consider a linear system in which the input is the wave elevation at some position (x, y) on the average water surface, and in which the output is the hydrodynamic pressure. Then according to Eq. (4.93), the transfer function is

$$H_p(\omega) = \frac{p(x, y, z; \omega)}{\eta(x, y; \omega)} = \rho g e(kz), \tag{4.337}$$

which for deep water becomes

$$H_p(\omega) = \rho g \exp\{\omega^2 z/g\}. \tag{4.338}$$

(Note that $z \leq 0$.) Observe that $H_p(\omega)$ is real, and hence its inverse Fourier transform is an even function of t, and for deep water is given by[37]

$$h_p(t) = h_p(-t) = \tfrac{1}{2}\rho g(-g/\pi z)^{1/2} \exp\{gt^2/4z\}, \tag{4.339}$$

which is, evidently, not causal, except for the case of $z = 0$, which gives

$$h_p(t) = \rho g \delta(t). \tag{4.340}$$

Compared with the value at $t = 0$, $h_p(t)$ is reduced to 1% for $t = \pm 4.3\sqrt{-z/g}$, which increases with the submergence of the considered point (x, y, z). Thus a deeper point has a "more non-causal" impulse response function for its hydrodynamic pressure.

Referring also to Eqs. (4.316) and (4.319), we find it evident that $H_p(\omega)$ is an even function of ω also in the case of finite water depth. Moreover, $H_p(\omega)$ is real when ω is real. It follows that the corresponding inverse Fourier transform $h_p(t)$ is an even function of t, and hence it is a non-causal impulse response function.

Physically we may interpret this as follows. There is an effect on the hydrodynamic pressure from the previous and the following wave troughs also at the instant when a wave crest is passing. The deeper the considered location is, the more this non-instantaneous effect contributes. From this point of view the well-known fact that the amplitude of the hydrodynamic pressure decreases with the distance below the free surface seems very reasonable.

Problems

Problem 4.1: Deriving Dispersion Relation Including Capillarity

Consider an infinitesimal surface element of length Δx in the x direction and unit length in the y direction. The net capillary attraction force is

$$p_k \Delta x = \gamma \sin(\theta) - \gamma \sin(\theta + \Delta\theta),$$

where γ is the surface tension; $[\gamma] = $ N/m. Furthermore, θ and $\theta + \Delta\theta$ are the angles between the surface and the horizontal plane at the positions x and $x + \Delta x$, respectively.

Derive a modified dispersion relation, including capillarity, by inserting a linearisation of the above-mentioned surface tension force into the Bernoulli equation (4.12). Observe that a discussion of this dispersion relation, for the deep-water case, is the subject of Problem 3.2. [Hint: first show that the free-surface boundary condition (4.43) has to be replaced by

$$\left[-\omega^2\hat{\phi} + g\frac{\partial\hat{\phi}}{\partial z} - \frac{\gamma}{\rho}\frac{\partial^3\hat{\phi}}{\partial x^2\partial z}\right]_{z=0} = 0.$$

Then use this condition in combination with Eq. (4.83) – which already satisfies the Laplace equation as well as the sea-bed boundary condition (4.42) – to determine a new dispersion relation to replace Eq. (4.57).]

Problem 4.2: Vertical Functions from Bottom Boundary Condition

Use the boundary condition (4.42) for $z = -h$ together with Eq. (4.44) in the determination of the integration constants c_+ and c_- in Eq. (4.49) for $Z(z)$. One additional condition is required. Determine both integration constants, and write down an explicit expression for $Z(z)$ for both of the following two cases of a second condition:

(a) $Z(0) = 1$,
(b) $\int_{-h}^{0} |Z(z)|^2 \, dz = h$.

Problem 4.3: Vertical Functions for Evanescent Solutions

(a) Show that the equation

$$Z_n''(z) = -m_n^2 Z_n(z)$$

with the boundary condition

$$Z_n'(-h) = 0$$

has a particular solution

$$Z_n(z) = C_n \cos(m_n z + m_n h).$$

(b) Further, show that the boundary condition

$$Z_n'(0) = \frac{\omega^2}{g} Z_n(0)$$

is satisfied if

$$\omega^2/(gm_n) = -\tan(m_m h).$$

(c) Discuss possible solutions of the equation

$$\omega^2/(gm_n) = -\tan(m_m h)$$

as well as normalisation of the corresponding functions $Z_n(z)$.

Problem 4.4: Distance to Wave Origin

Assume that the ocean is calm and that suddenly a storm develops. Waves are being generated, and swells propagate away from the storm region. Far away, at a distance l from the storm centre, swells of frequency $f = 1/T = \omega/2\pi$ are recorded a certain time τ after the start of the storm. We assume deep water between the storm centre and the place where the swells are recorded. Derive an expression for the "waiting time" τ in terms of l, f and g, where g is the acceleration of gravity. The starting point of the derivation should be the dispersion relationship

$$\omega^2 = gk$$

for waves on deep water.

In the South Pacific, at Tuvalu, a long swell was registered as follows:

16 Nov 1991   at 0300 GMT   T = 22.2 s
                at 1400 GMT   T = 20.0 s
17 Nov 1991   at 0100 GMT   T = 18.2 s

Find the distance to the swell source. Approximately on which date did this storm occur? Assuming the swell was travelling southward, determine the latitude at which the swell was formed, given that Tuvalu is located at 8°30′ S (1′ corresponds to a nautical mile = 1852 m. Or, 90°, the distance from equator to the north pole is 10^7 m).

Problem 4.5: Reflection of Plane Wave at Vertical Wall

(a) Show that a gravitational wave for which the elevation has a complex amplitude

$$\hat{\eta}_i = \eta_0 \exp(-ik_x x - ik_y y) \tag{1}$$

may propagate on an infinite lake of constant depth h. That is, show that this wave satisfies the partial differential equation

$$\left[\frac{\partial^2}{\partial x^2} + \frac{\partial^2}{\partial y^2} + k^2 \right] \hat{\eta} = 0.$$

(b) Moreover, it is assumed that the boundary conditions at the free water surface $z = 0$ and at the bottom $z = -h$ are satisfied (although a proof of this statement is not required in the present problem). Express k_x and k_y by the angular repetency k and the angle β between the direction of propagation and the x axis.

(c) Assume that the incident wave (1) in the region $x < 0$ is reflected at a fixed vertical wall in the plane $x = 0$. Use the additional boundary condition at the wall to determine the resultant wave

$$\hat{\eta} = \hat{\eta}_i + \hat{\eta}_r,$$

where $\hat{\eta}_r$ represents the reflected wave. Check that the reflection angle equals the angle of incidence.

Problem 4.6: Cross Waves in a Wave Channel

An infinitely long wave channel in the region $-d/2 < x < d/2$ is bounded by fixed plane walls, the planes $x = -d/2$ and $x = d/2$. The channel has constant water depth h.

Show that a harmonic wave of the type

$$\hat{\eta} = (A\exp\{-ik_xx\} + B\exp\{ik_xx\})\exp\{-ik_yy\}$$

may propagate in the y direction. Use the boundary conditions to express possible values of k_x in terms of d. Find the limiting value of the angular frequency, below which no cross wave can propagate.

Derive expressions for the phase velocity $v_p = \omega/k_y$ and the group velocity $v_g = d\omega/dk_y$ for the cross waves which may propagate along the wave channel.

Problem 4.7: Kinetic Energy for Progressive Wave

Derive an expression for the time-average of the stored kinetic energy per unit volume associated with a progressive wave

$$\hat{\phi} = -(g/i\omega)Ae(kz)e^{-ikx}$$

in terms of $\rho, g, k, \omega, |A|$ and z.

Problem 4.8: Kinetic Energy Density

Prove that a plane wave

$$\hat{\eta} = Ae^{-kx} + Be^{ikx}$$

on water of constant depth h is associated with a kinetic energy density per unit horizontal area equal to

$$\langle E_k \rangle = (\rho g/4)(|A|^2 + |B|^2)$$

(averaged over the time and over the horizontal plane). Also prove that the time-averaged kinetic energy per unit volume is

$$e_k = \frac{\rho}{2}\left[\frac{e(kz)\,gk}{\omega}\right]^2 (|A|^2 + |B|^2) - \frac{\rho g^2 k^2}{4\omega^2 \cosh^2(kh)}\,|\hat{\eta}|^2.$$

Problem 4.9: Propagation Velocities on Intermediate Water Depth

Find numerical values for the group velocity and phase velocity for a wave of period $T = 9$ s on water of depth $h = 8$ m, and also for the case when $h = 6$ m. This problem involves numerical (or graphical) solution of the transcendental equation

$$\omega^2 = gk\tanh(kh)$$

(the dispersion relation). The acceleration of gravity is $g = 9.81$ m/s^2.

Problem 4.10: Evanescent Waves

The complex amplitude $\hat{\phi}$ of a velocity potential satisfying the Laplace equation may be written

$$\hat{\phi} = X(x)\,Z(z),$$

where $X(x)$ and $Z(z)$ have to satisfy the corresponding differential equations

$$X''(x) = -\lambda\,X(x),$$
$$Z''(z) = \lambda\,Z(z).$$

Further, $Z(z)$ has to satisfy the boundary conditions

$$Z'(-h) = 0, \quad -\omega^2 Z(0) + g\,Z'(0) = 0.$$

Solutions for $\hat{\phi}$ corresponding to negative values of λ, $\lambda = -m^2$, represent so-called evanescent waves.

Consider a propagating wave travelling in an infinite basin of water depth h, where some discontinuities at $x = x_1$ and $x = x_2$ $(x_2 > x_1)$ introduce evanescent waves. Show that an evanescent wave in the x region $-\infty < x < x_1$ or $x_2 < x < \infty$ does not transport any time-average power. Is it possible that real energy transport is associated with an evanescent wave in the region $x_1 < x < x_2$?

Problem 4.11: Slowly Varying Water Depth and Channel Width

A regular wave

$$\eta = \eta(x,t) = A_o\cos(\omega t - k_0 x)$$

is travelling in deep water. Its amplitude is $A_o = 1$ m and its period is $T = 9$ s.

(a) Give mathematical expressions and numerical values for the angular frequency ω and the angular repetency (wave number) k_0. Also determine the wavelength

λ_0. Give mathematical expressions and numerical values for the phase velocity v_{p0}, the group velocity v_{g0} and the maximum water particle velocity $v_{0,\max}$.

Now assume that this wave is moving toward a coast line which is normal to the x axis. The water depth varies slowly with x, $h = h(x)$, which means that we may neglect reflection (or partial reflection) of the wave. If we also neglect energy losses (e.g., caused by friction at the sea bed), the wave-energy transport J (wave power per unit width of the wave front) stays constant as the wave travels into shallower water (Note: for simplicity we assume that the water depth does not vary with the y coordinate.)

(b) At some position $x = x_1$ the water depth is $h = h(x_1) = 8$ m. Here the phase velocity is $v_{p1} = 8.27$ m/s and the group velocity is $v_{g1} = 7.24$ m/s. Give relative numerical values of the wavelength λ_1 and the wave amplitude A_1 at this location; that is, state the numerical values of (λ_0/λ_1) and (A_1/A_0).

(c) Note that for a 9 s wave in 8 m depth, neither the deep-water nor the shallow-water approximations give accurate values for v_{g1} and v_{p1}. If you had used those approximations, how many percent too small or too large would the computed values be for v_{g1} and v_{p1}?

A tapered horizontal wave channel with vertical walls has its entrance (mouth) at the position x_1 (see Figure 4.14). The entrance width is b_1 ($b_1 = 30$ m). As an approximation we assume that the wave continues as a plane wave into the channel, but the narrower width available to the wave results in an increased wave amplitude. If the channel width changes slowly with x, $b = b(x)$, we may neglect partial reflections of the wave. Neglecting also other energy losses,

(d) derive an expression for the wave amplitude A_2 at x_2 in terms of A_1, b_1 and b_2 where b_2 is the channel width at x_2 ($x_2 > x_1$). The water depth is still 8 m at x_2. The bottom of the vertical channel wall is at the same depth, of course. The top of the vertical wall is 3.5 m above the mean water level. Find the numerical value of b_2 when it is given that $A_2 = 3.5$ m. (Remember that in the deep sea, the wave amplitude is $A_0 = 1$ m.)

(e) If the water depth at x_2 had been 6 m instead of 8 m, while the depth at the entrance remains 8 m, find for this case the wave amplitude A_2' at x_2' in terms

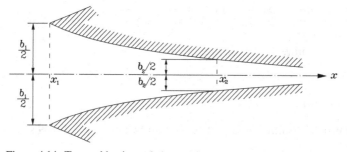

Figure 4.14: Tapered horizontal channel for wave concentration.

of A_1, b_1, b_2, v_{g1} and v_{g2}. In a 6-m depth the 9-s wave has a group velocity of $v_{g2} = 6.60$ m/s. With this channel (where the depth decreases slowly with increasing x), what would be the width b_2' at x_2' if $A_2' = 3.5$ m?

(f) If 5% of the energy was lost (in friction and other processes) when the wave travelled from deep water to the position x_1, how would this influence the answers under point (b)? (Will the numerical values be increased, decreased or unchanged?) If changed, find the new values.

(g) Moreover, if there is also energy loss when the wave travels from x_1 to x_2, how would that influence the answers under point (d)? Assuming that this energy loss is 20% (and that 5% of the deep-water wave energy was lost before the entrance of the channel), obtain new numerical values for the answers under (d).

Problem 4.12: Depth Function $D(kh)$

(a) Show that the depth function $D(kh)$ defined by Eq. (4.80) has a maximum x_0 for $kh = x_0$, where x_0 is a root of the transcendental equation

$$x_0 \tanh(x_0) = 1.$$

(b) Further, show that

$$2k \int_{-h}^{0} \{e(kz)\}^2 \, dz = D(kh),$$

where

$$e(kz) = \frac{\cos h(kh + kz)}{\cos h(kh)}.$$

(See also Problem 4.2.)

(c) Show that the depth function may be written as

$$D(kh) = \left(1 + \frac{4khe^{-2kh}}{1 - e^{-4kh}}\right) \frac{1 - e^{-2kh}}{1 + e^{-2kh}}.$$

This expression is recommended for use in computer programmes which are run with large values of kh.

Problem 4.13: Transmission and Reflection at a Barrier

We consider the following two-dimensional problem. In a wave channel, which is infinitely long in the x direction and which has a constant water depth h, there is at $x = 0$ a barrier, a stiff plate of negligible thickness, occupying the region $x = 0$, $-h_1 > z > -h_2$ ($h \geq h_2 > h_1 \geq 0$). A wave propagating in the positive x direction originates from $x = -\infty$, and it is partially reflected and partially transmitted at the barrier.

In the region $x < 0$ the complex amplitude of the velocity potential may be written as

$$\hat{\phi}(x, z) = A_0 Z_0(z) e^{-ikx} + \sum_{n=0}^{\infty} b_n Z_n(z) \exp\{m_n x\},$$

where the first term on the right-hand side represents the incident wave. The orthogonal and complete set of vertical eigenfunctions $\{Z_n(z)\}$ is defined by Eq. (4.73), and m_n $(n \geq 1)$ is a positive solution of Eq. (4.74), while $m_0 = ik$. See also Eqs. (4.77), (4.78) and (4.115). For the region $x > 0$ we correspondingly write

$$\hat{\phi}(x, z) = \sum_{n=0}^{\infty} a_n Z_n(z) \exp\{-m_n x\}.$$

Although the coefficients $\{a_n\}$ and $\{b_n\}$ are unknown, we consider A_0 to be known. We define the complex reflection and transmission coefficients by

$$\Gamma = b_0/A_0, \quad T = a_0/A_0,$$

respectively. Obviously $\hat{\phi}(x, z)$ satisfies the Laplace equation and the homogeneous boundary conditions (4.42) at $z = -h$ and (4.43) at $z = 0$.

Next we have to apply continuity and boundary conditions at the plane $x = 0$. Firstly, $\partial \hat{\phi}/\partial x$ (representing the horizontal component of the fluid velocity) has to be continuous there (for $0 > z > -h$), and moreover it has to vanish on the barrier ($-h_1 < z < -h_2$). Finally, $\hat{\phi}$, representing the hydrodynamic pressure, see Eq. (4.39), has to be continuous above and below the barrier (i.e., on $0 > z > -h_1$ and on $-h_2 > z > h$). On this basis show that $a_n = -b_n$ for $n \geq 1$, whereas $a_0 = A_0 - b_0$. The latter result means that $\Gamma + T = 1$, and energy conservation means that $|\Gamma|^2 + |T|^2 = 1$. Using this, show that $\Gamma T^* + \Gamma^* T = 0$. Moreover, show that the set of coefficients $\{b_n\}$ has to satisfy an equation of the type

$$\sum_{m=0}^{\infty} B_{mn} b_m = c_n A_0 \quad \text{for } n = 0, 1, 2, 3, \ldots,$$

which may be solved approximately by truncating the infinite sum and then using a computer for numerical solution. Derive an expression for B_{mn} and for c_n. [Hint: use the orthogonality condition – Eqs. (4.70) and (4.76) – i.e.,

$$\int_{-h}^{0} Z_m(z) Z_m(z) \, dz = h \delta_{mn}.$$

If the integral is taken only over the interval $-h_1 > z > -h_2$ it takes another value, which we may denote by $h D_{mn}$. Likewise we define $h E_{mn}$ as the integral taken over the remaining two parts to the interval $0 > z > -h$. Thus $D_{mn} + E_{mn} = \delta_{mn}$. Observe that both D_{mn} and E_{mn} have to enter (explicitly or implicitly) into the expression for B_{mn}, in order for B_{mn} to be influenced by the boundary condition on the barrier as well as by the condition of continuous hydrodynamic pressure above and below the barrier.]

Problem 4.14: Energy-Absorbing Wall

Assume that a wave

$$\hat{\eta}_i = A \exp\{-ik(x \cos \beta + y \sin \beta)\} \quad (|\beta| < \pi/2)$$

is incident on a vertical wall $x = 0$ where there is a boundary condition

$$\hat{p} = Z_n \hat{v}_x = Z_n \frac{\partial \hat{\phi}}{\partial x} \quad (x = 0).$$

Here the constant parameter Z_n is a distributed surface impedance (of dimension Pa s/m = N s/m^3). Show that the time-average absorbed power per unit of the surface of the wall is $\frac{1}{2}\text{Re}\{Z_n\}|\hat{v}_x|^2$. Further show that a wave

$$\eta_r = \Gamma A \exp\{ik(x \cos \beta - y \sin \beta)\}$$

is reflected from the wall. Assume infinite water depth and derive an expression for Γ in terms of ω, ρ and Z_n. Show that $|\Gamma| = 1$ if $\text{Re}\{Z_n\} = 0$ and that $|\Gamma| < 1$ if $\text{Re}\{Z_n\} > 0$.

Problem 4.15: Wave-Energy Transport with Two Waves

A harmonic wave on water of constant depth h is composed of two plane waves propagating in different directions. Let the wave elevation be given by

$$\eta = Ae^{-ikx} + Be^{-ik(x \cos \beta + y \sin \beta)}$$

Derive an expression for the intensity

$$\vec{I} = \overline{p(t)\vec{v}(t)}$$

which, by definition, is a time-independent vector. In general it depends on x and y. Show that, in spite of this dependence there is nowhere an accumulation of energy. Discuss the cases $\beta = 0$, $\beta = \pi$ and $\beta = \pi/2$.

Derive expressions for the space-averaged wave-energy transport $\langle J_x \rangle$ and $\langle J_y \rangle$. Pay special attention to the case of $\cos\beta = 1$.

CHAPTER FIVE

Wave-Body Interactions

The subject of this chapter is, as with Chapter 3, a discussion on the interaction between waves and oscillating bodies. Now, however, the discussion is limited to the case of interaction with water waves, which we discussed in some detail in Chapter 4. Let us start by studying body oscillations and wave forces on bodies. Next let us consider the phenomenon of wave generation by oscillating bodies, and let us discuss more general relationships between the two kinds of wave-body interactions. Some particular consideration is given to two-dimensional cases and to axisymmetric cases. Whereas most of the analysis is carried out in the frequency domain, some studies in the time domain are also made.

5.1 Six Modes of Body Motion: Wave Forces and Moments

We choose a coordinate system with the z axis through the centre of gravity of a body immersed in water. Further, we choose a reference point $(z = z_0)$ on the z axis, for instance the centre of gravity or (as in Figure 5.1) $z = 0$. The body has a wet surface S, which separates it from the water. Consider a surface element dS at position $\vec{s}$ (Figure 5.1). The vector $\vec{s}$ originates in the chosen reference point. Let $\vec{U}$ be the velocity of the reference point. The surface element dS has velocity

$$\vec{u} = \vec{U} + \vec{\Omega} \times \vec{s}, \tag{5.1}$$

where $\vec{\Omega}$ is the angular velocity vector corresponding to rotation about the reference point. Thus, with no rotation $\vec{u} = \vec{U}$ and with no translation, $\vec{u} = \vec{\Omega} \times \vec{s}$. For a given body the motion of each point $\vec{s}$ (of the wet body surface S) is characterised by the time-dependent vectors $\vec{U}$ and $\vec{\Omega}$.

The velocity potential ϕ must satisfy the boundary condition (4.17) $\vec{n} \cdot \nabla \phi = \vec{n} \cdot \vec{u}$ (or $\partial \phi / \partial n = u_n$) everywhere on S:

$$\frac{\partial \phi}{\partial n} = \vec{n} \cdot \nabla \phi = u_n = \vec{U} \cdot \vec{n} + \vec{\Omega} \times \vec{s} \cdot \vec{n} = \vec{U} \cdot \vec{n} + \vec{\Omega} \cdot \vec{s} \times \vec{n}. \tag{5.2}$$

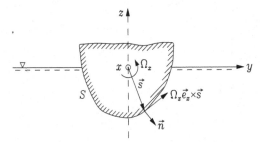

Figure 5.1: Body oscillating in water. Vector $\vec{s}$ gives the position of a point of wet surface S, where the unit normal is $\vec{n}$.

Here, u_n is the normal component of velocity $\vec{u}$ of the wet-surface element dS of the body.

5.1.1 Six Modes of Motion

We introduce six-dimensional generalised vectors, a velocity vector **u** with components

$$(u_1, u_2, u_3) \equiv (U_x, U_y, U_z) = \vec{U}, \tag{5.3}$$
$$(u_4, u_5, u_6) \equiv (\Omega_x, \Omega_y, \Omega_z) = \vec{\Omega}, \tag{5.4}$$

and a normal vector **n** with components

$$(n_1, n_2, n_3) \equiv (n_x, n_y, n_z) = \vec{n}, \tag{5.5}$$
$$(n_4, n_5, n_6) \equiv \vec{s} \times \vec{n}. \tag{5.6}$$

Thus $n_4 = (\vec{s} \times \vec{n})_x = s_y n_z - s_z n_y$, and so on. Note that the components numbered 1 to 3 have different dimension from the remaining components numbered 4 to 6. Thus u_1, u_2 and u_3 have SI units of m/s, whereas u_4, u_5 and u_6 have SI units of rad/s.

The motion of the rigid body is characterised by six components corresponding to six degrees of freedom or modes of (oscillatory) motion. For a ship like, elongated body (directed parallel to the x axis, as indicated in Figure 5.2), the modes are named as shown in Table 5.1. For an axisymmetric body or another not-elongated body, the mode numbers 1 and 2 (and 4 and 5) are ambiguous. They may be

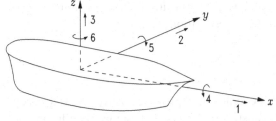

Figure 5.2: A rigid body has six modes of motion: surge (1), sway (2), heave (3), roll (4), pitch (5) and yaw (6).

Table 5.1. Six modes of oscillatory motion of
a body

Mode No.	Component	Mode Name
1	$u_1 = U_x$	surge
2	$u_2 = U_y$	sway
3	$u_3 = U_z$	heave
4	$u_4 = \Omega_x$	roll
5	$u_5 = \Omega_y$	pitch
6	$u_6 = \Omega_z$	yaw

arbitrarily interchanged. However, we may remove this ambiguity when there is
an incident wave where the propagation direction defines the x direction.

Note that in the two-dimensional case (see Section 5.8), that is, when there
is no variation in the y direction, there are only three modes of motion, namely
surge, heave and pitch.

With

$$u_n = \vec{u} \cdot \vec{n} = \vec{U} \cdot \vec{n} + \vec{\Omega} \times \vec{s} \cdot \vec{n} = \vec{U} \cdot \vec{n} + \vec{\Omega} \cdot \vec{s} \times \vec{n}$$
$$= (u_1 n_1 + u_2 n_2 + u_3 n_3) + (u_4 n_4 + u_5 n_5 + u_6 n_6) \qquad (5.7)$$

the boundary condition (5.2) becomes

$$\frac{\partial \phi}{\partial n} = \sum_{j=1}^{6} u_j n_j = \mathbf{u}^T \mathbf{n} \quad \text{on } S. \qquad (5.8)$$

Here the superscript T denotes transpose. Thus $\mathbf{u}^T = (u_1, u_2, u_3, u_4, u_5, u_6)$ is
a line vector, whereas $\mathbf{u}$ is the corresponding column vector. Moreover, $\mathbf{n} =
(n_1, n_2, n_3, n_4, n_5, n_6)^T$. These generalised vectors are six dimensional.

Note that in the above equations the variables $\vec{U}$, $\vec{\Omega}$, $u_j(t)$ and $\phi(x, y, z, t)$
may, in the case of harmonic oscillations and waves, be replaced by their complex
amplitudes $\hat{U}$, $\hat{\Omega}$, $\hat{u}_j(t)$ and $\hat{\phi}(x, y, z, t)$. Thus

$$\frac{\partial \hat{\phi}}{\partial n} = \sum_{j=1}^{6} \hat{u}_j n_j = \hat{\mathbf{u}}^T \mathbf{n} \quad \text{on } S. \qquad (5.9)$$

Let us, for a while, consider the radiation problem. When the body oscillates,
a radiated wave ϕ_r is generated which is a superposition (linear combination) of
radiated waves caused by each of the six oscillation modes,

$$\hat{\phi}_r = \sum_{j=1}^{6} \varphi_j \hat{u}_j, \qquad (5.10)$$

where $\varphi_j = \varphi_j(x, y, z)$ is a complex coefficient of proportionality. The frequency-
dependent complex amplitude $\hat{\phi}_r$ must satisfy the body-boundary condition

$$\frac{\partial \varphi_j}{\partial n} = n_j \quad \text{on } S, \qquad (5.11)$$

because the complex amplitude $\hat{\phi}_r$ of the radiated velocity potential must satisfy the boundary condition (5.9). Whereas ϕ_r has SI unit m^2/s, φ_1, φ_2 and φ_3 have unit m, and φ_4, φ_5 and φ_6 have unit m^2 (or, more precisely, m^2/rad). The coefficient φ_j may be interpreted as the complex amplitude of the radiated velocity potential which is due to body oscillation in mode j with unit velocity amplitude ($\hat{u}_j = 1$). Note that, in the present linear theory, viscosity is neglected. Coefficients φ_j are independent of the oscillation amplitude, because boundary condition (5.2) as well as the Laplace equation (4.35) are linear.

Coefficient φ_j must satisfy the same homogeneous equations as $\hat{\phi}_r$, namely the Laplace equation

$$\nabla^2 \varphi_j = 0, \tag{5.12}$$

the sea-bed boundary condition

$$\left[\frac{\partial \varphi_j}{\partial z}\right]_{z=-h} = 0 \tag{5.13}$$

and the free-surface boundary condition

$$\left[-\omega^2 \varphi_j + g \frac{\partial \varphi_j}{\partial z}\right]_{z=0} = 0. \tag{5.14}$$

[See Eqs. (4.35), (4.42) and (4.43)]. Moreover, φ_j (and $\hat{\phi}_r$) must satisfy a radiation condition at infinite distance. This will be discussed later (also see Section 4.6).

Remember that in this linear theory it is consistent to take the boundary condition (5.14) at the mean free surface $z = 0$ instead of at the real actual free surface $z = \eta$. Similarly, when we use boundary condition (5.11), we may consider S to be the mean wet surface of the body instead of the actual wet surface.

5.1.2 Hydrodynamic Force Acting on a Body

Next, let us derive expressions for the force and the moment which act on the body, in terms of a given velocity potential ϕ.

Firstly, let us consider the vertical (or "heave") force component F_z. The vertical force on the element dS (Figure 5.3) is $p(-n_z)\,dS = -pn_3\,dS$. Here p is the hydrodynamic pressure. Integrating over the wet surface S gives the total heave

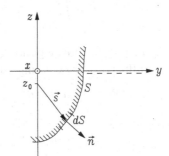

Figure 5.3: Surface element dS of wet surface S of a rigid body with unit normal $\vec{n}$ and with position $\vec{s}$ relative to the chosen reference point $(x, y, z) = (0, 0, z_0)$ of the body.

force

$$F_3 \equiv F_z = - \iint_S p n_3 \, dS. \tag{5.15}$$

In terms of complex amplitudes we have, using Eq. (4.39), that is, $\hat{p} = -i\omega\rho\hat{\phi}$,

$$\hat{F}_3 = \hat{F}_z = i\omega\rho \iint_S \hat{\phi} n_3 \, dS. \tag{5.16}$$

Analogous expressions apply for the horizontal force components, that is, surge force F_1 and sway force F_2.

Next, let us consider the moment about the x axis. The force moment acting on surface element dS is

$$dM_x = s_y dF_z - s_z dF_y = (-pn_z s_y + pn_y s_z) \, dS. \tag{5.17}$$

Integrating over the whole wet surface S gives

$$M_x = - \iint_S p(s_y n_z - s_z n_y) \, dS = - \iint_S p(\vec{s} \times \vec{n})_x \, dS. \tag{5.18}$$

Thus, using Eq. (5.6), that is, $n_4 = (\vec{s} \times \vec{n})_x$, we find that the roll moment is

$$M_x = - \iint_S p n_4 \, dS. \tag{5.19}$$

In terms of complex amplitudes,

$$\hat{M}_x = i\omega\rho \iint_S \hat{\phi} n_4 \, dS. \tag{5.20}$$

Similar expressions may be obtained for pitch moment M_y and yaw moment M_z.

We now define a generalised force vector having six components:

$$\mathbf{F} \equiv (F_1, F_2, F_3, F_4, F_5, F_6) \equiv (F_x, F_y, F_z, M_x, M_y, M_z) = (\vec{F}, \vec{M}). \tag{5.21}$$

The three first components have SI unit N (newton), and the three remaining ones have SI unit N m. For component j ($j = 1, 2, \ldots, 6$) we have

$$F_j = - \iint_S p n_j \, dS, \tag{5.22}$$

$$\hat{F}_j = i\omega\rho \iint_S \hat{\phi} n_j \, dS \tag{5.23}$$

for an arbitrary, given potential $\hat{\phi}$.

If the body is oscillating, surface element dS receives an instantaneous power

$$\begin{aligned}
dP(t) &= -\vec{u} \cdot \vec{n} \, p \, dS \\
&= -(\vec{U} \cdot \vec{n} + \vec{\Omega} \times \vec{s} \cdot \vec{n}) \, p \, dS = -(\vec{U} \cdot \vec{n} + \vec{\Omega} \cdot \vec{s} \times \vec{n}) \, p \, dS \\
&= -(u_1 n_1 + u_2 n_2 + u_3 n_3 + u_4 n_4 + u_5 n_5 + u_6 n_6) \, p \, dS. \tag{5.24}
\end{aligned}$$

Integrating over the wet surface, we get the power received by the oscillating body,

$$P(t) = \vec{F}(t) \cdot \vec{U}(t) + \vec{M}(t) \cdot \vec{\Omega}(t) = \sum_{j=1}^{6} F_j u_j, \qquad (5.25)$$

where we have used Eqs. (5.1), (5.7), (5.21) and (5.22).

5.1.3 Excitation Force

If the body is fixed, a non-vanishing potential $\hat{\phi}$ is the result of an incident wave, only. In such a case F_j is called the "excitation force" (or for $j = 4, 5, 6$, the "excitation moment"). Then with no body motion, no radiated wave is generated. The generalised vector of the *excitation force* is written

$$\mathbf{F}_e \equiv (F_{e,1}, F_{e,2}, \dots, F_{e,6}) = (\vec{F}_e, \vec{M}_e), \qquad (5.26)$$

and using Eq. (5.23), we have

$$F_{e,j} = i\omega\rho \iint_S (\hat{\phi}_0 + \hat{\phi}_d) n_j \, dS, \qquad (5.27)$$

where $\hat{\phi}_0$ represents the undisturbed incident wave and $\hat{\phi}_d$ the diffracted wave. The latter wave is induced when the former wave does not satisfy the homogeneous boundary condition (4.18) on the fixed wet surface S. The boundary condition

$$-\frac{\partial \hat{\phi}_d}{\partial n} = \frac{\partial \hat{\phi}_0}{\partial n} \quad \text{on } S \qquad (5.28)$$

has to be satisfied for the diffraction problem. Note that $\hat{\phi}_d$ as well as $\hat{\phi}_0$ satisfy the homogeneous boundary conditions (4.42) on the sea bed, $z = -h$, and (4.43) on the free water surface, $z = 0$.

If the diffraction term $\hat{\phi}_d$ is neglected in Eq. (5.27), the resulting force is termed the Froude-Krylov force, which is considered in more detail in Section 5.6. It may represent a reasonable approximation to the excitation force, in particular if the extension of the immersed body is very small compared with the wavelength. It may be computationally convenient to use such an approximation because it is not then required to solve the boundary-value problem for finding the diffraction potential $\hat{\phi}_d$.

Except for very simple geometries it is not possible to find solutions to the diffraction problem in terms of elementary functions. However, for the case of a wave

$$\hat{\eta}_0 = A e^{-ikx} \qquad (5.29)$$

incident upon a vertical wall at $x = 0$, the diffracted wave is simply the totally reflected wave

$$\hat{\eta}_d = A e^{ikx} \qquad (5.30)$$

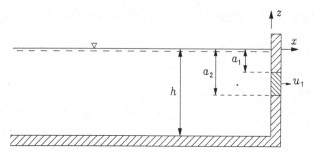

Figure 5.4: Piston in a vertical end wall of a basin with water of depth h. The piston, which is able to oscillate in surge (horizontal motion), occupies, in this two-dimensional problem, the strip $-a_2 < z < -a_1$.

(cf. Problem 4.5). Then, see Eq. (4.92), the velocity potential is given by

$$\hat{\phi}_0 + \hat{\phi}_d = -\frac{g}{i\omega} e(kz)(\hat{\eta}_0 + \hat{\eta}_d) = -\frac{2g}{i\omega} e(kz) A \cos(kx). \tag{5.31}$$

Using Eq. (5.27), and noting that, on S, $x = 0$ and $n_1 = -1$, we now find the surge component of the excitation force (per unit width in the y direction)

$$\hat{F}'_{e,1} = -i\omega\rho \int_{-a_2}^{-a_1} (\hat{\phi}_0 + \hat{\phi}_d)\, dz = 2\rho g A \int_{-a_2}^{-a_1} e(kz)\, dz \tag{5.32}$$

for the striplike piston shown in Figure 5.4. (The dash is used to denote a quantity per unit width. $\hat{F}'_{e,1}$ has SI unit N/m.) Inserting the integrand from Eq. (4.54) and carrying out the integration give (cf. Problem 5.2)

$$\hat{F}'_{e,1} = 2\rho g A \frac{\sinh(kh - ka_1) - \sinh(kh - ka_2)}{k \cosh(kh)}. \tag{5.33}$$

Setting $a_2 = h$ and $a_1 = 0$, we find the surge excitation force per unit width of the whole vertical wall:

$$\hat{F}'_{e,1} = 2\rho g A \frac{\sinh(kh)}{k \cosh(kh)} = 2\rho A \left(\frac{\omega}{k}\right)^2, \tag{5.34}$$

where we have made use of the dispersion relationship (4.57).

If the incident wave is as given by Eq. (5.29), then, according to Eqs. (4.92) and (5.28), $\hat{\phi}_0$ and $\hat{\phi}_d$ are proportional to the complex elevation amplitude A at the origin $(x, y) = (0, 0)$. Hence, also all excitation force components $\hat{F}_{e,j}$ are proportional to A. It should be remembered that the excitation force on a body is the wave force when the body is not moving. Thus, for instance, the surge velocity u_1 (see Figure 5.4) was zero for the situation analysed above. In the next section (subsection 5.2.3), however, we shall consider the case when $u_1 \neq 0$.

Although the excitation force in the above example is in phase with the incident wave elevation at the origin, there may, in the general case, be a non-zero phase difference θ_j between these two variables such that

$$\frac{\hat{F}_{e,j}}{A} = \frac{|\hat{F}_{e,j}|}{|A|} \exp(i\theta_j). \tag{5.35}$$

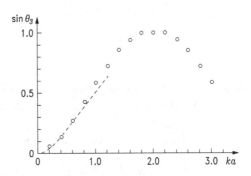

Figure 5.5: Phase angle θ_3 between the heave excitation force and the undisturbed incident wave at the body centre for a semi-submerged floating sphere of radius a. The circle points are numerically computed by Greenhow.[38] The curve is computed by Kyllingstad[39] by using a second-order scattering method.

As we shall see later (in Section 5.6), for a floating body which is very much smaller than the wavelength we have $\theta_1 \approx \pi$ and $\theta_3 \approx 0$ for the surge and heave modes, respectively. For a floating semisubmerged sphere on deep water, numerically computed values for θ_3 are given by the graph[38,39] in Figure 5.5 (provided the sphere is located such that its vertical axis coincides with the z axis). It can be seen from the graph that $\theta_3 < 0.1$ for $ka < 0.25$. For the semisubmerged sphere the amplitude ratio $|\hat{F}_{e,j}|/|A|$ may be obtained by utilising matter presented later [Eqs. (5.200) and (5.279) in combination with the ϵ curve in Figure 5.6].

5.2 Radiation from an Oscillating Body

5.2.1 The Radiation Impedance Matrix

In the following paragraphs let us consider a case when a body is oscillating in the absence of an incident wave. We shall study the forces acting on the body which are due to the wave which is radiated as a result of the body's oscillation. Let the body oscillate in a single mode j with complex velocity amplitude $\hat{u}_j$. The radiated wave is associated with a velocity potential ϕ_r given by

$$\hat{\phi}_r = \varphi_j \hat{u}_j, \tag{5.36}$$

where $\varphi_j = \varphi_j(x, y, z)$ is a coefficient of proportionality, as introduced in Eq. (5.10). The radiated wave reacts with a force on the body. Component j' of the force is

$$\hat{F}_{r,j'} = i\omega\rho \iint_S \varphi_j \hat{u}_j n_{j'} dS \tag{5.37}$$

according to Eqs. (5.23) and (5.36).

Whether j denotes a translation mode ($j = 1, 2, 3$) or a rotation mode ($j = 4, 5, 6$), $\hat{u}_j$ is a constant under the integration. This reflects the fact that the body is rigid. Hence, we may write

$$\hat{F}_{r,j'} = -Z_{j'j}\hat{u}_j \tag{5.38}$$

where

$$Z_{j'j} = -i\omega\rho \iint\limits_{S} \varphi_j n_{j'} \, dS \tag{5.39}$$

is an element of the so-called *radiation impedance* matrix. The SI units are $\text{N s/m} = \text{kg/s}$ for $Z_{q'q}$ $(q', q = 1, 2, 3)$, N s m (or N s m/rad) for $Z_{p'p}$ $(p', p = 4, 5, 6)$ and N s for Z_{qp} and Z_{pq}.

Using the wet-surface boundary condition (5.11), we have

$$Z_{j'j} = -i\omega\rho \iint\limits_{S} \varphi_j \frac{\partial\varphi_{j'}}{\partial n} \, dS. \tag{5.40}$$

Whereas φ_j is complex, $\partial\varphi_j/\partial n$ is real on S, because n_j is real; compare boundary condition (5.11). Hence, we may in the integrand replace $\partial\varphi_{j'}/\partial n$ by $\partial\varphi_{j'}^*/\partial n$ if we so wish. Thus, we have the alternative formula

$$Z_{j'j} = -i\omega\rho \iint\limits_{S} \varphi_j \frac{\partial\varphi_{j'}^*}{\partial n} \, dS. \tag{5.41}$$

Note that φ_j has to satisfy boundary conditions elsewhere – see Eqs. (5.13) and (5.14). This means, for instance, that the radiation-impedance matrix of a body does not have the same value if it is placed in a wave channel as if it is placed in open sea.[40,41] Note that a homogeneous boundary condition such as Eq. (4.18), namely $\partial\varphi_j/\partial n = 0$, has to be satisfied on the vertical walls of the wave channel.

We may interpret $-Z_{j'j}$ as the j' component of the reaction force which is due to wave radiation from mode j oscillating with unit amplitude ($\hat{u}_j = 1$). In fact, it is equal to the j component of the reaction force which is due to wave radiation from mode j' oscillating with unit amplitude ($\hat{u}_{j'} = 1$). This follows from the "reciprocity" relation

$$Z_{jj'} = Z_{j'j}, \tag{5.42}$$

which will be proved below. Thus, the 6×6 radiation-impedance matrix is symmetric.

The following derivation proves the reciprocity relation. Using Eqs. (5.40), (4.240) and (4.249), we obtain

$$Z_{j'j} - Z_{jj'} = -i\omega\rho \, I(\varphi_j, \varphi_{j'}) = 0 \tag{5.43}$$

when we utilise the fact that φ_j and $\varphi_{j'}$ satisfy the same radiation condition.

For certain body geometries some of the elements of the radiation impedance matrix vanish. If $y = 0$ is a plane of symmetry (which is typical for a ship hull), then n_2, n_4 and n_6 in Eq. (5.39) are odd functions of y, whereas φ_1, φ_3 and φ_5 are even functions. Hence $Z_{21} = Z_{23} = Z_{25} = Z_{41} = Z_{43} = Z_{45} = Z_{61} = Z_{63} = Z_{65} = 0$. Moreover, if $x = 0$ is a plane of symmetry, then n_1, n_5 and n_6 are odd functions of x, whereas φ_2, φ_3 and φ_4 are even functions. Hence $Z_{12} = Z_{13} = Z_{14} = Z_{52} = Z_{53} = Z_{54} = Z_{62} = Z_{63} = Z_{64} = 0$. From this observation, while noting Eq. (5.42), it follows that if both $y = 0$ and $x = 0$ are planes of symmetry, then the only non-vanishing off-diagonal elements of the radiation impedance matrix are $Z_{15} = Z_{51}$ and $Z_{24} = Z_{42}$.

Because ω is real, it is convenient to split $Z_{j'j}$ into real and imaginary parts

$$Z_{j'j} = R_{j'j} + i X_{j'j} = R_{j'j} + i\omega m_{j'j}, \tag{5.44}$$

where we term $R_{j'j}$ the *radiation resistance* matrix, $X_{j'j}$ the *radiation reactance* matrix and $m_{j'j}$ the *added-mass* matrix. Note that $R_{j'j}$ is called the "added damping coefficient matrix" by many writers. Equation (5.44) is a generalisation of the scalar Eqs. (3.33) and (3.37).

5.2.2 Energy Interpretation of the Radiation Impedance

We have previously, in Eq. (3.29), defined a radiation resistance by considering the radiated power. We shall later, compare Eqs. (5.177) and (5.188), in a rather general way, relate radiation resistance and reactance to power and energy associated with wave radiation. At present let us, however, just consider a diagonal element of the radiation-impedance matrix. Using Eq. (5.41) we have

$$\begin{aligned} \frac{1}{2} Z_{jj} |\hat{u}_j|^2 &= \frac{1}{2} Z_{jj} \hat{u}_j \hat{u}_j^* \\ &= \frac{1}{2} \iint_S (-i\omega\rho\, \varphi_j \hat{u}_j) \frac{\partial}{\partial n} (\varphi_j^* \hat{u}_j^*) \, dS. \end{aligned} \tag{5.45}$$

Further, using Eqs. (5.36), (4.10) and (4.39), we find that this becomes

$$\begin{aligned} \frac{1}{2} Z_{jj} |\hat{u}_j|^2 &= \frac{1}{2} \iint_S (-i\omega\rho\hat{\phi}_r) \frac{\partial \hat{\phi}_r^*}{\partial n} \, dS \\ &= \frac{1}{2} \iint_S \hat{p}\, \hat{v}_n^* \, dS = \mathcal{P}_r. \end{aligned} \tag{5.46}$$

Referring to Eqs. (3.17) and (3.20), we see that the real part of $\mathcal{P}_r$ is the (time-average) power P_r delivered to the fluid from the body oscillating in mode j:

$$P_r = \mathrm{Re}\{\mathcal{P}_r\} = \tfrac{1}{2} R_{jj} |\hat{u}_j|^2. \tag{5.47}$$

Here R_{jj} is the diagonal element of the radiation-resistance matrix

$$R_{jj'} = \mathrm{Re}\{Z_{jj'}\}. \tag{5.48}$$

The real part P_r of the radiated "complex power" $\mathcal{P}_r$ [see Eq. (2.92)] represents the radiated "active power". The imaginary part

$$\mathrm{Im}\{\mathcal{P}_r\} = \tfrac{1}{2} X_{jj} |\hat{u}_j|^2 \tag{5.49}$$

represents the radiated "reactive power". The radiation-reactance matrix is

$$X_{jj'} = \text{Im}\{Z_{jj'}\}. \tag{5.50}$$

We may note that (see Problem 5.7 or Subsection 7.2.2)

$$P_r = \sum_{j'=1}^{6} \sum_{j=1}^{6} Z_{j'j}\hat{u}_{j'}\hat{u}_j^* \tag{5.51}$$

is a generalisation of Eq. (5.46), which is valid when all except one of the oscillation modes have a vanishing amplitude.

5.2.3 Wavemaker in a Wave Channel

Let us consider an example referring to Figure 5.4, when the piston has a surge velocity with complex amplitude $\hat{u}_1 \neq 0$. To be slightly more general we shall, for this example, write the inhomogeneous boundary condition (5.11) as

$$\frac{\partial \varphi_1}{\partial x} = c(z) \quad \text{for } x = 0, \tag{5.52}$$

where $c(z)$ is a given function. For the case shown in Figure 5.4 we must set $c(z) = 1$ for $-a_2 < z < -a_1$ and $c(z) = 0$ elsewhere. If we set $c(z) = 1$ for $-h < z < 0$, this corresponds to the whole vertical wall oscillating as a surging piston. Another case would be to set $c(z) = 1 + z/h$ for $-h < z < 0$. This corresponds to an oscillating flap, hinged at $z = -h$. Such a flap is typical as a wavemaker in a wave channel.

The given oscillation at $x = 0$ generates a wave which propagates in the negative x direction toward $x = -\infty$ (Figure 5.4). Hence it is a requirement that our solution of the boundary-value problem is finite in the region $x < 0$, and that it satisfies the radiation condition at $x = -\infty$. Referring to Eq. (4.115), which is a solution to the boundary-value problem given by Eqs. (4.35), (4.42) and (4.43), we can immediately write down the following solution:

$$\varphi_1 = c_0 \exp(ikx)Z_0(z) + \sum_{n=1}^{\infty} c_n \exp(m_n x)Z_n(z) = \sum_{n=0}^{\infty} X_n(x) Z_n(z), \tag{5.53}$$

which (for $j = 1$) satisfies the Laplace equation (5.12), the homogeneous boundary conditions (5.13) and (5.14), the finiteness condition for $-\infty < x < 0$ and the radiation condition for $x = -\infty$. [Concerning the orthogonal functions $Z_n(z)$ used in Eq. (5.53), it may be referred to equations (4.70)–(4.79).] For convenience, we have introduced $m_0 = ik$ and

$$X_n(x) = c_n \exp(m_n x). \tag{5.54}$$

In Eq. (5.53) the terms corresponding to $n \geq 1$ represent evanescent waves which are non-negligible only near the wave generator. Thus in the far-field region $-x \gg 1/m_1$ we have asymptotically

$$\varphi_1 \sim c_o \exp(ikx)Z_0(z). \tag{5.55}$$

In order to determine the unknown constants c_n in Eq. (5.53), we use boundary condition (5.52) at $x = 0$:

$$c(z) = \left[\frac{\partial \varphi_1}{\partial x}\right]_{x=0} = \sum_{n=0}^{\infty} X'_n(0)\, Z_n(z). \qquad (5.56)$$

We multiply by $Z_m^*(z)$, integrate from $z = -h$ to $z = 0$ and use orthogonality condition (4.70). This gives

$$\int_{-h}^{0} c(z)\, Z_m^*(z)dz = \sum_{n=0}^{\infty} X'_n(0) \int_{-h}^{0} Z_m^*(z)\, Z_n(z)\, dz = X'_m(0)h. \qquad (5.57)$$

That is,

$$X'_n(0) = \frac{1}{h} \int_{-h}^{0} c(z)\, Z_n^*(z)\, dz. \qquad (5.58)$$

According to Eq. (5.54) we have

$$X'_0(0) = ikc_0, \quad X'_n(0) = m_n c_n, \qquad (5.59)$$

and hence

$$c_0 = \frac{1}{ikh} \int_{-h}^{0} c(z)\, Z_0^*(z)\, dz, \qquad (5.60)$$

$$c_n = \frac{1}{m_n h} \int_{-h}^{0} c(z)\, Z_n^*(z)\, dz. \qquad (5.61)$$

Thus, c_n is a kind of "Fourier" coefficient for the expansion of the velocity-distribution function $c(z)$ in the function "space" spanned by the complete function set $\{Z_n(z)\}$.

The velocity potential of the radiated wave is

$$\hat{\phi}_r = \varphi_1 \hat{u}_1, \qquad (5.62)$$

where φ_1 is given by Eq. (5.53). The wave elevation of the radiated wave is [cf. Eq. (4.41)]

$$\hat{\eta}_r = \frac{i\omega}{-g} [\hat{\phi}_r]_{z=0} = \frac{i\omega}{-g} \hat{u}_1 \sum_{n=0}^{\infty} c_n Z_n(0)\, \exp(m_n x). \qquad (5.63)$$

In the far-field region, that is, for $-x \gg 2h/\pi > 1/m_1$, the evanescent waves are negligible, and we have the asymptotic solution

$$\hat{\eta}_r \sim A_r^- \exp(ikx) = a_1^- \hat{u}_1 \exp(ikx), \qquad (5.64)$$

where the far-field coefficient A_r^- for the wave radiated in the negative x direction is given by

$$a_1^- = \frac{A_r^-}{\hat{u}_1} = \frac{i\omega}{-g} c_0 Z_0(0) = \frac{i\omega}{-g} c_0 \sqrt{\frac{2kh}{D(kh)}}, \qquad (5.65)$$

where Eqs. (4.54) and (4.79) have been used in the last step. Note that a_1^- is real, because $c(z)$ and $Z_0(z)$ are real, and hence [cf. Eq. (5.60)] also ic_0 is real.

If Figure 5.4 represents a wave channel of width d, the radiation-impedance of the wavemaker is, according to Eq. (5.41), with $\partial/\partial n = -\partial/\partial x$,

$$Z_{11} = i\omega\rho\, d \int_{-h}^{0} \left[\varphi_1 \frac{\partial \varphi_1^*}{\partial x} \right]_{x=0} dz. \tag{5.66}$$

Considering Figure 5.4 as representing a two-dimensional problem, we find that the radiation impedance per unit width is

$$Z_{11}' = Z_{11}/d = i\omega\rho \int_{-h}^{0} \left[c_0 Z_0(z) + \sum_{n=1}^{\infty} c_n Z_n(z) \right] c^*(z)\, dz$$

$$= i\omega\rho \left[c_0(-ikh)c_0^* + \sum_{n=1}^{\infty} c_n(m_n h)c_n^* \right], \tag{5.67}$$

where we have used Eqs. (5.52), (5.53), (5.60) and (5.61) in Eq. (5.66). Hence,

$$Z_{11}' = \frac{Z_{11}}{d} = \omega k\rho h |c_0|^2 + i\omega\rho h \sum_{n=1}^{\infty} m_n |c_n|^2. \tag{5.68}$$

The radiation resistance is

$$R_{11} = \text{Re}\{Z_{11}\} = \omega k\rho |c_0|^2 hd. \tag{5.69}$$

The added mass is

$$m_{11} = \frac{1}{\omega} \text{Im}\{Z_{11}\} = \rho h d \sum_{n=1}^{\infty} m_n |c_n|^2 \tag{5.70}$$

[cf. Eqs. (5.48) and (5.44)]. Because all m_n are positive we see that all contributions to the radiation reactance ωm_{11} are positive. We also see that the radiation resistance is associated with the far field (corresponding to the propagating wave), whereas the radiation reactance and hence the added mass are associated with the near field (corresponding to the evanescent waves).

It might be instructive to consider the complex energy transport, $\mathcal{J}(x)$, associated with the radiated wave. Referring to Eqs. (2.92) and (4.130), we obtain the complex-wave energy transport (which in this case is in the negative x direction):

$$\mathcal{J}(x) = \int_{-h}^{0} \frac{1}{2} \hat{p}(-\hat{v}_x)^*\, dz = -\frac{1}{2} i\omega\rho \int_{-h}^{0} \hat{\phi}_r \left(-\frac{\partial \hat{\phi}_r^*}{\partial x} \right) dz$$

$$= \frac{i\omega\rho}{2} \sum_{n=0}^{\infty} \sum_{l=0}^{\infty} X_n(x) X_l'^*(x) \int_{-h}^{0} Z_n(z) Z_l^*(z)\, dz\, \hat{u}_1 \hat{u}_1^*$$

$$= \frac{1}{2} i\omega\rho h \sum_{n=0}^{\infty} X_n(x) X_n'^*(x) |\hat{u}_1|^2, \tag{5.71}$$

where we have utilised Eqs. (5.53), (5.62) and (4.70). Use of Eqs. (5.59) and (5.71) gives

$$
\mathcal{J}(x) = \left[\frac{1}{2} \omega k h \rho |c_0|^2 + \frac{1}{2} i \omega \rho h \sum_{n=1}^{\infty} m_n |c_n|^2 \exp(2m_n x) \right] |\hat{u}_1|^2. \tag{5.72}
$$

We see that, because of orthogonality condition (4.70), there are no cross terms from different vertical eigenfunctions in the product.

The (active) power transport per unit width of wave frontage is

$$
J = \mathrm{Re}\{\mathcal{J}(x)\} = \tfrac{1}{2} \omega k h \rho |c_0|^2 |\hat{u}_1|^2. \tag{5.73}
$$

Moreover, we see that there is a reactive power $\mathrm{Im}\{\mathcal{J}(x)\}$ which exists only in the near field. This demonstrates that the radiation reactance and hence the added mass are related to reactive energy transport in the near-field region of the wavemaker. The active energy transport J in a plane wave in an ideal (loss free) liquid is of course independent of x. We see that the radiated power, in accordance with Eq. (3.29), is

$$
\tfrac{1}{2} R_{11} |\hat{u}_1|^2 = P_r = J d = \mathrm{Re}\{\mathcal{J}(x)\} d = \mathcal{J}(-\infty)\, d. \tag{5.74}
$$

Correspondingly the radiation impedance Z_{11}, as given by Eq. (5.68), satisfies

$$
\tfrac{1}{2} Z_{11} |\hat{u}_1|^2 = \mathcal{J}(0) d, \tag{5.75}
$$

in accordance with Eq. (5.46).

Imagine now that we have a slightly flexible wavemaker, such that for a particular value of k (and hence of ω) a horizontal oscillation can be realised where

$$
c(z) = e(kz). \tag{5.76}
$$

Using Eq. (4.79) in Eq. (5.60) we then find that

$$
c_0 = \frac{i}{k} \sqrt{\frac{D(kh)}{2kh}}, \tag{5.77}
$$

and from Eq. (4.70) that $c_n = 0$ for $n \geq 1$. From Eq. (5.69) we then find that

$$
R_{11} = \omega k \rho \, \frac{D(kh)}{2k^3}\, d = \frac{\omega \rho \, D(kh) d}{2k^2}, \tag{5.78}
$$

and from Eq. (5.70) that $m_{11} = 0$. In such a case we have wave generation without added mass and without any evanescent wave. The reason is that the far-field solution alone [cf. Eq. (5.55)] matches the wavemaker boundary condition (5.52) on the wavemaker's wet surface, when $c(z)$ is chosen as in Eq. (5.76).

In contrast, if we are able to realise

$$
c(z) = \frac{\cos(m_1 z + m_1 h)}{\cos(m_1 h)}, \tag{5.79}
$$

we generate only one evanescent mode and no progressive wave. (Note that in this case the lower and upper parts of the vertical end wall have to oscillate in opposite phases.) For this example, the radiation resistance vanishes at the particular frequency for which Eq. (5.79) is satisfied. This shows that it is, at least in principle, possible to make oscillations in the water without producing a propagating wave. This knowledge may be of use if it is desired that a body oscillating in the sea with a particular frequency shall not generate undesirable waves.

Finally, let us consider the case of a stiff surging vertical plate. That is, we choose $c(z) = 1$ for $-h < z < 0$. Then, in general, $c_n \neq 0$ for all n. In particular, we have

$$c_0 = \frac{i\omega^2}{gk^3h} \sqrt{\frac{2kh}{D(kh)}} \tag{5.80}$$

as shown in Problem 5.3. From Eqs. (5.65) and (5.69) we then find that the far-field coefficient is

$$a_1^- = \frac{2\omega^3}{g^2k^2 D(kh)} \tag{5.81}$$

and that the radiation resistance is

$$R_{11} = \frac{2\omega^5 \rho d}{g^2k^4 D(kh)}. \tag{5.82}$$

If a wave

$$\hat{\eta}_0 = A\exp(-ikx) \tag{5.83}$$

originating from $x = -\infty$ is incident upon the stiff vertical plate at $x = 0$, we have, in accordance with Eq. (5.34), a surge excitation force given by

$$\hat{F}_{e,1} = 2A\frac{\rho\omega^2 d}{k^2}. \tag{5.84}$$

Let us now, as a prelude to sections on reciprocity relations (Sections 5.4 and, in particular, 5.8), establish that the following relations hold in the present example, with interaction of a surging vertical plate with a wave on one of its sides (the left-hand side, as shown in Figure 5.4). The radiation resistance and the excitation force may be expressed in terms of the far-field coefficients as

$$R_{11} = \frac{\rho g^2 D(kh)d}{2\omega} |a_1^-|^2, \tag{5.85}$$

$$\hat{F}_{e,1} = A\frac{\rho g^2 D(kh)d}{\omega} a_1^-, \tag{5.86}$$

respectively. Using these relations in Eq. (3.45), we see that at optimum oscillation the maximum power absorbed by the body is

$$P_{max} = \frac{|\hat{F}_{e,1}|^2}{8R_{11}} = \frac{\rho g^2 D(kh)d}{4\omega} |A|^2 = Jd, \tag{5.87}$$

which, according to Eq. (4.136), is the complete energy transport of the incident wave in the wave channel of width d. In this case the absorbing body generates a wave which, in the far-field region, cancels the reflected wave. Thus $A_r^- = a_1^- \hat{u}_1 = -A$ or

$$\hat{u}_1 = \hat{u}_{1,\mathrm{opt}} = -A/a_1^- = -Aa_1^-/|a_1^-|^2 = \hat{F}_{e,1}/2R_{11}, \tag{5.88}$$

in accordance with Eq. (3.46). Note that a_1^- is real in the present example. See Eq. (5.65) and observe that c_0 is purely imaginary. The results of Eqs. (5.87) and (5.88) are examples of matter discussed in a more general and systematic way in Chapter 6 (see Section 6.1 and Subsection 6.4.3).

5.2.4 Examples of Other Body Geometries

For a general geometry it is complicated to solve the radiation problem for an oscillating body; that is, it is complicated to solve the boundary-value problem represented by Eqs. (5.11)–(5.14). A simple example was discussed above in Subsection 5.2.3. In the present subsection some numerically obtained results are presented for the radiation impedance and also for the excitation force for some particular bodies. These quantities are examples of so-called hydrodynamic parameters.

Let us at first consider a sphere of radius a semisubmerged on water of infinite depth. With this body geometry it is obvious that rotary modes cannot generate any wave in an ideal fluid. Moreover, all off-diagonal elements of the radiation impedance vanish, because both $x = 0$ and $y = 0$ are planes of symmetry. Thus the only non-vanishing elements are $Z_{11} = Z_{22}$ and Z_{33}, which we write as

$$Z_{jj} = (\omega \rho\, 2\pi a^3/3)(\epsilon_{jj} + i\,\mu_{jj}), \tag{5.89}$$

where ϵ and μ are non-dimensionalised radiation resistance and added mass, respectively. These parameters are shown by curves shown in Figure 5.6, which is based on previous numerical results.[42,43] We may note that the radiation resistance tends to zero as ka approaches infinity or zero. The added mass is finite in both

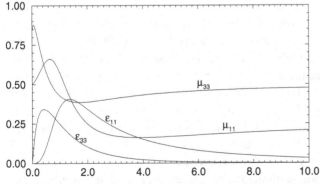

Figure 5.6: Graphs showing non-dimensionalised surge and heave coefficients of radiation resistance and of added mass versus ka for a semisubmerged sphere of radius a on deep water. Here $k = \omega^2/g$ is the angular repetency. The numerical values are from Havelock[42] and Hulme.[43] The radiation impedance for surge ($j = 1$) and heave ($j = 3$) is given by $Z_{jj} = (\omega \rho\, 2\pi a^3/3)(\epsilon_{jj} + i\mu_{jj})$.

these limits. Observe that

$$\frac{R_{jj}}{\omega m_{jj}} = \frac{\epsilon_{jj}}{\mu_{jj}} = \mathcal{O}\{(ka)^n\} \quad \text{as } ka \to 0, \tag{5.90}$$

where $n = 2$ for the surge mode and $n = 1$ for the heave mode. It can be shown (cf. Problem 5.10) that

$$\epsilon_{33} \to (3\pi/4)ka \quad \text{as } ka \to 0. \tag{5.91}$$

When $ka \ll 1$, the diameter is much shorter than the wavelength. Also for body geometries other than a sphere, it is generally so that the radiation reactance dominates over the radiation resistance when the extension of the body is small compared with the wavelength.

As the next example we present in Figure 5.7 the excitation-force coefficient

$$f_3 \equiv F_{e,3}/A = \kappa_3 \rho g \pi a^2 \tag{5.92}$$

and the radiation-impedance

$$Z_{33} = (\omega \rho 2\pi a^3/3)(\epsilon_{33} + i\mu_{33}) \tag{5.93}$$

for the heave mode of a floating truncated vertical cylinder of radius a, and of draft b. The cylinder axis coincides with the z axis ($x = y = 0$). The numerical values for these hydrodynamic parameters have been computed by using a method described by Eidsmoen.[44] We observe from Figure 5.7 that, for low frequencies ($ka \to 0$), the heave excitation force has a magnitude as we could expect by simply applying Archimedes' law, which neglects effects of wave interference and of wave diffraction. For high frequencies ($ka \to \infty$) the force tends to zero, which was to be expected because of the mentioned effects, but also because of the decrease of hydrodynamic pressure with increasing submergence below the free water surface.

From the lower left-hand graph we see that for large frequencies the curve for the phase approaches a straight line which (in radians) has a steepness equal to 1. This means that the heave excitation force is, for $ka \to \infty$, in phase with the incident wave elevation at $x = -a$ (i.e., where the wave first hits the cylinder), whereas for $ka \to 0$ it is in phase with the (undisturbed) incident wave elevation at $x = 0$. We observe that the curves for ϵ_{33} and μ_{33} in Figure 5.7 are qualitatively similar to the corresponding curves in Figure 5.6 for the semisubmerged sphere. Note, however, that ϵ_{33} is, at larger frequencies, significantly smaller for the vertical cylinder. The limiting values for $ka \to 0$ are also different. It can be shown (cf. Problem 5.10) that for finite water depth h, $\epsilon_{33} \to 3\pi a/8h$, whereas for infinite water depth $\epsilon_{33} \to 3\pi ka/4$ as $ka \to 0$. The reason for the different appearance of the two ϵ_{33} curves for $ka \to 0$ is that infinite water depth was assumed with the semisubmerged sphere results in Figure 5.6. For the vertical cylinder $\epsilon_{33} \to 0.078$ as $ka \to 0$.

The final example for which numerical results are presented in this subsection is the ISSC (International Ship and Offshore Structures Congress) tension-leg platform (TLP)[45] in water with a depth of $h = 450$ m. The platform consists of four cylindrical columns (each of radius 8.435 m) and four pontoons of rectangular

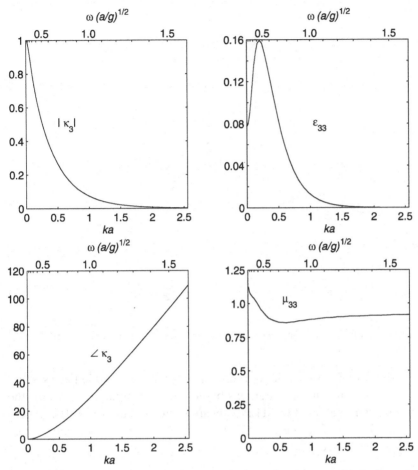

Figure 5.7: Non-dimensional graphs for the hydrodynamic parameters for the heave mode of a floating truncated vertical cylinder of radius a and of draft $b = 1.88a$ on water of depth $h = 15a$. In each graph the horizontal scales give ka (lower scale) and $\omega(a/g)^{1/2}$ (upper scale). The four graphs give, in dimensionless values, the amplitude $|\kappa_3|$ and phase $\angle\kappa_3$ (in degrees) of the excitation force, and the real and imaginary parts of the radiation impedance $\epsilon_{33} + i\mu_{33}$.

cross section (7.5 m wide and 10.5 m high) connecting the columns. The distance between adjacent columns is $2L = 86.25$ m. The draft of the TLP is 35.0 m, corresponding to the lowermost surface of the columns as well as of the pontoons. This is a large structure with a geometry which is not simple. To numerically solve the hydrodynamic boundary-value problems as given in Subsection 5.1.1 or in the first part of Section 4.2, we have applied the computer programme WAMIT.[46] The wet surface of the immersed TLP is approximated to 512 plane panels (see Figure 5.8). The structure is symmetric with respect to the planes $x = 0$ and $y = 0$, and the width and length are equal. The computer programme WAMIT provides values for the fluid velocity, for the hydrodynamic pressure, for the body parameters (e.g., immersed volume, centre of gravity, and hydrostatic coefficients), for drift forces and for motion response, in addition to the hydrodynamic parameters

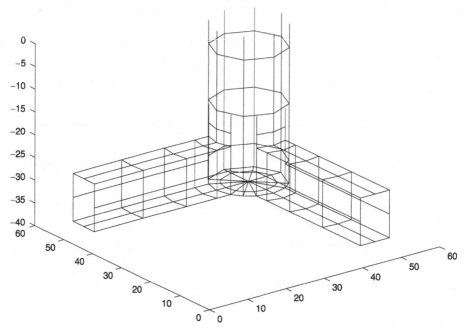

Figure 5.8: One quadrant of the ISSC TLP as approximated by 128 panels.[46] The scales on the indicated coordinate system are in metres.

(i.e., excitation forces and moments, radiation resistance and added masses and inertia moments). Dimensionless graphs based on the computed results for the hydrodynamic parameters of the TLP body are shown in Figure 5.9. The dimensionless hydrodynamic parameters $A_{jj'}$, $B_{jj'}$ and X_j are defined as follows from the radiation impedance,

$$Z_{jj'} = \omega \rho L^k (B_{jj'} + i A_{jj'}),\tag{5.94}$$

and the excitation force,

$$\hat{F}_{e,j} = f_j A = \rho g A L^m X_j,\tag{5.95}$$

where $L = 43.125$ m. Further, $m = 2$ for $j = 1$ and $j = 3$, whereas $m = 3$ for $j = 5$, and $k = 3$ for $(j, j') = (1, 1)$ and $(j, j') = (3, 3)$, whereas $k = 4$ for $(j, j') = (1, 5)$ and $k = 5$ for $(j, j') = (5, 5)$. The excitation force computed here applies for the case when the incident wave propagates in the x direction, and A is the wave elevation at the origin $(x, y) = (0, 0)$.

We observe that the curves in Figure 5.9 have a somewhat wavy nature. This is due to constructive and destructive wave interference associated with different parts of the TLP body. Let us consider, for instance, the surge mode. When the distance $2L$ between adjacent columns is an odd multiple of half wavelengths, corresponding to values of dimensionless frequency $\Omega = 1.25, 2.17, 2.80, \ldots$, the surge excitation force contribution to one pair of columns is cancelled by the contribution from the remaining pair. Similarly, generated waves from the two pairs cancel

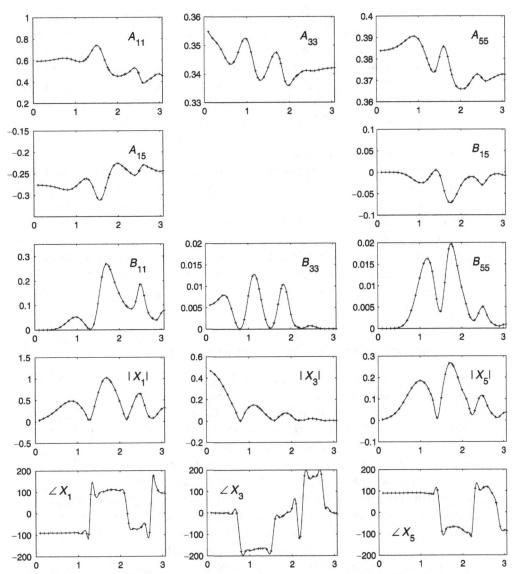

Figure 5.9: Hydrodynamic parameters for the ISSC TLP. The horizontal scale gives the non-dimensional frequency $\Omega = \omega(L/g)^{1/2}$, where $L = 43.125$ m is half of the distance between centres of adjacent TLP columns, and $g = 9.81$ m/s^2. The eight graphs in the three upper rows give the radiation impedance – cf. Eq. (5.94) – and the six graphs in the two lower rows the excitation force – cf. Eq. (5.95) – in dimensionless quantities, for the modes surge ($j = 1$), heave ($j = 3$) and pitch ($j = 5$). In the lowest row the phase is given in degrees. The curves are obtained by using cubic-spline interpolation between computed results, shown as dots in the diagram.

each other. This explains why $|X_1|$ and B_{11} are so small at these frequencies. If, however, the distance $2L$ corresponds to a multiple of wavelengths, which happens for $\Omega = 1.77, 2.51, \ldots$, then we expect constructive interference, which explains why $|X_1|$ and B_{11} are large at these frequencies. We see similar behaviour with the pitch mode. However, with the heave mode, maxima and minima for $|X_3|$ and B_{33} occur at other frequencies. Graphs similar to those shown in Figure 5.9 have been published[47] for a TLP body with six columns. They show, as we could expect, even more interference effects than disclosed by the wavy nature of the curves in Figure 5.9 for a four-column TLP.

A submerged body of the pontoon's cross section and of total length $8L$ displaces a water mass of 0.339 (in normalised units). It is interesting to note that this value is rather close to the added-mass value A_{33} for heave, as given in Figure 5.9. A deeply submerged horizontal cylinder of infinite length has an added mass for heave, as well as for surge, that equals the mass of displaced water. If, however, the cylinder is only slightly below the water surface, the added mass may depend strongly on frequency and in some cases it may even be negative. Numerical results for the added mass, and also for the radiation resistance have been published by McIver and Evans.[19]

5.3 Impulse Response Functions in Hydrodynamics

We previously discussed two kinds of wave-body interaction in the frequency domain. At first (in Section 5.1) we introduced the excitation force, that is, the wave force on the body when the body is held fixed ($u_j = 0$ for $j = 1, \ldots, 6$); see Eqs. (5.27) and (5.33). Then (in Section 5.2.1) we introduced the hydrodynamic radiation parameters: radiation impedance $Z_{jj'}$, see Eq. (5.40), and its real and imaginary parts, radiation resistance $R_{jj'}$ and radiation reactance $X_{jj'}$, respectively. Also the added mass $m_{jj'} = X_{jj'}/\omega$ was introduced; See Eq. (5.44).

To the two kinds of interaction there correspond two kinds of boundary-value problems, which we call the excitation problem and the radiation problem. They have different inhomogeneous boundary conditions on the wet body surface S, namely as given by Eqs. (5.28) and (5.11), respectively.

Following the introduction of generalised six-dimensional vectors (in Section 5.1), we may define a six-dimensional column vector $\hat{\mathbf{F}}_e$ for the excitation force's complex amplitude. Similarly we have a vector $\hat{\mathbf{u}}$, which represents the complex amplitudes of the six components of the velocity of the oscillating body. Correspondingly, we denote the radiation-impedance matrix by $\mathbf{Z} = \mathbf{Z}(\omega)$, which is of dimension 6×6. The reaction force which is due to the oscillating body's radiation is given by

$$\hat{\mathbf{F}}_r = -\mathbf{Z}\,\hat{\mathbf{u}} = -\mathbf{Z}(\omega)\,\hat{\mathbf{u}}, \qquad (5.96)$$

which is an alternative way of writing Eq. (5.38). We may interpret $-\mathbf{Z}$ as the transfer function of a linear system, where $\hat{\mathbf{u}}$ is the input and $\hat{\mathbf{F}}_r$ the output. Similarly to this definition of a linear system for the radiation problem, we may define the following linear system for the excitation problem:

$$\hat{\mathbf{F}}_e = \mathbf{f}(\omega)\, A. \tag{5.97}$$

Here the system's input is $A = \hat{\eta}_0(0, 0)$, which is the complex elevation amplitude of the (undisturbed) incident wave at the origin, $(x, y) = (0, 0)$. Further, the transfer function is the six-dimensional column vector $\mathbf{f}$, which we shall call the excitation-force coefficient vector.

In the following let us generalise to situations in which the oscillation need not be sinusoidal. Then, referring to Eqs. (2.136), (2.161) and (2.166), we have the Fourier transforms of the reaction force $\mathbf{F}_{r,t}(t)$ caused by radiation, and the excitation force $\mathbf{F}_{e,t}(t)$

$$\mathbf{F}_r(\omega) = -\mathbf{Z}(\omega)\mathbf{u}(\omega), \tag{5.98}$$

$$\mathbf{F}_e(\omega) = \mathbf{f}(\omega)A(\omega). \tag{5.99}$$

Here $\mathbf{u}(\omega)$ is the Fourier transform of $\mathbf{u}_t(t)$, and $A(\omega)$ is the Fourier transform of $a(t) \equiv \eta_0(0, 0, t)$, the wave elevation of the undisturbed incident wave at the origin $(x, y) = (0, 0)$. (A subscript t is used to denote the inverse Fourier transforms, which are functions of time.)

The inverse Fourier transforms of transfer functions $\mathbf{f}(\omega)$ and $\mathbf{Z}(\omega)$ correspond to time-domain impulse response functions, introduced into ship hydrodynamics by Cummins.[48] Below let us first discuss the causal linear system given by Eq. (5.98). Later let us discuss the system given by Eq. (5.99); we shall see that this system may be non-causal.

5.3.1 The Kramers-Kronig Relations in Hydrodynamic Radiation

We consider a body which oscillates and thereby generates a radiated wave on otherwise calm water. Let the body's oscillation velocity (in the time domain) be given by $\mathbf{u}_t(t)$. The reaction force $\mathbf{F}_{r,t}(t)$ is given by the convolution product

$$\mathbf{F}_{r,t}(t) = -\mathbf{z}(t) * \mathbf{u}_t(t) \tag{5.100}$$

in the time domain. [See Eq. (2.128), and remember that convolution is commutative. Note, however, that the matrix multiplication implied here is not commutative.] The impulse response matrix $-\mathbf{z}(t)$ is the negative of the inverse Fourier transform of the radiation impedance

$$\mathbf{z}(t) = \mathcal{F}^{-1}\{\mathbf{Z}(\omega)\}. \tag{5.101}$$

Note that this system is causal; that is,

$$\mathbf{z}(t) = 0 \quad \text{for } t < 0, \tag{5.102}$$

which means that there is no output $\mathbf{F}_{r,t}(t)$ before a non-vanishing $\mathbf{u}_t(t)$ is applied as input. For this reason, when we compute the convolution integral in Eq. (5.100), we may apply Eq. (2.177), instead of the more general Eq. (2.128), where the range of integration is from $t = -\infty$ to $t = +\infty$. The fact that $\mathbf{z}(t)$ is a causal function also has some consequence for its Fourier transform

$$\mathbf{Z}(\omega) = \mathbf{R}(\omega) + i\omega\mathbf{m}(\omega), \tag{5.103}$$

namely that the radiation-resistance matrix $\mathbf{R}(\omega)$ and the added-mass matrix $\mathbf{m}(\omega)$ are related by Kramers-Kronig relations. Apparently, Kotik and Mangulis[49] were the first to apply these relations for a hydrodynamic problem. Later, some consequences of the relations were discussed by, for example, Greenhow.[50] Here let us derive the relations in the following way. At first we note that $\mathbf{m}(\omega)$ does not, in general, vanish in the limit $\omega \to \infty$. We wish to remove this singularity by considering the following related transfer function:

$$\mathbf{H}(\omega) = \mathbf{m}(\omega) - \mathbf{m}(\infty) + [\mathbf{R}(\omega)]/i\omega \tag{5.104}$$

Note that $\mathbf{R}(\omega) \to 0$ as $\omega \to \infty$. We may assume that $\mathbf{R}(\omega)$ tends sufficiently fast to zero as $\omega \to 0$ to make $\mathbf{H}(\omega)$ non-singular at $\omega = 0$. The Kramers-Kronig relations (2.204) and (2.206) give for the real and imaginary parts of $\mathbf{H}(\omega)$

$$\mathbf{m}(\omega) - \mathbf{m}(\infty) = \frac{2}{\pi} \int_0^\infty \frac{-\mathbf{R}(y)}{\omega^2 - y^2} \, dy, \tag{5.105}$$

$$\mathbf{R}(\omega) = \frac{2\omega^2}{\pi} \int_0^\infty \frac{\mathbf{m}(y) - \mathbf{m}(\infty)}{\omega^2 - y^2} \, dy. \tag{5.106}$$

By using Eqs. (5.98), (5.103) and (5.104), we find that the reaction force caused by radiation is given by

$$\mathbf{F}_r(\omega) = -\mathbf{Z}(\omega)\mathbf{u}(\omega) = \mathbf{F}'_r(\omega) - i\omega\mathbf{m}(\infty)\mathbf{u}(\omega), \tag{5.107}$$

where

$$\mathbf{F}'_r(\omega) = -i\omega\mathbf{H}(\omega)\mathbf{u}(\omega) = -\mathbf{K}(\omega)\mathbf{u}(\omega). \tag{5.108}$$

Here we have introduced an alternative transfer function

$$\mathbf{K}(\omega) = i\omega\mathbf{H}(\omega) = \mathbf{Z}(\omega) - i\omega\mathbf{m}(\infty)$$
$$= \mathbf{R}(\omega) + i\omega[\mathbf{m}(\omega) - \mathbf{m}(\infty)]. \tag{5.109}$$

The corresponding inverse Fourier transforms are

$$\mathbf{F}_{r,t}(t) = \mathbf{F}'_{r,t}(t) - \mathbf{m}(\infty)\dot{\mathbf{u}}_t(t) \tag{5.110}$$

with

$$\mathbf{F}'_{r,t}(t) = -\mathbf{h}(t) * \dot{\mathbf{u}}_t(t) = -\mathbf{k}(t) * \mathbf{u}_t(t), \tag{5.111}$$

where

$$\mathbf{h}(t) = \frac{1}{2\pi} \int_{-\infty}^\infty \mathbf{H}(\omega) e^{i\omega t} \, d\omega,$$

$$\mathbf{k}(t) = \frac{1}{2\pi} \int_{-\infty}^\infty \mathbf{K}(\omega) e^{i\omega t} \, d\omega. \tag{5.112}$$

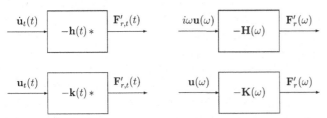

Figure 5.10: Block diagrams of linear systems in the time (left) and frequency (right) domains, where the acceleration (upper) or the velocity (lower) vectors of the oscillating body is considered as input and where a certain part [cf. Eqs. (5.104) and (5.108)] of the radiation reaction force vector is, in both cases, considered as the output.

Note that

$$\mathbf{k}(t) = \frac{d}{dt}\mathbf{h}(t). \tag{5.113}$$

The linear systems corresponding to the two alternative transfer functions $\mathbf{H}(\omega)$ and $\mathbf{K}(\omega)$, or equivalently to the convolutions in Eq. (5.111), are represented by block diagrams in Figure 5.10.

Because of the condition of causality, we have $\mathbf{h}(t) = 0$ and $\mathbf{k}(t) = 0$ for $t < 0$. Moreover, using Eqs. (2.184), (2.185), (5.104) and (5.109), we find for $t > 0$

$$\mathbf{h}(t) = \frac{2}{\pi} \int_0^\infty [\mathbf{m}(\omega) - \mathbf{m}(\infty)] \cos(\omega t)\, d\omega$$

$$= \frac{2}{\pi} \int_0^\infty \frac{\mathbf{R}(\omega)}{\omega} \sin(\omega t)\, d\omega, \tag{5.114}$$

$$\mathbf{k}(t) = \frac{2}{\pi} \int_0^\infty \mathbf{R}(\omega) \cos(\omega t)\, d\omega = 2\mathcal{F}^{-1}\{\mathbf{R}(\omega)\}$$

$$= -\frac{2}{\pi} \int_0^\infty \omega[\mathbf{m}(\omega) - \mathbf{m}(\infty)] \sin(\omega t)\, d\omega. \tag{5.115}$$

Note that as a consequence of the principle of causality, all information contained in the two real matrix functions $\mathbf{R}(\omega)$ and $\mathbf{m}(\omega)$ is contained alternatively in the single real matrix function $\mathbf{k}(t)$ together with the constant matrix $\mathbf{m}(\infty)$. A function of the type $\mathbf{k}(t)$ has been applied in a time-domain analysis of a heaving body, for instance by Count and Jefferys.[51]

As an example let us consider the heave mode of the floating truncated vertical cylinder for which the radiation resistance was given in Figure 5.7. The corresponding impulse response function $k_3(t)$ for radiation is given by the curve in the left-hand graph of Figure 5.11.

5.3.2 Non-causal Impulse Response for the Excitation Force

The impulse responses corresponding to radiation impedance $\mathbf{Z}(\omega)$ and related transfer functions $\mathbf{H}(\omega)$ and $\mathbf{K}(\omega)$, as described above, are causal because their inputs (velocity or acceleration of the oscillating body) are the actual causes of their responses (reaction forces from the radiated wave which is generated by the oscillating body).

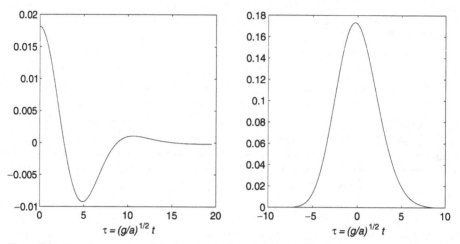

Figure 5.11: Impulse-response functions for the heave mode of a floating truncated vertical cylinder of radius a and of draft $b = 1.88a$ on water of depth $h = 15a$. The scales are dimensionless. The dimensionless time on the horizontal scale is $t(g/a)^{1/2}$. The curve in the left graph gives the dimensionless impulse-response function $k_3/(\rho g a^2)$ for the radiation problem. The curve in the right graph gives the dimensionless impulse-response function $f_{t,3}/[\pi \rho(ga)^{3/2}]$ for the excitation force [cf. Eqs. (5.115) and (5.117)].

In contrast, the impulse response functions related to the excitation forces are not necessarily causal, because the output excitation force, as well as the input incident wave elevation at the origin (or at some other specified reference point), has a distant primary cause such as a storm or an oscillating wavemaker. The linear system, whose output is the excitation force, has the undisturbed incident wave elevation at the origin of the body as input. However, this wave may hit part of the body and exert a force, before the arrival of the wave at the origin.[52] Note, however, that this is not sufficient[31] for a complete explanation of the non-causality of the impulse response function $\mathbf{f}_t(t)$ defined below.

The linear system associated with the excitation problem is represented by Eq. (5.99) or its inverse Fourier transform

$$\mathbf{F}_{e,t}(t) = \mathbf{f}_t(t) * a(t)$$
$$= \int_{-\infty}^{\infty} \mathbf{f}_t(t - \tau)\, a(\tau)\, d\tau = \int_{-\infty}^{\infty} \mathbf{f}_t(\tau)\, a(t - \tau)\, d\tau, \qquad (5.116)$$

where

$$\mathbf{f}_t(t) = \frac{1}{2\pi} \int_{-\infty}^{\infty} \mathbf{f}(\omega)\, e^{i\omega t}\, d\omega \qquad (5.117)$$

is the inverse Fourier transform of the excitation-force-coefficient vector $\mathbf{f}(\omega)$, and where $a(t)$ is the wave elevation which is due to the (undisturbed) incident wave at the reference point $(x, y) = (x_0, y_0)$ of the body. (We may choose this to be the origin, i.e., $x_0 = 0$ and $y_0 = 0$.) Note that, as a result of the non-causality of $\mathbf{f}_t(t)$, we could apply neither Eq. (2.184) nor Eq. (2.185) to compute the inverse Fourier transform in Eq. (5.117). Rather we had to apply Eq. (2.170). When $\mathbf{f}_t(t) \neq 0$ for $t < 0$ it appears from Eq. (5.116) that in order to compute $\mathbf{F}_{e,t}(t)$, future information is required on the wave elevation $a(t)$. For this reason it may be desirable to predict the incident wave a certain time length t_1 into the future, where t_1 is sufficiently large to make $\mathbf{f}_t(t)$ negligible for $t < -t_1$.

The statement that $\mathbf{f}_t(t)$ is not, in general, causal is easy to demonstrate by considering the example of Figure 5.4, where the surge excitation-force coefficient per unit width is

$$f_1'(\omega) = \hat{F}_{e,1}'/A = 2\rho(\omega/k)^2 \tag{5.118}$$

according to Eq. (5.34). Note that k is a function of ω as determined through the dispersion relationship (4.57). Thus $f_1'(\omega)$ is real and an even function of ω. Hence, the inverse Fourier transform $f_{t,1}(t)$ is an even, and hence non-causal, function of time. Let us, for mathematical simplicity, consider the surge excitation force on an infinitesimal horizontal strip, corresponding to $z = -a_1$ and $z + \Delta z = -a_2$ in Figure 5.4, when the water is deep. Then with $k = \omega^2/g$ and $e(kz) = e^{kz}$ [see Eqs. (4.58) and (4.54)], we have from Eq. (5.32)

$$\Delta f_1'(\omega) = \Delta F_{e,1}'/A = 2\rho g \exp(\omega^2 z/g)\Delta z. \tag{5.119}$$

By taking the inverse Fourier transform [see Erdélyi,[37] formula (1.4.11)], we have the corresponding impulse response

$$\Delta f_{t,1}'(t) = \rho g(-g/\pi z)^{1/2} \exp(gt^2/4z)\Delta z, \tag{5.120}$$

which is evidently non-causal, except for the case of $z = 0$, for which

$$\Delta f_{t,1}'(t) = 2\rho g \delta(t)\Delta z \tag{5.121}$$

(see Subsection 4.9.2 and Problem 5.8).

Taking the inverse Fourier transform of the heave excitation force given in Figure 5.7 for a floating truncated vertical cylinder, we find for this example the excitation impulse-response function $f_{3,t}$ as shown in the right-hand graph of Figure 5.11. Looking at the curves in this figure, we may say that the causal radiation impulse response "remembers" roughly fifteen to twenty time units $\sqrt{a/g}$ back into the past. In contrast, the non-causal excitation impulse "remembers" only approximately eight time units $\sqrt{a/g}$ back into the past, but it also "requires" information on the incident wave elevation approximately seven time units $\sqrt{a/g}$ into the future.

5.4 Reciprocity Relations

In this section let us state and prove some relations (so-called reciprocity relations) which may be very useful when we study the interaction between a wave and an oscillating body.

We have already proved Eq. (5.42), which states that the radiation impedance is symmetric, $Z_{jj'} = Z_{j'j}$, or in the matrix notation

$$\mathbf{Z} = \mathbf{Z}^T, \tag{5.122}$$

where $\mathbf{Z}^T$ denotes the transpose of matrix $\mathbf{Z}$. It follows that also the radiation resistance matrix and the added mass matrix are symmetric:

$$\mathbf{R} = \mathbf{R}^T, \qquad \mathbf{m} = \mathbf{m}^T. \tag{5.123}$$

Reciprocity relations of this type are well known in several branches of physics. An analogy is, for instance, the symmetry of the electric impedance matrix for a linear electric network (see, e.g., Section 2.13 in Goldman[53]).

Another analogy, from the subject of statics, is the relationship between a set of forces applied to a body of a linear elastic material and the resulting linear displacements of the points of force attack, where each displacement is in the same direction as the corresponding force. In this case a reciprocity relation may be derived by considering a force F applied statically to two different points a and b. Applied to point a, the displacement is δ_{aa} at point a and δ_{ba} at point b, and the work done is $(F/2)\delta_{aa}$. If applied to point b, the displacement is δ_{ab} at point a and δ_{bb} at point b, and the work done is $(F/2)\delta_{bb}$. Further, if, in addition, an equally large force F is afterward applied to point a, the displacements are $\delta_{ab} + \delta_{aa}$ and $\delta_{bb} + \delta_{ba}$ because the principle of superposition is applicable for the linear system. The work done is

$$W_{ba} = (F/2)\delta_{bb} + (F/2)\delta_{aa} + F\delta_{ba}. \tag{5.124}$$

However, if a force F is first applied to point a and then a second force F is applied to point b, the work would be

$$W_{ab} = (F/2)\delta_{aa} + (F/2)\delta_{bb} + F\delta_{ab}. \tag{5.125}$$

The work done is stored as elastic energy in the material, and this energy must be independent of the order of applying the two forces, that is, $W_{ba} = W_{ab}$. Hence,

$$\delta_{ab} = \delta_{ba}, \tag{5.126}$$

which proves the following reciprocity relation from the subject of statics: The displacement at point b caused by a force at point a equals the displacement at point a if the same force is applied to point b (see, e.g., p. 423 in Timoshenko and Gere[54]).

Several kinds of reciprocity relations relating to hydrodynamics have been formulated and proved by Newman.[30] Some of these relations are considered in the following subsections.

5.4.1 Radiation Resistance in Terms of Far-Field Coefficients

The complex amplitude of the velocity potential of the wave radiated from an oscillating body may, according to Eq. (5.10), be written as

$$\hat{\phi}_r(x, y, z) = \boldsymbol{\varphi}^T(x, y, z)\hat{\mathbf{u}} = \hat{\mathbf{u}}^T\boldsymbol{\varphi}(x, y, z), \tag{5.127}$$

where we have introduced the column vectors $\boldsymbol{\varphi}(x, y, z)$ and $\hat{\mathbf{u}}$, which are six dimensional if the wave is radiated from a single body oscillating in all its six degrees of freedom [see Eqs. (5.3) and (5.4)]. Note that, because $\hat{\phi}_r$ satisfies the radiation condition of an outgoing wave at infinity, we have [see Eqs. (4.222) or (4.266)] the far-field approximations

$$\hat{\phi}_r \sim A_r(\theta)\, e(kz)\, (kr)^{-1/2} e^{-ikr}, \tag{5.128}$$

$$\boldsymbol{\varphi} \sim \mathbf{a}(\theta)\, e(kz)\, (kr)^{-1/2} e^{-ikr}, \tag{5.129}$$

where

$$A_r(\theta) = \mathbf{a}^T(\theta)\, \hat{\mathbf{u}} = \hat{\mathbf{u}}^T \mathbf{a}(\theta) \tag{5.130}$$

is the far-field coefficient for the radiated wave. Note that each component φ_j of the column vector satisfies the radiation condition, and that $a_j(\theta)$ is the far-field coefficient for this component. Using relation (4.282) we have the following Kochin functions for the radiated wave:

$$\begin{bmatrix} H_r(\theta) \\ \mathbf{h}(\theta) \end{bmatrix} = \sqrt{2\pi} \begin{bmatrix} A_r(\theta) \\ \mathbf{a}(\theta) \end{bmatrix} e^{i\pi/4}. \tag{5.131}$$

Here $\mathbf{h}(\theta)$ and $\mathbf{a}(\theta)$ are column vectors composed of components $h_j(\theta)$ and $a_j(\theta)$, respectively.

As a further preparation for proof of reciprocity relations involving radiation parameters, let us introduce the following matrix version of Eq. (4.240):

$$\mathbf{I}(\boldsymbol{\varphi}^*, \boldsymbol{\varphi}^T) = \iint\limits_S \left(\boldsymbol{\varphi}^* \frac{\partial \boldsymbol{\varphi}^T}{\partial n} - \frac{\partial \boldsymbol{\varphi}^*}{\partial n} \boldsymbol{\varphi}^T \right) dS. \tag{5.132}$$

Because $\boldsymbol{\varphi}$ satisfies the radiation condition, we may apply Eq. (4.249), which means that

$$\mathbf{I}(\boldsymbol{\varphi}, \boldsymbol{\varphi}^T) = 0. \tag{5.133}$$

However, the integral does not vanish when we take the complex conjugate of one of the two functions entering into the integral. Thus, the complex conjugate of Eqs. (4.254) and (4.255) may in matrix notation be rewritten as

$$\mathbf{I}(\boldsymbol{\varphi}^*, \boldsymbol{\varphi}^T) = -i \frac{D(kh)}{k} \int_0^{2\pi} \mathbf{a}^*(\theta)\, \mathbf{a}^T(\theta)\, d\theta$$

$$= -\lim_{r\to\infty} 2 \iint\limits_{S_\infty} \frac{\partial \boldsymbol{\varphi}^*}{\partial r} \boldsymbol{\varphi}^T\, dS. \tag{5.134}$$

Note that here $\boldsymbol{I}(\boldsymbol{\varphi}^*, \boldsymbol{\varphi}^T)$, which should not be confused with the identity matrix $\mathbf{I}$, is a square matrix, which is of dimension 6×6 for the case of one (three-dimensional) single body oscillating in all its degrees of freedom.

As Eq. (5.36) has been generalised to Eq. (5.127), it follows that expression (5.41) for the radiation-impedance may be rewritten in matrix notation as

$$\mathbf{Z} = -i\omega\rho \iint\limits_S \frac{\partial \boldsymbol{\varphi}^*}{\partial n} \boldsymbol{\varphi}^T\, dS. \tag{5.135}$$

The complex conjugate of this matrix is

$$\mathbf{Z}^* = i\omega\rho \iint\limits_S \frac{\partial\boldsymbol{\varphi}}{\partial n}\boldsymbol{\varphi}^\dagger \, dS, \tag{5.136}$$

where $\boldsymbol{\varphi}^\dagger = (\boldsymbol{\varphi}^T)^*$ is the complex-conjugate transpose of $\boldsymbol{\varphi}$. Note, however, that $\mathbf{Z}^T = \mathbf{Z}$ [see Eq. (5.122)], and hence

$$\mathbf{Z}^* = \mathbf{Z}^\dagger = i\omega\rho \iint\limits_S \boldsymbol{\varphi}^* \frac{\partial\boldsymbol{\varphi}^T}{\partial n} \, dS. \tag{5.137}$$

The real part of $\mathbf{Z}$ is the radiation-resistance matrix

$$\mathbf{R} = \frac{1}{2}(\mathbf{Z} + \mathbf{Z}^*) = \frac{1}{2}(\mathbf{Z} + \mathbf{Z}^\dagger) =$$

$$= -\frac{i}{2}\,\omega\rho \iint\limits_S \left(\frac{\partial\boldsymbol{\varphi}^*}{\partial n}\boldsymbol{\varphi}^T - \boldsymbol{\varphi}^*\frac{\partial\boldsymbol{\varphi}^T}{\partial n}\right) dS =$$

$$= \frac{i}{2}\omega\rho\,\mathbf{I}(\boldsymbol{\varphi}^*, \boldsymbol{\varphi}^T), \tag{5.138}$$

where Eq. (5.132) has been used. (It has been assumed that ω is real.) Now using Eq. (5.134) in Eq. (5.138), we have

$$\mathbf{R} = -i\omega\rho \lim_{r\to\infty} \iint\limits_{S_\infty} \frac{\partial\boldsymbol{\varphi}^*}{\partial r}\boldsymbol{\varphi}^T \, dS. \tag{5.139}$$

It is interesting to compare this with Eq. (5.135). If the integral is taken over S_∞ in the far field, instead of over the wet surface S of the body, we arrive at the radiation-resistance matrix instead of the radiation-impedance matrix. This indicates that the radiation reactance and hence the added mass are somehow related to the near-field region of the wave-generating oscillating body.

The radiation-resistance matrix may be expressed in terms of the far-field coefficient vector $\mathbf{a}(\theta)$ of the radiated wave. From Eqs. (5.131), (5.134) and (5.138) we obtain

$$\mathbf{R} = \frac{\omega\rho\,D(kh)}{2k} \int_0^{2\pi} \mathbf{a}^*(\theta)\,\mathbf{a}^T(\theta)\,d\theta = \frac{\omega\rho\,D(kh)}{4\pi k} \int_0^{2\pi} \mathbf{h}^*(\beta)\,\mathbf{h}^T(\beta)\,d\beta. \tag{5.140}$$

We have here been able to express the radiation-resistance matrix in terms of far-field quantities of the radiated wave. The physical explanation of this is that the energy radiated from the oscillating body must be retrieved without loss in the far-field region of the assumed ideal fluid. It is possible to derive Eq. (5.140) simply by using this energy argument as a basis for the derivation (see Problem 5.7).

Whereas the radiation resistance is related to energy in the far-field region, the radiation reactance, and hence the added mass, is somehow related to the near-field region, as we remarked above when comparing Eqs. (5.135) and (5.139). We shall return to this matter in a more quantitative fashion in Subsection 5.5.4.

5.4.2 The Excitation Force: the Haskind Relation

Some important relations in which the excitation force is expressed in terms of radiation parameters are derived in the following paragraphs. Using wet-surface conditions (5.11) and (5.28), we may rewrite expression (5.27) for the excitation force as

$$\hat{F}_{e,j} = i\omega\rho \iint\limits_S \left\{ (\hat{\phi}_0 + \hat{\phi}_d)\frac{\partial \varphi_j}{\partial n} - \varphi_j \frac{\partial}{\partial n}(\hat{\phi}_0 + \hat{\phi}_d) \right\} dS$$

$$= i\omega\rho\, I\{(\hat{\phi}_0 + \hat{\phi}_d), \varphi_j\}$$

$$= i\omega\rho\, I\{\hat{\phi}_0, \varphi_j\} + i\omega\rho\, I\{\hat{\phi}_d, \varphi_j\}, \tag{5.141}$$

where integral (4.240) has been used. Furthermore, because $\hat{\phi}_d$ and φ_j both satisfy the radiation condition, which means that $I(\hat{\phi}_d, \varphi_j) = 0$ according to Eq. (4.249), we obtain

$$\hat{F}_{e,j} = i\omega\rho\, I(\hat{\phi}_0, \varphi_j). \tag{5.142}$$

In order to calculate the excitation force from formula (5.27), we need to know the diffraction potential at wet surface S. Alternatively, the excitation force may be calculated from Eq. (5.142), where we need to know not the diffraction potential ϕ_d, but the radiation potential φ_j, either at wet surface S or at control surface S_∞ in the far-field region [see Eq. (4.243)]. This latter method of calculation of the excitation force was first pointed out by Haskind [55,56] for the case of a single body. Equation (5.142) is one way of formulating the so-called Haskind relation.

Using column vector $\boldsymbol{\varphi}$, which we introduced in Eq. (5.127), we may, as in Eq. (5.97), write the excitation force as a column vector

$$\hat{\mathbf{F}}_e = \mathbf{f}A = i\omega\rho\, \mathbf{I}(\hat{\phi}_0, \boldsymbol{\varphi}) \tag{5.143}$$

as an alternative to Eq. (5.142). We have also here introduced the excitation-force-coefficient vector $\mathbf{f}$.

In another version of the Haskind relation, the excitation force may also be expressed in terms of the radiated wave's far-field coefficients or of its Kochin functions. We assume now that the incident plane wave is given as

$$\hat{\eta}_0 = A\, \exp\{-ik(x\cos\beta + y\sin\beta)\} = Ae^{-ikr(\beta)}. \tag{5.144}$$

Here β determines the angle of incidence. See Eqs. (4.102) and (4.277). Note that the frequency-dependent excitation-force-coefficient vector $\mathbf{f} = \hat{\mathbf{F}}_e/A$ is a function also of β, $\mathbf{f} = \mathbf{f}(\beta)$. Applying now Eq. (4.281) with $\phi_i = \hat{\phi}_0$ (i.e., $A_i = A$ and $\beta_i = \beta$) and with $\phi_j = \hat{\phi}_r$ (i.e., $A_j = 0$), we have

$$I(\hat{\phi}_0, \hat{\phi}_r) = \frac{g\,D(kh)}{i\omega k} H_r(\beta \pm \pi)\, A. \tag{5.145}$$

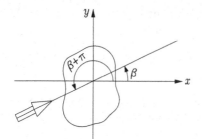

Figure 5.12: Excitation force experienced by an oscillating body caused by a wave with angle of incidence β is related to the body's ability to radiate a wave in the opposite direction, $\theta = \beta + \pi$.

Furthermore, using this together with Eqs. (5.127), (5.130) and (5.131) in Eq. (5.143), we may write the excitation-force vector as

$$\mathbf{F}_e = \mathbf{f}(\beta)\, A = i\omega\rho\, \mathbf{I}(\phi_0, \boldsymbol{\varphi}) = \frac{\rho g\, D(kh)}{k}\, \mathbf{h}(\beta \pm \pi)\, A, \tag{5.146}$$

or for the excitation-force-coefficient vector,

$$\mathbf{f}(\beta) = \frac{\rho g\, D(kh)}{k}\, \mathbf{h}(\beta \pm \pi) = \frac{\rho g\, D(kh)}{k} \sqrt{2\pi}\, \mathbf{a}(\beta \pm \pi)\, e^{i\pi/4}. \tag{5.147}$$

Reciprocity relation (5.147), which is another form of the Haskind relation, means that a body's ability to radiate a wave *into* a certain direction is related to the excitation force which the body experiences when a plane wave is incident *from* just that direction (see Figure 5.12). If we, in an experiment, measure the excitation-force coefficients for various directions of wave incidence β, we can also, by application of reciprocity relation (5.147), obtain the corresponding far-field coefficients for waves radiated in the opposite directions.

5.4.3 Reciprocity Relation Between Radiation Resistance and Excitation Force

Let us, finally, remark that by using the Haskind relation in the form of Eq. (5.147) together with reciprocity relation (5.140) between the radiation resistance and the far-field coefficients, we find

$$\mathbf{R} = \frac{\omega k}{4\pi\rho g^2\, D(kh)} \int_{-\pi}^{\pi} \mathbf{f}(\beta)\, \mathbf{f}^{\dagger}(\beta)\, d\beta = \frac{\omega k}{4\pi\rho g^2\, D(kh)} \int_{-\pi}^{\pi} \mathbf{f}^{*}(\beta)\, \mathbf{f}^{\mathrm{T}}(\beta)\, d\beta. \tag{5.148}$$

Here we have also utilised the fact that $\mathbf{R}$ is a symmetrical matrix; that is, it equals its own transpose. Remembering that $\mathbf{f}(\beta) = \hat{\mathbf{F}}_e(\beta)/A$, we may rewrite Eq. (5.148) as

$$\mathbf{R} = \frac{\omega k}{4\pi\rho g^2\, D(kh)|A|^2} \int_{-\pi}^{\pi} \hat{\mathbf{F}}_e(\beta)\, \hat{\mathbf{F}}_e^{\dagger}(\beta)\, d\beta. \tag{5.149}$$

Introducing the wave-energy transport J per unit frontage, see Eq. (4.136), we may also write the radiation-resistance matrix as

$$\mathbf{R} = \frac{k}{16\pi J} \int_{-\pi}^{\pi} \hat{\mathbf{F}}_e(\beta)\, \hat{\mathbf{F}}_e^{\dagger}(\beta)\, d\beta. \tag{5.150}$$

For instance, if the body is axisymmetric, heave component $F_{e,3}$ of the excitation force is independent of β (the angle of wave incidence) and we have

$$R_{33} = (k/8J)|\hat{F}_{e,3}|^2. \tag{5.151}$$

We may note that the coefficient in front of the integral in Eqs. (5.149) and (5.150) may also be written as $k/(8\pi\rho g v_g |A|^2)$ if Eq. (4.110) for the group velocity v_g is used.

5.5 Several Bodies Interacting with Waves

(The text in this section is partly taken from *Applied Ocean Research*, Vol. 2, J. Falnes, "Radiation impedance matrix and optimum power absorption for inter-acting oscillators in surface waves", pp. 75–80, 1980, with permission from Elsevier Science.) So far, in this chapter we have studied the interaction between waves and a body which may oscillate in six independent modes, as introduced in Section 5.1. There are two different kinds of interaction, as represented by two hydrodynamic parameters: the excitation force and the radiation impedance.

Let us now consider the interaction between waves and an arbitrary number of oscillating bodies, partly or totally submerged in water (see Figure 5.13). A systematic study of many hydrodynamically interacting bodies was made inde-pendently by Evans[57] and Falnes,[21] following Budal's analysis[58] of wave-energy absorption by such a system of several bodies.

At first in the following subsection we shall introduce the excitation force and the radiation impedance in a phenomenological way. Later we shall relate these phenomenologically defined parameters with hydrodynamic theory.

Each oscillating body has, in general, six degrees of freedom, corresponding to three translatory modes and three rotary modes. We chose a coordinate system

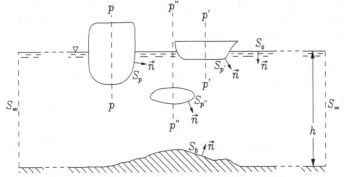

Figure 5.13: System of interacting oscillating bodies contained within an imaginary cylindrical control surface S_∞ at water depth h. Vertical lines through (the average position of) the centre of gravity of the bodies are indicated by $p - p$, $p' - p'$ and $p'' - p''$. Wet surfaces of oscillating bodies are indicated by S_p, $S_{p'}$ and $S_{p''}$. Fixed surfaces, including the sea bed, are given by S_b, and S_0 denotes the external free water surface. The arrows indicate unit normals pointing into the fluid region.

with the z-axis pointing upward and with the plane $z = 0$ coinciding with the average free water surface. We shall consider each body as six *oscillators*. Thus, if the number of oscillating bodies is N, there are $6N$ oscillators. Parameters pertaining to a particular oscillator will be denoted by a subscript

$$i = 6(p-1) + j, \tag{5.152}$$

where $j = 1, 2, \ldots, 6$ is the mode number, and where p is the number of the body.

With an assumed angular frequency ω, the state of an oscillator is given by amplitude and phase. These two real quantities are conveniently incorporated into a single complex quantity, the complex amplitude. Thus, we shall specify the state of oscillator number i by the complex velocity amplitude $\hat{u}_i$, or alternatively by $\hat{u}_{pj}$. This is a translatory velocity for the modes surge, sway and heave ($j = 1, 2, 3$), and an angular velocity for the modes roll, pitch and yaw ($j = 4, 5, 6$).

5.5.1 Phenomenological Discussion

When all oscillators are non-moving, an excitation force, represented by its complex amplitude $\hat{F}_{e,i}$, acts on oscillator number i. The excitation force is primarily caused by a given incoming wave, but diffraction effects caused by the fixed bodies are included in the quantity $\hat{F}_{e,i}$. When oscillator number i' oscillates with a complex velocity amplitude $\hat{u}_{i'}$, it radiates a wave which acts on oscillator number i with an additional force having a complex amplitude $-Z_{ii'}\hat{u}_{i'}$. In linear theory the complex coefficient $-Z_{ii'}$ is a factor of proportionality, but it depends on ω and on the geometry of the problem. Using the principle of superposition, we find that the total force acting on oscillator number i becomes

$$\hat{F}_{t,i} = \hat{F}_{e,i} - \sum_{i'} Z_{ii'}\hat{u}_{i'}, \tag{5.153}$$

where the sum is taken over all oscillators, including also oscillator number i.

The set of complex amplitudes of the total force, the excitation force and the oscillator velocity are conveniently assembled into column vectors, $\hat{\mathbf{F}}_t$, $\hat{\mathbf{F}}_e$ and $\hat{\mathbf{u}}$, respectively. Considering Eq. (5.153) as a set of equations for all oscillators, we may write it in matrix form as

$$\hat{\mathbf{F}}_t = \hat{\mathbf{F}}_e - \mathbf{Z}\hat{\mathbf{u}}, \tag{5.154}$$

where $\mathbf{Z}$ is a square matrix with elements $Z_{ii'}$.

We call $\mathbf{Z}$ the radiation-impedance matrix in analogy with term usage in theories for antennas and for acoustic transducers. The complex matrix $\mathbf{Z}$ may be decomposed into its real part $\mathbf{R}$, the radiation-resistance matrix, and imaginary part $\mathbf{X}$, the radiation-reactance matrix. Thus,

$$\mathbf{Z} = \mathbf{R} + i\mathbf{X} = \mathbf{R} + i\omega\mathbf{m}, \tag{5.155}$$

where ω is the angular frequency and $\mathbf{m}$ is the hydrodynamic added-mass matrix.

In general $\mathbf{Z}$ is a $6N \times 6N$ matrix. It may be partitioned into 6×6 matrices $\mathbf{Z}_{pp'}$, which are defined as follows. Assume that body number p' is oscillating with complex velocity amplitudes given by the generalised six-dimensional vector $\hat{\mathbf{u}}_{p'}$. This results in a contribution $-\mathbf{Z}_{pp'}\hat{\mathbf{u}}_{p'}$ to the generalised six-dimensional vector representing the complex amplitudes of the radiation force on body number p. For the particular case of $p' = p$, $\mathbf{Z}_{pp'} = \mathbf{Z}_{pp}$ represents the ordinary radiation-impedance matrix for body number p, but modified because of the presence of all the other bodies. The radiation-impedance matrix for the whole system of the N oscillating bodies is, thus, a partitioned matrix of the type

$$\mathbf{Z} = \begin{bmatrix} \ddots & \vdots & & \vdots & \\ \cdots & \mathbf{Z}_{pp} & \cdots & \mathbf{Z}_{pp'} & \cdots \\ & \vdots & \ddots & \vdots & \\ \cdots & \mathbf{Z}_{p'p} & \cdots & \mathbf{Z}_{p'p'} & \cdots \\ & \vdots & & \vdots & \ddots \end{bmatrix}. \tag{5.156}$$

In this way matrix $\mathbf{Z}$ is composed of N^2 matrices.

If $p' \neq p$, and if the distance $d_{pp'}$ between the two bodies' vertical reference axes $p - p$ and $p' - p'$ (see Figure 5.13) is sufficiently large, we expect that matrix $\mathbf{Z}_{pp'}$ depends on the distance $d_{pp'}$ approximately as

$$\mathbf{Z}_{pp'} \sim (\mathbf{Z}_{pp'})_0 \sqrt{d_0/d_{pp'}} \, \exp\{-ikd_{pp'}\}, \tag{5.157}$$

where $(\mathbf{Z}_{pp'})_0$ is independent of $d_{pp'}$. Compare this with approximation (4.222) or (4.246) and remember that in linear theory, the wave-force contribution on a body is proportional to the corresponding wave hitting the body. We expect that approximation (5.157) might be useful when spacing $d_{pp'}$ is large in comparison with the wavelength and with the largest horizontal extension (diameter) of bodies p and p'.

5.5.2 Hydrodynamic Formulation

We still assume an incompressible ideal fluid and irrotational motion. Potential theory then applies. We also assume harmonic time variation and use linearised theory. In general, the velocity potential

$$\phi = \phi_0 + \phi_d + \phi_r \tag{5.158}$$

is composed of three main contributions. Here ϕ_0 represents a given incident wave, which results in a diffracted wave ϕ_d when all bodies are fixed. If the bodies are oscillating, a radiated wave ϕ_r is, in addition, set up.

Note that, in general, ϕ_0 is the given velocity potential in case all the N bodies are absent. It could, for instance, represent a plane wave with complex amplitude as given by Eqs. (4.92) and (4.102), or alternatively by Eq. (4.83). In the former case it is a propagating incident wave. In the latter case it is a plane wave which includes also a wave which is partly reflected from, for example, a straight coast line. If the coast topography is more irregular, the mathematical description of ϕ_0 would be more complicated. The term ϕ_d in Eq. (5.158) thus represents diffraction which is due to the immersed N bodies only.

The velocity potential associated with the radiated wave has a complex amplitude

$$\hat{\phi}_r = \sum_i \varphi_i \hat{u}_i, \tag{5.159}$$

where the sum is taken over all oscillators. Here the complex coefficient φ_i represents the velocity potential which is due to the unit velocity amplitude of oscillator number i, when all other oscillators are non-moving. For one single body, we introduced the six-dimensional column vector $\boldsymbol{\varphi}$ and its transpose $\boldsymbol{\varphi}^T$ in connection with Eqs. (5.127) and (5.132). If we extend the dimension to $6N$, the vector includes all φ_i as its components. Then we may write Eqs. (5.159) as

$$\hat{\phi}_r = \boldsymbol{\varphi}^T \hat{\mathbf{u}} = \hat{\mathbf{u}}^T \boldsymbol{\varphi}. \tag{5.160}$$

For each of the bodies we may define corresponding six-dimensional column vectors $\boldsymbol{\varphi}_p$ and $\mathbf{u}_p$. Then we may write

$$\hat{\phi}_r = \sum_{p=1}^{N} \boldsymbol{\varphi}_p^T \hat{\mathbf{u}}_p = \sum_{p=1}^{N} \hat{\mathbf{u}}_p^T \boldsymbol{\varphi}_p. \tag{5.161}$$

At large horizontal distance r_p from the reference axis $p - p$ of body number p, we have, according to Eq. (4.265), an asymptotic expression

$$\boldsymbol{\varphi}_p \sim \mathbf{b}_p(\theta_p) e(kz)(kr_p)^{-1/2} \exp\{-ikr_p\}. \tag{5.162}$$

Here $\mathbf{b}_p(\theta_p)$ is a six-dimensional vector composed of body number p's six far-field coefficients referred to the local origin (the point where the axis $p - p$ crosses the plane of the mean water surface). See also Figure 4.11 (with subscript p instead of i) for a definition of the local coordinates r_p and θ_p.

All terms in Eq. (5.158) satisfy the Laplace equation (4.16) in the fluid domain and the usual homogeneous boundary conditions (4.29) and (4.18) or (4.19) on the free surface, $z = 0$, and on the sea bed. The radiation condition of outgoing waves at infinity has to be satisfied for velocity potentials ϕ_d, ϕ_r and all φ_i. The inhomogeneous boundary condition (4.17) on the wet surfaces S_p of the oscillating bodies is satisfied, provided

$$\frac{\partial}{\partial n}(\hat{\phi}_0 + \hat{\phi}_d) = 0 \quad \text{on all } S_p \tag{5.163}$$

$$\frac{\partial \varphi_i}{\partial n} = \begin{cases} n_{pj} & \text{on } S_p \\ 0 & \text{on } S_{p'} \ (p' \neq p) \end{cases}. \tag{5.164}$$

Here $\partial/\partial n$ is the normal derivative in the direction of the outward unit normal $\vec{n}$

to the surface of the body. Remembering Eq. (5.152), we find that n_{pj} is defined as the x, y or z component of $\vec{n}$, when $j = 1, 2$ or 3, respectively. Furthermore, n_{pj} is the x, y or z component of vector $\vec{s}_p \times \vec{n}$ when $j = 4, 5$ or 6, respectively. Here $\vec{s}_p$ is the position vector referred to a selected point on reference axis $p - p$ of body number p, for instance the centre of gravity or the centre of buoyancy. Note that Eqs. (5.164) and (5.163) represent a generalisation of the corresponding boundary conditions (5.11) and (5.28) for the single body.

5.5.3 Radiation-Impedance and Radiation-Resistance Matrices

According to Eq. (5.22) the force on oscillator number i caused by potential $\varphi_{i'}$ (i.e., caused by unit velocity amplitude of oscillator number i') is

$$-Z_{ii'} = -\iint_{S_p} p_{i'} n_{pj} \, dS_p = i\omega\rho \iint_S \varphi_{i'} \frac{\partial \varphi_i}{\partial n} \, dS, \, j, \tag{5.165}$$

where ρ is the water density and $p_{i'}$ represents the hydrodynamic pressure corresponding to $\varphi_{i'}$. The last integral, which represents a generalisation of Eq. (5.40), is, by virtue of Eq. (5.164), to be taken over S_p or over the totality of wet body surfaces

$$S = \sum_{p=1}^{N} S_p. \tag{5.166}$$

Because $\partial \varphi_i / \partial n$ is real everywhere on S, according to boundary condition (5.164), we may use φ_i or φ_i^* as we wish in the integrand above. Note that Eq. (5.165) is valid for $i' = i$ as well as for $i' \neq i$. When $i' \neq i$, oscillator i' may pertain to any of the oscillating bodies, including body number p.

Matrix $\mathbf{Z}_{pp'}$ representing the radiation interaction between bodies p and p' [see Eq. (5.156)] may be written in terms of the (generally six-dimensional) column vectors $\boldsymbol{\varphi}_p$ and $\boldsymbol{\varphi}_{p'}$ as

$$\mathbf{Z}_{pp'} = -i\omega\rho \iint_S \frac{\partial \boldsymbol{\varphi}_p}{\partial n} \boldsymbol{\varphi}_{p'}^T \, dS. \tag{5.167}$$

In the integrand here we may replace $\boldsymbol{\varphi}_p$ by $\boldsymbol{\varphi}_p^*$.

The radiation-impedance matrix for the total system of N oscillating bodies may be written as

$$\mathbf{Z} = -i\omega\rho \iint_S \frac{\partial \boldsymbol{\varphi}}{\partial n} \boldsymbol{\varphi}^T \, dS. \tag{5.168}$$

If in the integrand we replace $\partial \boldsymbol{\varphi}/\partial n$ by $\partial \boldsymbol{\varphi}^*/\partial n$, we get the alternative expression

$$\mathbf{Z} = -i\omega\rho \iint_S \frac{\partial \boldsymbol{\varphi}^*}{\partial n} \boldsymbol{\varphi}^T \, dS, \tag{5.169}$$

which is a generalisation of Eq. (5.135).

We next show that radiation-impedance matrix $\mathbf{Z}$ is symmetric; that is,

$$\mathbf{Z} = \mathbf{Z}^T. \tag{5.170}$$

From Eqs. (5.168) and (4.240) – see also Eq. (5.132) – it follows that

$$\mathbf{Z} - \mathbf{Z}^T = -i\omega\rho \iint\limits_{S} \left(\frac{\partial \boldsymbol{\varphi}}{\partial n} \boldsymbol{\varphi}^T - \boldsymbol{\varphi} \frac{\partial \boldsymbol{\varphi}^T}{\partial n} \right) dS = i\omega\rho\, \mathbf{I}(\boldsymbol{\varphi}, \boldsymbol{\varphi}^T) = 0, \tag{5.171}$$

where the last equality follows from Eq. (4.249) – see also Eq. (5.133) – because all components of vector $\boldsymbol{\varphi}$ satisfy the same radiation condition at infinity. We may note that reciprocity relation (5.170) is a generalisation of Eq. (5.42) or (5.122). Taking a look at Eq. (5.156), we see that Eq. (5.170) means

$$\mathbf{Z}_{p'p} = \mathbf{Z}_{pp'}^T. \tag{5.172}$$

Note that matrix $\mathbf{Z}_{p'p}$ is not necessarily symmetric if $p' \neq p$. Moreover, from Eq. (5.170) it follows that

$$\mathbf{Z}^* = (\mathbf{Z}^T)^* = \mathbf{Z}^\dagger. \tag{5.173}$$

Splitting the radiation impedance matrix into real and imaginary parts, as in Eq. (5.155), we may write the radiation-resistance matrix as

$$
\begin{aligned}
\mathbf{R} &= \frac{1}{2}(\mathbf{Z} + \mathbf{Z}^*) = \frac{1}{2}(\mathbf{Z} + \mathbf{Z}^\dagger) \\
&= -\frac{1}{2}i\omega\rho \iint\limits_{S} \left(\frac{\partial \boldsymbol{\varphi}^*}{\partial n} \boldsymbol{\varphi}^T - \boldsymbol{\varphi}^* \frac{\partial \boldsymbol{\varphi}^T}{\partial n} \right) dS \\
&= \frac{i}{2}\omega\rho\, \mathbf{I}(\boldsymbol{\varphi}^*, \boldsymbol{\varphi}^T) = -\frac{i}{2}\omega\rho\, \mathbf{I}(\boldsymbol{\varphi}, \boldsymbol{\varphi}^\dagger).
\end{aligned} \tag{5.174}
$$

Here we have used Eqs. (5.169), (5.170) and (5.173). Note that Eq. (5.174) is an extension of Eq. (5.138) from the case of a 6×6 matrix to the case of a $6N \times 6N$ matrix. In an analogous way we may extend Eqs. (5.139) and (5.140). Thus, we may also write the $6N \times 6N$ radiation-resistance matrix as

$$\mathbf{R} = -i\omega\rho \lim_{r \to \infty} \iint\limits_{S_\infty} \frac{\partial \boldsymbol{\varphi}^*}{\partial r} \boldsymbol{\varphi}^T \, dS \tag{5.175}$$

or as

$$
\begin{aligned}
\mathbf{R} &= \frac{\omega\rho\, D(kh)}{2k} \int_0^{2\pi} \mathbf{a}^*(\theta)\, \mathbf{a}^T(\theta)\, d\theta \\
&= \frac{\omega\rho\, D(kh)}{4\pi k} \int_0^{2\pi} \mathbf{h}^*(\theta)\, \mathbf{h}^T(\theta)\, d\theta,
\end{aligned} \tag{5.176}
$$

where $\mathbf{a}(\theta)$ and $\mathbf{h}(\theta)$ are $6N$-dimensional column vectors consisting of all far-field coefficients $a_i(\theta)$ for the radiated wave and of the corresponding Kochin functions $h_i(\theta)$, respectively. Note that the far-field coefficients $a_i(\theta)$ are referred to the common (global) origin. We may observe that becasue $\mathbf{R}$ is real, we may take the complex conjugate of the integrands without changing the resulting integral in Eq (5.176).

We shall now prove the following relationship between radiation-resistance matrix $\mathbf{R}$ and the radiated power P_r:

$$P_r = \tfrac{1}{2}\hat{\mathbf{u}}^T \mathbf{R}\hat{\mathbf{u}}^* = \tfrac{1}{2}\hat{\mathbf{u}}^\dagger \mathbf{R}\hat{\mathbf{u}}. \tag{5.177}$$

The last equality follows from the fact that P_r is real ($P_r^* = P_r$) and that $\mathbf{R}$ is a real matrix. Because $P_r \geq 0$, $\mathbf{R}$ is a positive semidefinite matrix (see Section 6.4).

In the far-field region ($r \to \infty$) the curvature of the wave front is negligible. The radiated power transport per unit width of the wave front is then

$$J(r,\theta) = \frac{\rho g^2 D(kh)}{4\omega}|\hat{\eta}_r(r,\theta)|^2 = \frac{\omega\rho\, D(kh)}{4}|\hat{\phi}_r(r,\theta,0)|^2, \tag{5.178}$$

where we have used Eqs. (4.136) and (4.41). In the far-field region ($r \to \infty$) we may use the asymptotic approximation (4.222)

$$\hat{\phi}_r \sim A_r(\theta)e(kz)(kr)^{-1/2}e^{-ikr}, \tag{5.179}$$

where

$$A_r(\theta) = \hat{\mathbf{u}}^T \mathbf{a}(\theta) = \mathbf{a}^T(\theta)\hat{\mathbf{u}} \tag{5.180}$$

is the far-field coefficient for the radiated wave, in accordance with Eq. (5.160). Thus the power radiated through the cylindrical control surface S_∞ of radius r ($r \to \infty$) is

$$P_r = \int_0^{2\pi} J_r(r,\theta)r\,d\theta = \frac{\omega\rho\, D(kh)}{4k}\int_0^{2\pi} A_r^*(\theta)A_r(\theta)d\theta$$

$$= \frac{\omega\rho\, D(kh)}{4k}\int_0^{2\pi} \hat{\mathbf{u}}^\dagger \mathbf{a}^*(\theta)\mathbf{a}^T(\theta)\hat{\mathbf{u}}\,d\theta. \tag{5.181}$$

[Also cf. Eq. (4.227).] Remembering that $\hat{\mathbf{u}}$ may be taken outside the last integral, we find that Eq. (5.177) results if use is made of Eq. (5.176).

Components $a_i(\theta)$ of vector $\mathbf{a}(\theta)$ may be arranged in groups $\mathbf{a}_p(\theta)$ pertaining to body p, $(p = 1, 2, \ldots, N)$. The real part of matrices $\mathbf{Z}_{pp'}$ on the right-hand side of Eq. (5.156) may thus, according to Eqs. (5.167), (5.174) and (5.176), be written as

$$\mathbf{R}_{pp'} = \frac{i}{2}\omega\rho\mathbf{I}(\boldsymbol{\varphi}_p^*, \boldsymbol{\varphi}_{p'}^T) = \frac{\omega\rho\, D(kh)}{2k}\int_0^{2\pi} \mathbf{a}_p^*(\theta)\mathbf{a}_{p'}^T(\theta)d\theta. \tag{5.182}$$

In analogy with Eqs. (4.265) and (4.266), we may use far-field coefficient vectors $\mathbf{b}_p(\theta_p)$ or $\mathbf{a}_p(\theta)$ referred to the local origin $(x_p, y_p, 0)$ or the global origin $(0, 0, 0)$, respectively. See Figure 4.11, which, with subscript i replaced by p, defines angles θ_p and α_p and distances x_p, y_p and d_p. Vertical axis $p - p$ in Figure 5.13 is given by $(x, y) = (x_p, y_p)$. Note that vector $\mathbf{b}_p(\theta_p)$ also appears in Eq. (5.162). According to Eq. (4.271) we have

$$\mathbf{a}_p(\theta) = \mathbf{b}_p(\theta)\exp\{ikd_p\cos(\alpha_p - \theta)\}$$

$$= \mathbf{b}_p(\theta)\exp\{ik(x_p\cos\theta + y_p\sin\theta)\}. \tag{5.183}$$

If we use this, or Eq. (4.302), in Eq. (5.182), we get

$$\mathbf{R}_{pp'} = \frac{\omega\rho\,D(kh)}{2k} \int_0^{2\pi} \mathbf{b}_p^*(\theta)\mathbf{b}_{p'}^T(\theta)\exp\{ik(r_p - r_{p'})\}d\theta. \tag{5.184}$$

The distance difference $r_p - r_{p'}$ entering in the exponent may, according to Eq. (4.303), be expressed in various ways, as

$$\begin{aligned} r_p - r_{p'} &= (x_{p'} - x_p)\cos\theta + (y_{p'} - y_p)\sin\theta \\ &= d_{p'}\cos(\alpha_{p'} - \theta) - d_p\cos(\alpha_p - \theta) \\ &= d_{p'p}\cos(\alpha_{p'p} - \theta). \end{aligned} \tag{5.185}$$

Using the last expression we have

$$\mathbf{R}_{pp'} = \frac{\omega\rho\,D(kh)}{2k} \int_0^{2\pi} \mathbf{b}_p^*(\theta)\mathbf{b}_{p'}^T(\theta)\exp\{ikd_{p'p}\cos(\alpha_{p'p} - \theta)\}\,d\theta. \tag{5.186}$$

5.5.4 Radiation-Reactance and Added-Mass Matrices

We note that, according to Eq. (5.155), the radiation-reactance matrix may be written as

$$\mathbf{X} = \omega\mathbf{m} = -i(\mathbf{Z} - \mathbf{R}) = \omega\rho\left(\lim_{r\to\infty}\iint_{S_\infty}\frac{\partial\varphi^*}{\partial r}\varphi^T dS - \iint_S \frac{\partial\varphi^*}{\partial n}\varphi^T dS\right), \tag{5.187}$$

where we have used Eqs. (5.169) and (5.175). This indicates that the reactance matrix $\mathbf{X}$, and hence the added-mass matrix $\mathbf{m}$, is somehow related to the radiated wave in the near-field region. Equation (2.90) shows how the reactance of a simple mechanical oscillator is related to the difference between the kinetic and potential energy. In the following paragraphs let us relate the radiation-reactance matrix to the difference between kinetic and potential energy in the near-field region. We note that in the far-field region, as also in a plane wave, this difference vanishes [cf. Eqs. (4.121) and (4.127)]. To be explicit, let us show below that

$$\frac{1}{4}\hat{\mathbf{u}}^T\mathbf{m}\hat{\mathbf{u}}^* = \frac{1}{4\omega}\hat{\mathbf{u}}^T\mathbf{X}\hat{\mathbf{u}}^* = W_k - W_p, \tag{5.188}$$

where $W_k - W_p$ is the difference between the kinetic energy and the potential energy associated with radiated wave $\hat{\phi}_r$ as given by Eq. (5.160). Negative added mass is possible if $W_p > W_k$.[19]

The (time-average) kinetic energy is [see Eq. (2.85)]

$$W_k = \frac{1}{4}\rho\iiint_V \hat{v}\cdot\hat{v}^* dV = \frac{1}{4}\rho\iiint_V \nabla\hat{\phi}_r^*\cdot\nabla\hat{\phi}_r dV, \tag{5.189}$$

where the integral is taken over the fluid volume shown in Figure 5.13. This volume is bounded by a closed surface consisting of the sea bed S_b, the control surface S_∞, the free water surface S_0 and the wave-generating surface S [defined by

Eq. (5.166)]. Because $\hat{\phi}_r$ obeys the Laplace equation (4.35) we have

$$\nabla \cdot (\hat{\phi}_r^* \nabla \hat{\phi}_r) \equiv \hat{\phi}_r^* \nabla^2 \hat{\phi}_r + \nabla \hat{\phi}_r^* \cdot \nabla \hat{\phi}_r = \nabla \hat{\phi}_r^* \cdot \nabla \hat{\phi}_r. \tag{5.190}$$

Hence, because the integrand is the divergence of vector $\hat{\phi}_r^* \nabla \hat{\phi}_r$, we may, by using Gauss' divergence theorem, replace volume integral (5.189) by the following closed-surface integral:

$$W_k = \frac{\rho}{4} \oiint (-\vec{n}) \cdot \hat{\phi}_r^* \nabla \hat{\phi}_r dS = -\frac{\rho}{4} \oiint \hat{\phi}_r^* \frac{\partial \hat{\phi}_r}{\partial n} dS, \tag{5.191}$$

where the unit normal $\vec{n}$ is pointing into the fluid (see Figure 5.13). Note that $\partial/\partial n = -\partial/\partial z$ on S_0 and $\partial/\partial n = -\partial/\partial r$ on S_∞. Furthermore, we observe that the integrand vanishes on the sea bed S_b as a result of boundary condition (4.18).

By using Eqs. (4.119), (4.41) and (4.43) we obtain the (time-average) potential energy

$$W_p = \iint_{S_0} E_p dS = \frac{\rho g}{4} \iint_{S_0} \hat{\eta}\hat{\eta}^* dS = \frac{\omega^2 \rho}{4g} \iint_{S_0} \hat{\phi}_r^* \hat{\phi}_r dS$$

$$= \frac{\rho}{4} \iint_{S_0} \hat{\phi}_r^* \frac{\partial \hat{\phi}_r}{\partial z} dS = -\frac{\rho}{4} \iint_{S_0} \hat{\phi}_r^* \frac{\partial \hat{\phi}_r}{\partial n} dS. \tag{5.192}$$

Subtraction between Eqs. (5.191) and (5.192) gives

$$W_k - W_p = \frac{\rho}{4} \iint_{S_\infty} \hat{\phi}_r^* \frac{\partial \hat{\phi}_r}{\partial r} dS - \frac{\rho}{4} \iint_{S} \hat{\phi}_r^* \frac{\partial \hat{\phi}_r}{\partial n} dS. \tag{5.193}$$

Observing from Eq. (5.160) that we may write $\hat{\phi}_r^* = \hat{u}^\dagger \boldsymbol{\varphi}^*$ and $\hat{\phi}_r = \boldsymbol{\varphi}^T \hat{u}$ and, further, that $\hat{u}^\dagger$ and $\hat{u}$ may be taken outside the integral, we find that a combination of Eqs. (5.187) and (5.193) shows that

$$\lim_{r \to \infty} (W_k - W_p) = \frac{1}{4\omega} \hat{u}^\dagger \mathbf{X} \hat{u} = \frac{1}{4} \hat{u}^\dagger \mathbf{m} \hat{u}. \tag{5.194}$$

Because $(W_k - W_p)$ is a real scalar and $\mathbf{X}$ (and $\mathbf{m}$) is a real symmetric matrix, the right-hand sides of Eqs. (5.187) and (5.194) must equal their own conjugates as well as their own transposes. Hence, we obtain Eq. (5.188) if we take the complex conjugate of Eq. (5.194) and assume that the limit operation $(r \to \infty)$ is implied, because the radius r of control cylinder S_∞ has to be sufficiently large in order to contain the complete near-field region inside S_∞.

An alternative expression for the added-mass matrix is obtained as follows. Use of Eqs (5.155), (5.173) and (5.169) gives

$$\mathbf{m} = \frac{\mathbf{X}}{\omega} = \frac{1}{2i\omega}(\mathbf{Z} - \mathbf{Z}^*) = \frac{1}{2i\omega}(\mathbf{Z} - \mathbf{Z}^\dagger)$$

$$= -\frac{\rho}{2} \iint_{S} \left(\frac{\partial \boldsymbol{\varphi}^*}{\partial n} \boldsymbol{\varphi}^T + \boldsymbol{\varphi}^* \frac{\partial \boldsymbol{\varphi}^T}{\partial n} \right) dS$$

$$= -\frac{\rho}{2} \iint_{S} \frac{\partial}{\partial n}(\boldsymbol{\varphi}^* \boldsymbol{\varphi}^T) dS. \tag{5.195}$$

In this way the added-mass matrix is expressed as an integral over the wave-generating surface S (the totality of wet body surfaces).

5.5.5 Excitation Force Vector: the Haskind Relation

For each of the N bodies we may define an excitation force vector $\mathbf{F}_{e,p}$, which has, in general, six components, as in Eq. (5.27) or (5.143). Note that boundary condition (5.28) or (5.163) for ϕ_d has to be satisfied on the wet surface, not only of body p, but also of all other bodies. Thus $\mathbf{F}_{e,p}$ represents the wave force on body p when *all* bodies are in a non-oscillating state.

For the total system of N bodies, we define a column vector $\mathbf{F}_e$ (in general of dimension $6N$) for the excitation forces:

$$\mathbf{F}_e = \left(\mathbf{F}_{e,1}^T \cdots \mathbf{F}_{e,p}^T \cdots \mathbf{F}_{e,N}^T\right)^T. \tag{5.196}$$

Here $\mathbf{F}_{e,p}$ is the (generally six-dimensional) excitation-force vector for body p. In accordance with Eq. (5.27) we may write

$$\hat{\mathbf{F}}_{e,p} = i\omega\rho \iint\limits_{S_p} (\hat{\phi}_0 + \hat{\phi}_d)\mathbf{n}_p dS, \tag{5.197}$$

where $\mathbf{n}_p$ is the (six-dimensional) unit normal vector as defined by Eqs. (5.5) and (5.6), but now for body p. It should be noted that the diffraction potential $\hat{\phi}_d$ should satisfy boundary condition (5.163), not only on S_p but also on the remaining part of S.

Because all components φ_i of vector $\boldsymbol{\varphi}$ satisfy boundary condition (5.164) and also the same radiation condition at infinity as $\hat{\phi}_d$, we may generalise Eqs. (5.141) and (5.143) to

$$\hat{\mathbf{F}}_e = i\omega\rho\mathbf{I}[(\hat{\phi}_0 + \hat{\phi}_d), \boldsymbol{\varphi}] = i\omega\rho\mathbf{I}(\hat{\phi}_0, \boldsymbol{\varphi}) \tag{5.198}$$

also for the case with N bodies. In particular, we have for body p

$$\hat{\mathbf{F}}_{e,p} = i\omega\rho\mathbf{I}(\hat{\phi}_0, \boldsymbol{\varphi}_p). \tag{5.199}$$

If the incident wave is plane in accordance with Eq. (5.144), we may also generalise Eqs. (5.146) and (5.147) to the case of N bodies. Thus, the excitation-force vector is

$$\hat{\mathbf{F}}_e(\beta) = \mathbf{f}_g(\beta)A = \frac{\rho g}{k}\,D(kh)\,\mathbf{h}(\beta \pm \pi)A, \tag{5.200}$$

where

$$\mathbf{f}_g(\beta) = \frac{\rho g}{k}\,D(kh)\,\mathbf{h}(\beta \pm \pi) = \frac{\rho g}{k}\,D(kh)\,\sqrt{2\pi}\,\mathbf{a}(\beta \pm \pi)e^{i\pi/4} \tag{5.201}$$

is, in general, a $6N$-dimensional column vector consisting of the excitation force coefficients referred to the global origin $(x, y) = (0, 0)$. Note that the excitation force depends on the angle of incidence β of the incident wave.

Similarly, reciprocity relations (5.148)–(5.150), where the radiation-resistance matrix is expressed in terms of excitation-force vectors, may be directly generalised to the $6N$-dimensional case.

Note that in Eq. (5.200) A is the elevation of the undisturbed incident wave at the (global) origin $(x, y) = (0, 0)$. Considering body p's excitation-force vector $\mathbf{F}_{e,p}$, we have from Eqs. (5.200) and (5.201) that

$$\hat{\mathbf{F}}_{e,p}(\beta) = \mathbf{f}_{g,p}(\beta) A = \frac{\rho g\, D(kh)}{k} \sqrt{2\pi}\, \mathbf{a}_p(\beta \pm \pi) e^{i\pi/4} A. \tag{5.202}$$

Using relation (5.183) between the local and global far-field coefficients, we find

$$\hat{\mathbf{F}}_{e,p}(\beta) = \mathbf{f}_p(\beta) A_p = \frac{\rho g\, D(kh)}{k} \sqrt{2\pi}\, \mathbf{b}_p(\beta \pm \pi) e^{i\pi/4} A_p, \tag{5.203}$$

where

$$A_p = A \exp\{-ik(x_p \cos\beta + y_p \sin\beta)\} \tag{5.204}$$

is [see Eq. (5.144)] the wave elevation corresponding to the undisturbed incident wave at body p's local origin $(x, y) = (x_p, y_p)$. In Eqs. (5.202) and (5.203) the excitation-force coefficients $\mathbf{f}_{g,p}(\beta)$ and $\mathbf{f}_p(\beta)$ for body p are referred to the global origin $(x, y) = (0, 0)$ and to the local origin $(x, y) = (x_p, y_p)$, respectively.

We may note that by using Eqs. (4.107) and (4.110) for the phase and group velocities, we may write the coefficient $\rho g\, D(kh)/k$, appearing in Eqs. (5.200)–(5.203), as $2v_g v_p \rho$. Note that Eq. (7.180) is an extension of Eq. (5.203).

5.5.6 Wide-Spacing Approximation

If the distance between two different bodies (p and p', say) is sufficiently large, the diffraction and radiation interactions between the two bodies may be well represented by the wide-spacing approximation. The requirement is that the one body is not influenced by the other body's near-field contribution to the diffracted or radiated waves. Here let us illustrate the wide-spacing approximation for the radiation problem. We shall apply asymptotic approximation (5.162) for both bodies. That means we have to assume equal water depth h at both bodies.

The interaction between the two bodies is then represented by the 6×6 matrix $\mathbf{Z}_{pp'}$. We consider the force

$$\Delta' \hat{\mathbf{F}}_p = -\mathbf{Z}_{pp'} \hat{\mathbf{u}}_{p'} \tag{5.205}$$

on body p which is due to the oscillation of body p' with velocity $\mathbf{u}_{p'}$. Using far-field approximation (5.162) with p' instead of p and also Eqs. (5.127) and (4.41), we find the corresponding wave elevation:

$$\Delta' \hat{\eta}_r(r_{p'}, \theta_{p'}) = -\frac{i\omega}{g} \boldsymbol{\varphi}_{p'}^T(r_{p'}, \theta_{p'}, 0) \hat{\mathbf{u}}_{p'} \tag{5.206}$$

$$\sim -\frac{i\omega}{g} \mathbf{b}_{p'}^T(\theta_{p'})(kr_{p'})^{-1/2} \exp\{-ikr_{p'}\} \hat{\mathbf{u}}_{p'}. \tag{5.207}$$

If $kd_{pp'} \to \infty$, this wave may be considered as a plane wave when it arrives at body p, where $(r_{p'}, \theta_{p'}) = (d_{pp'}, \alpha_{pp'})$; see Figure 4.13. Using also Eq. (5.203) yields

$$\Delta'\hat{\mathbf{F}}_p \sim \frac{\rho g\, D(kh)}{k} \sqrt{2\pi}\, \mathbf{b}_p(\alpha_{pp'} \pm \pi) e^{i\pi/4} \Delta'\hat{\eta}_r(d_{pp'}, \alpha_{pp'}). \tag{5.208}$$

Inserting here from approximation (5.207) and comparing with Eq. (5.205) results in the following asymptotic approximation for $\mathbf{Z}_{pp'}$ as $kd_{pp'} \to \infty$:

$$\mathbf{Z}_{pp'} \sim \frac{i\omega\rho\, D(kh)}{k} \sqrt{2\pi}\, \mathbf{b}_p(\alpha_{p'p}) \mathbf{b}_{p'}^T(\alpha_{pp'})(kd_{pp'})^{-1/2} \exp\left\{-ikd_{pp'} + \frac{\pi}{4}\right\}. \tag{5.209}$$

Note that $\mathbf{b}_p(\alpha_{p'p})$ represents the action on body p' from body p, and vice versa for $\mathbf{b}_{p'}(\alpha_{pp'})$. The result agrees with the expected approximation (5.157). We also easily see that the general reciprocity relationship (5.170) is satisfied by approximation (5.209).

5.6 The Froude-Krylov Force and Small-Body Approximation

We may, in accordance with Eq. (5.197), decompose the excitation-force vector $\mathbf{F}_{e,p}$ for body p into two parts:

$$\hat{\mathbf{F}}_{e,p} = \hat{\mathbf{F}}_{FK,p} + \hat{\mathbf{F}}_{d,p}, \tag{5.210}$$

where [with $\mathbf{n}_p = (\vec{n}_p, \vec{s}_p \times \vec{n}_p)$ – see Eqs. (5.5)–(5.6) –]

$$\hat{\mathbf{F}}_{FK,p} = i\omega\rho \iint\limits_{S_p} \hat{\phi}_0 \mathbf{n}_p\, dS \tag{5.211}$$

is the Froude-Krylov force vector, and where

$$\hat{\mathbf{F}}_{d,p} = i\omega\rho \iint\limits_{S_p} \hat{\phi}_d \mathbf{n}_p\, dS, \tag{5.212}$$

is the diffraction force vector. Alternatively, we have from the Haskind relationship [cf. Eq. (5.199)]

$$\hat{\mathbf{F}}_{e,p} = i\omega\rho\, \mathbf{I}(\hat{\phi}_0, \boldsymbol{\varphi}_p) = i\omega\rho \iint\limits_{S} \left(\hat{\phi}_0 \frac{\partial\boldsymbol{\varphi}_p}{\partial n} - \boldsymbol{\varphi}_p \frac{\partial\hat{\phi}_0}{\partial n} \right) dS. \tag{5.213}$$

The first term represents the Froude-Krylov force, because $\partial\boldsymbol{\varphi}_p/\partial n = 0$ on $S_{p'}$ when $p' \neq p$, and $\partial\boldsymbol{\varphi}_p/\partial n = \mathbf{n}_p$ on S_p; see Eq. (5.164). Hence, we may compute the diffraction force vector

$$\hat{\mathbf{F}}_{d,p} = -i\omega\rho \iint\limits_{S} \boldsymbol{\varphi}_p \frac{\partial\hat{\phi}_0}{\partial n}\, dS \tag{5.214}$$

by solving the radiation problem instead of solving the scattering (or diffraction) problem.

With given incident wave and a given geometry for body p, the Froude-Krylov force vector $\hat{\mathbf{F}}_{FK,p}$ is obtained by direct integration. For the diffraction force to be computed it is necessary to solve a boundary-value problem corresponding to either scattering (diffraction) or radiation. The latter kind of problem is frequently, but not always, the easier to solve.

5.6.1 The Froude-Krylov Force and Moment

Next we shall consider the translational components of the Froude-Krylov force. For

$$i = 6(p-1) + q \quad (q = 1, 2, 3) \tag{5.215}$$

we write the three components of the Froude-Krylov force as

$$\hat{F}_{FK,i} = \hat{F}_{FK,pq} = i\omega\rho \iint_{S_p} \hat{\phi}_0 n_{pq} dS = i\omega\rho \iint_{S_p} \hat{\phi}_0 \vec{e}_q \cdot \vec{n}_p dS. \tag{5.216}$$

Here $\vec{e}_q$ ($q = 1, 2, 3$) is the unit vector in the direction x, y or z, for surge, sway or heave, respectively. The unit normal components n_{pq} are as given by Eq. (5.5), but for body p. By making the union of the wet surface S_p and the water plane area S_{wp} of body p (see Figure 5.14), we may decompose this integral into one integral over the closed surface $S_p + S_{wp}$ and one integral over S_{wp}:

$$\hat{F}_{FK,pq} = i\omega\rho \oiint_{S_p+S_{wp}} \hat{\phi}_0 \vec{e}_q \cdot \vec{n}_p dS - i\omega\rho \iint_{S_{wp}} \hat{\phi}_0 \delta_{q3} dS. \tag{5.217}$$

Note that the last term vanishes except for the heave mode ($q = 3$), because $n_q = \delta_{q3}$ on S_{wp}. We apply Gauss' theorem to the closed-surface integral and convert it to an integral over the volume V_p of displaced water:

$$i\omega\rho \oiint_{S_p+S_{wp}} \hat{\phi}_0 \vec{e}_q \cdot \vec{n}_p dS = i\omega\rho \iiint_{V_p} (\nabla\hat{\phi}_0) \cdot \vec{e}_q dV = i\omega\rho \iiint_{V_p} \frac{\partial\hat{\phi}_0}{\partial x_q} dV \tag{5.218}$$

($x_1 = x$, $x_2 = y$, $x_3 = z$). This may be obtained directly from Gauss' divergence theorem applied to the divergence $\nabla \cdot (\hat{\phi}_0 \vec{e}_q)$ where $\vec{e}_q$ is a constant (unit) vector.

Figure 5.14: Floating body with water plane area S_{wp}, submerged volume V_p (volume of displaced water), and wet surface S_p with unit normal $\vec{n}_p$. The vertical unit vector is indicated by $\vec{e}_3$.

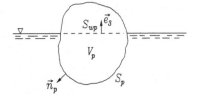

Thus we have

$$\hat{F}_{FK,pq} = i\omega\rho \iiint_{V_p} \frac{\partial \hat{\phi}_0}{\partial x_q} \, dV + \delta_{q3}\rho g \iint_{S_{wp}} \hat{\eta}_0 \, dS \tag{5.219}$$

where also Eq. (4.41) has been used. Note that $\hat{\phi}_0$ and $\hat{\eta}_0$ are the velocity potential and the wave elevation of the undisturbed incident wave, that is, as the wave would have been in the absence of all the N bodies. Observing that [see Eq. (4.38)] $i\omega\partial\hat{\phi}_0/\partial x_q = i\omega\hat{v}_{0q} = \hat{a}_{0q}$, we may alternatively write Eq. (5.219) as

$$\hat{F}_{FK,pq} = \rho \iiint_{V_p} \hat{a}_{0q} \, dV + \delta_{q3}\rho g \iint_{S_{wp}} \hat{\eta}_0 \, dS, \tag{5.220}$$

where $\hat{a}_{0q}$ represents the q component of the fluid acceleration which is due to the incident wave. In the particular case of a purely propagating incident plane wave, $\hat{\phi}_0$ is given by Eqs. (4.92) and (4.102), and then we have the undisturbed fluid particle acceleration given by the components

$$\hat{a}_{0z} = \hat{a}_{03} = i\omega\hat{v}_{03} = -gkAe'(kz)\exp\{-ikr(\beta)\}$$
$$= -\omega^2 A\frac{\sinh(kz+kh)}{\sinh(kh)}\exp\{-ikr(\beta)\} \tag{5.221}$$

and

$$(\hat{a}_{0x}, \hat{a}_{0y}) = (\hat{a}_{01}, \hat{a}_{02}) = i\omega(\hat{v}_{01}, \hat{v}_{02}) = a_{0H}(\cos\beta, \sin\beta), \tag{5.222}$$

where

$$a_{0H} = ikg\,Ae(kz)\exp\{-ikr(\beta)\}. \tag{5.223}$$

Compare Eqs. (4.94), (4.95) and (4.277).

Finally we consider the rotational modes of the Froude-Krylov "force" (i.e., Froude-Krylov moments of roll, pitch and yaw) for body p. For

$$i = 6(p-1) + 3 + q \quad (q = 1, 2, 3) \tag{5.224}$$

set $F_i = M_{pq}$; thus, according to Eqs. (5.6) and (5.20), the Froude-Krylov moment is given by

$$\hat{M}_{FK,pq} = \hat{F}_{FK,i} = i\omega\rho \iint_{S_p} \hat{\phi}_0 n_{pq} \, dS = i\omega\rho \iint_{S_p} \hat{\phi}_0 \vec{e}_q \cdot \vec{s}_p \times \vec{n}_p \, dS, \tag{5.225}$$

where $\vec{s}_p$ is the vector from the body's reference point (x_p, y_p, z_p) to the surface element dS of the wet surface S_p (cf. Figures 5.3 and 5.13). On the water plane area S_{wp} (see Figure 5.14) we have $\vec{n}_p = \vec{e}_3$ and $\vec{e}_q \cdot (\vec{s}_p \times \vec{e}_3) = (y - y_p)\delta_{q1} - (x - x_p)\delta_{q2}$. Thus,

$$\hat{M}_{FK,pq} = i\omega\rho \oiint_{S_p+S_{wp}} \hat{\phi}_0 \vec{e}_q \times \vec{s}_p \cdot \vec{n}_p \, dS$$
$$-i\omega\rho \iint_{S_{wp}} \hat{\phi}_0[(y - y_p)\delta_{q1} - (x - x_p)\delta_{q2}] \, dS. \tag{5.226}$$

Note that for the yaw mode ($j = 6$ or $q = 3$) there is no contribution from the last term. Applying Gauss' theorem on the closed-surface integral, we have

$$\hat{M}_{FK,pq} = i\omega\rho \iiint\limits_{V_p} \nabla \cdot (\hat{\phi}_0 \vec{e}_q \times \vec{s}_p) \, dV$$

$$-i\omega\rho \iint\limits_{S_{wp}} \hat{\phi}_0[(y - y_p)\delta_{q1} - (x - x_p)\delta_{q2}] \, dS. \qquad (5.227)$$

In the volume integral here, the integrand may be written in various ways:

$$\nabla \cdot (\hat{\phi}_0 \vec{e}_q \times \vec{s}_p) = \vec{s}_p \cdot \nabla \times (\hat{\phi}_0 \vec{e}_q) - \hat{\phi}_0 \vec{e}_q \cdot \nabla \times \vec{s}_p = \vec{s}_p \cdot \nabla \times (\hat{\phi}_0 \vec{e}_q)$$

$$= \vec{s}_p \cdot (\nabla \hat{\phi}_0) \times \vec{e}_q = \vec{e}_q \cdot \vec{s}_p \times \nabla \hat{\phi}_0 = (\vec{s}_p \times \nabla \hat{\phi}_0)_q$$

$$= (\vec{s}_p \times \hat{v}_0)_q = (\vec{s}_p \times \hat{a}_0)_q \frac{1}{i\omega}. \qquad (5.228)$$

Here $\nabla \hat{\phi}_0 = \hat{v}$ represents the fluid velocity $\vec{v}_0$ corresponding to the undisturbed velocity potential ϕ_0, and $\vec{a}_0$ is the corresponding fluid acceleration.

5.6.2 The Diffraction Force

Later (in next subsection) let us consider the Froude-Krylov force, the diffraction force and the excitation force for a small body. Then we shall make use of Eqs. (5.220), (5.227) and (5.228). We shall then also approximate the following equation for the diffraction force, which is still exact (within potential theory). From Eq. (5.214) we write the diffraction force as

$$\hat{F}_{d,p} = -i\omega\rho \sum_{p'=1}^{N} \iint\limits_{S_{p'}} \varphi_p \frac{\partial \hat{\phi}_0}{\partial n} \, dS. \qquad (5.229)$$

On $S_{p'}$ we have

$$\frac{\partial \hat{\phi}_0}{\partial n} = \vec{n}_{p'} \cdot \nabla \hat{\phi}_0 = \vec{n}_{p'} \cdot \hat{v}_0 = \frac{1}{i\omega} \vec{n}_{p'} \cdot \hat{a}_0 = \frac{1}{i\omega} \sum_{q=1}^{3} n_{p'q} \hat{a}_{0q} \qquad (5.230)$$

and hence

$$\hat{F}_{d,p} = -\rho \sum_{q=1}^{3} \sum_{p'=1}^{N} \iint\limits_{S_{p'}} \varphi_p n_{p'q} \hat{a}_{0q} \, dS. \qquad (5.231)$$

Here the sum over p' shows that the diffracted waves, not only from body p, but also from all the other bodies ($p \neq p'$), contribute to the diffraction force on body p.

5.6.3 Small-Body Approximation for a Group of Bodies

If the horizontal and vertical extensions of body p are very much shorter than one wavelength, we may in Eqs. (5.220) and (5.227), as an approximation, take incident-wave quantities, such as $\hat{\phi}_0$, $\hat{\vec{v}}_0$ and $\hat{\vec{a}}_0$, outside the integral. Then (for $q = 1, 2, 3$) the Froude-Krylov force and moment components are given by

$$\hat{F}_{\text{FK},pq} \approx \rho \hat{a}_{0pq} V_p + \delta_{q3}\rho g \hat{\eta}_{0p} S_{wp},$$ (5.232)

$$\hat{M}_{\text{FK},pq} \approx \rho(\vec{\Gamma}_p \times \hat{\vec{a}}_{0p})_q + \rho g \hat{\eta}_{0p}(S_{yp}\delta_{1q} - S_{xp}\delta_{2q}),$$ (5.233)

where $\vec{a}_{0p}$ is the fluid acceleration caused by the undisturbed incident wave at the chosen reference point $(x, y, z) = (x_p, y_p, z_p)$ – for instance, the centre of mass or the centre of displaced volume – for body p. Furthermore, $\hat{\eta}_{0p}$ is the same wave's elevation at $(x, y) = (x_p, y_p)$. Moreover,

$$\vec{\Gamma}_p = \iiint\limits_{V_p} \vec{s}_p \, dV_p$$ (5.234)

and

$$S_{xp} = \iint\limits_{S_{wp}} (x - x_p) \, dS, \quad S_{yp} = \iint\limits_{S_{wp}} (y - y_p) \, dS$$ (5.235)

are moments of the displaced volume V_p and of the water plane area S_{wp} of body p. Furthermore, Eqs. (4.41) and (5.228) have been used to obtain Eq. (5.233).

The small-body approximation corresponds to replacing the velocity potential $\hat{\phi}_0$ and its derivatives by the first (zero-order) term of their Taylor expansions about the reference point (x_p, y_p, z_p). Thus, assuming that $\hat{\phi}_0$ is as given by Eq. (4.87) or by Eqs. (4.92) and (4.102), we find it evident that the error of the approximation is small, of the order of $\mathcal{O}\{ka\}$ as $ka \to 0$, where

$$a = \max(|\vec{s}_p|).$$ (5.236)

Hence we expect the approximation to be reasonably good if $ka \ll 1$, that is, if the linear extension of the body is very small compared with the wavelength. For this reason we call it the small-body approximation, or, alternatively, the long-wavelength approximation. The choice of the reference point is of little importance if $k \to 0$. However, in order for the approximation to be reasonably good also for cases when k is not close to zero, particular choices may be better than others. To assist in a good selection of the reference point, comparisons with experiments or with numerical computations may be useful.

Before approximating the diffraction force, we make the further assumption that, not only body p, but each of the N bodies has horizontal and vertical extensions very small compared with one wavelength. For each body p' we may expand $\hat{a}_{0q}$ in the integrand of Eq. (5.231) as a Taylor series around the reference point $(x_{p'}, y_{p'}z_{p'})$, and if we then neglect terms of order $\mathcal{O}\{ka\}$ as $ka \to 0$, the j component ($j = 1, 2, \ldots, 6$) of the diffraction force on body p may be approximated to

$$\hat{F}_{d,i} \approx \frac{1}{i\omega} \sum_{p'=1}^{N} \sum_{q=1}^{3} Z_{ii'}\hat{a}_{0p'q} = \sum_{p'=1}^{N} \sum_{q=1}^{3} Z_{ii'}\hat{v}_{0p'q},$$ (5.237)

where $i = 6(p-1) + j$, $i' = 6(p'-1) + q$, and

$$Z_{ii'} = -i\omega\rho \iint_{S_{p'}} \varphi_i n_{p'q}\, dS \tag{5.238}$$

is an element of the radiation-impedance matrix. See Eqs. (5.152), (5.39) and (5.165).

The excitation force is, in accordance with Eq. (5.210), given as the sum of the Froude-Krylov force and the diffraction force. Thus the j component of the excitation force on body p is given by

$$\hat{F}_{e,i} = \hat{F}_{FK,i} + \hat{F}_{d,i}, \tag{5.239}$$

where $\hat{F}_{d,i}$ is given by Eq. (5.237) and $\hat{F}_{FK,i}$ by Eq. (5.232) for a translational mode ($j = q = 1, 2, 3$) or by Eq. (5.233) for a rotational mode ($j = q + 3 = 4, 5, 6$).

5.6.4 Small-Body Approximation for a Single Body

In the remaining part of this section let us consider the case of only one body; thus we have $N = 1$, $p = p' = 1$, $i = j$ and $i' = q$. Hence, in this subsection let us omit the subscript p on $\hat{a}_{0pq}$, $\hat{v}_{0pq}$ and $\hat{\eta}_{0p}$. Then the sum over p' in Eq. (5.237) has only one term, and, hence, the diffraction force is given by

$$\hat{F}_{d,j} \approx \frac{1}{i\omega}\sum_{q=1}^{3} Z_{jq}\hat{a}_{0q} = \sum_{q=1}^{3} Z_{jq}\hat{v}_{0q}, \tag{5.240}$$

which, in view of Eq. (5.44), may be written as

$$\hat{F}_{d,j} \approx \sum_{q=1}^{3}(m_{jq}\hat{a}_{0q} + R_{jq}\hat{v}_{0q}). \tag{5.241}$$

As we observed in Subsection 5.2.4, the radiation impedance for a small body is usually dominated by the radiation reactance; that is,

$$|\omega m_{jq}| = |X_{jq}| \gg R_{jq}. \tag{5.242}$$

On deep water it can be shown that (see below)

$$|R_{33}/\omega m_{33}| = \mathcal{O}\{ka\} \tag{5.243}$$

(and even smaller for other jq combinations) as $ka \to 0$. Then the diffraction force is

$$\hat{F}_{d,j} \approx \sum_{q=1}^{3} m_{jq}\hat{a}_{0q}. \tag{5.244}$$

This is a consistent approximation when we neglect terms of order $\mathcal{O}\{ka\}$ relative to terms of $\mathcal{O}\{1\}$. Kyllingstad[59] has given a more consistent higher-order approximation than Eq. (5.241).

The free-surface boundary condition is, in the case of deep water,

$$0 = \left(-\frac{\omega^2}{g} + \frac{\partial}{\partial z} \right)\varphi_j = \left(-k + \frac{\partial}{\partial z} \right)\varphi_j. \tag{5.245}$$

When $k \to 0$ this condition becomes

$$\frac{\partial}{\partial z}\varphi_j = 0 \quad \text{on } z = 0. \tag{5.246}$$

That is, it corresponds to a stiff plate (or ice) on the water. In this case no surface wave can propagate on the surface as a result of the body's (slow!) oscillation. However, there is still a finite added mass in the limit $k \to 0$, because there is some finite amount of kinetic energy in the fluid; see Eq. (5.188). Small-body approximations (5.232)–(5.233) and reciprocity relation (5.149) show explicitly for a body floating on deep water (with $D = 1$) that when $ka \to 0$, R_{jq}/ω is of $\mathcal{O}\{k\}$ and of $\mathcal{O}\{a^4\}$ for $(j, q) = (3, 3)$ and even smaller for other possible jq combinations. Moreover, in agreement with the text which follows Eq. (5.91) [also see Eq. (5.90)], it is reasonable to assume that m_{jq} is of $\mathcal{O}\{k^0\}$ and of $\mathcal{O}\{a^3\}$. Hence, $|R_{jq}/\omega m_{jq}|$ is of $\mathcal{O}\{ka\}$ or smaller.

Neglecting terms of order $\mathcal{O}\{ka\}$, we have, for the translational modes, $j = q = 1, 2, 3$, the following approximation for the excitation force on a single small body:

$$\hat{F}_{e,q} \approx \rho g \hat{\eta}_0 S_w \delta_{q3} + \rho V \hat{a}_{0q} + \sum_{q=1}^{3} m_{jq} \hat{a}_{0q}. \tag{5.247}$$

For the rotational modes ($j = q + 3 = 4, 5, 6$) an approximation to the excitation moment of the small body is obtained by addition of the two approximations (5.233) and (5.244).

Let us consider the excitation force on a body having two mutually orthogonal vertical planes of symmetry (e.g., the planes $x = 0$ and $y = 0$). A special case of this is an axisymmetric body (the z axis being the axis of symmetry). Because in this case φ_3 is symmetric, whereas n_1 and n_2 are antisymmetric, it follows from definition (5.39) of the radiation impedance that $Z_{13} = 0$ and $Z_{23} = 0$, and hence $m_{13} = m_{23} = 0$. For $j = q = 1, 2, 3$ the above approximation for $F_{e,j}$ simplifies to

$$\hat{F}_{e,q} \approx \rho g \hat{\eta}_0 S_w \delta_{q3} + \rho V (1 + \mu_{qq}) \hat{a}_{0q}, \tag{5.248}$$

where

$$\mu_{jj} = m_{jj}/\rho V \tag{5.249}$$

is the non-dimensionalised added-mass coefficient. If the undisturbed velocity potential represents a purely propagating wave as given by Eqs. (4.92) and (4.102) and if the reference axis $p - p$ of the single body (see Figure 5.13) is the vertical line $(x, y) = (0, 0)$, then $\hat{\eta}_0 = A$. Inserting this into Eq. (5.248), we obtain for the heave excitation force

$$\hat{F}_{e,3} = \hat{F}_{\text{FK},3} + \hat{F}_{d,3} \approx \left[\rho g S_w + (\rho V + m_{33}) \frac{\hat{a}_{03}}{A} \right] A. \tag{5.250}$$

Using Eq. (5.221) gives

$$\hat{F}_{e,3} \approx \left[\rho g S_w - \omega^2 (\rho V + m_{33}) \frac{\sinh(kz_1 + kh)}{\sinh(kh)} \right] A, \tag{5.251}$$

where z_1 is the z coordinate of the reference point for the body. On deep water we now have

$$\hat{F}_{e,3} \approx [\rho g S_w - \omega^2 \rho V (1 + \mu_{33}) \exp\{kz_1\}] A, \tag{5.252}$$

where we have introduced the added-mass coefficient $\mu_{33} = m_{33}/\rho V$. For a small floating body with $z_1 \approx 0$ this gives

$$\hat{F}_{e,3} \approx [\rho g S_w - \omega^2 \rho V (1 + \mu_{33})] A. \tag{5.253}$$

Note that as $\omega \to 0$, $F_3/A \to \rho g S_w$, which is the hydrostatic (buoyancy) stiffness of the floating body (see Subsection 5.9.1.)

For small values of ω the first term $\rho g S_w$ dominates over the second term $-\omega^2 \rho V (1 + \mu_{33})$. On deep water where $\omega^2 = gk$ the second term is approximately linear in k (forgetting the usually small frequency dependence of μ_{33}). This corresponds to the tangent at $ka = 0$ of the curve indicated in Figure 5.15.

For most body shapes, heave excitation force $\hat{F}_{e,3}$ is closely in phase with the undisturbed wave amplitude A for those small values of ka where the small-body approximation is applicable. An exception is a body shape with a relatively small water plane area, as indicated in Figure 5.16. Then the second term of the formula may become the dominating term even for a frequency interval within the range of validity of the small-body approximation. Then there is a phase shift of π between $\hat{F}_3$ and A within a narrow frequency interval near the angular frequency $\{g S_w / V (1 + \mu_{33})\}^{1/2}$.

For a submerged body ($S_w = 0$) we have

$$\hat{F}_{e,3} \approx -\omega^2 \rho V (1 + \mu_{33}) \exp\{kz_1\} A. \tag{5.254}$$

Now $\hat{F}_3$ and A are in antiphase everywhere within the region of validity of the approximation. This antiphase may be simply explained by the fact that the

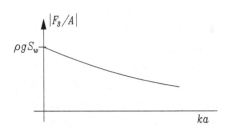

Figure 5.15: Typical variation of the amplitude of the heave force on a floating body for relatively small values of the angular repetency.

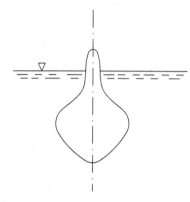

Figure 5.16: Floating body with a relatively small water plane area.

hydrodynamic pressure, which is in phase with the wave elevation [cf. Eq. (4.93)], is smaller below than above the submerged body. It should be observed that even if the heave excitation force in the small-body approximation is in antiphase with the wave elevation, the body's vertical motion is essentially in phase with the wave elevation, provided no other external forces are applied to the submerged body. This follows from the fact that the body's mechanical impedance is dominated by inertia. Then the acceleration and excursion are essentially in phase and antiphase, respectively, with the excitation force. (See also Sections 2.2 and 5.9.) For small values of $ka = \omega^2 a/g$ the above approximate formula (5.254) for $\hat{F}_{e,3}$ corresponds to the first increasing part of the force amplitude curve shown in Figure 5.17. The decreasing part of the curve for larger values of ka is due to the exponential factor in formula (5.254).

For a horizontal cylinder whose centre is submerged at least one diameter below the free water surface, the added-mass coefficients $\mu_{33} = \mu_{11}$ are close to 1, which is their exact value in an infinite fluid (without any free water surface).[19] A similar approximation applies for the added-mass coefficients μ_{11}, μ_{22} and μ_{33} for a sufficiently submerged sphere in which the added-mass coefficient is $\frac{1}{2}$ in an infinite fluid (cf. Newman,[24] p. 144).

5.7 Axisymmetric Oscillating System

A group of concentric axisymmetric bodies is an axisymmetric system. A case with three bodies is shown in Figure 5.18. Concerning the sea bed, we have

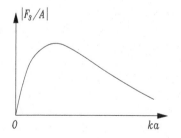

Figure 5.17: Typical variation of the amplitude of the heave force on a submerged body versus the angular repetency.

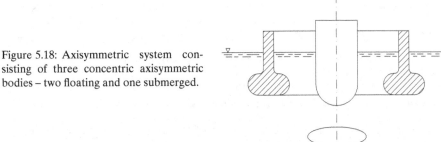

Figure 5.18: Axisymmetric system consisting of three concentric axisymmetric
bodies – two floating and one submerged.

to assume either deep water or a sea-bed structure which is axisymmetric and
concentric with the bodies. (The latter situation is true if the water depth is
constant and equal to h.) Because the reference axis ($p - p$ in Figure 5.13) is
the same for all bodies, we may choose it as the z axis. Correspondingly, the
global coordinates $(x, y, z) = (r \cos \theta, r \sin \theta, z)$ coincide with the local coordinates $(x_p, y_p, z) = (r_p \cos \theta_p, r_p \sin \theta_p, z)$ associated with each body p (for
all p).

For an axisymmetric body p, as shown in Figure 5.19, a wet-surface element
dS in position

$$\vec{s}_p = (x, y, z) = (r \cos \theta, r \sin \theta, z) \tag{5.255}$$

has a unit normal

$$\vec{n}_p = (n_{px}, n_{py}, n_{pz}) = (n_{pr} \cos \theta, n_{pr} \sin \theta, n_{pz}) = (n_{p1}, n_{p2}, n_{p3}). \tag{5.256}$$

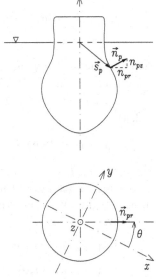

Figure 5.19: Side view and top view of a body symmetric
with respect to the z axis, with unit normal $\vec{n}_p$ on element
dS of the wet surface in position $\vec{s}_p$ from the reference point
$(0, 0, z_p) = (0, 0, 0)$.

Furthermore;

$$\vec{s}_p \times \vec{n}_p = (n_{p4}, n_{p5}, n_{p6}),$$ (5.257)

where

$$n_{p4} = yn_{pz} - zn_{py} = (rn_{pz} - zn_{pr})\sin\theta \equiv -n_{pM}\sin\theta$$
$$n_{p5} = zn_{px} - xn_{pz} = (zn_{pr} - rn_{pz})\cos\theta = n_{pM}\cos\theta$$
$$n_{p6} = xn_{py} - yn_{px} = rn_{pr}(\cos\theta\sin\theta - \sin\theta\cos\theta) = 0.$$ (5.258)

To summarise, for $j = 1, 6$ the components of the unit normal may be collected into the column vector

$$\mathbf{n}_p = (n_{pr}\cos\theta, n_{pr}\sin\theta, n_{pz}, -n_{pM}\sin\theta, n_{pM}\cos\theta, 0)^T,$$ (5.259)

where

$$n_{pM} = zn_{pr} - rn_{pz}, \quad n_{pr} = \sqrt{n_{px}^2 + n_{py}^2}.$$ (5.260)

We notice that for $j = 6$, boundary condition (5.164) is satisfied if $\varphi_{p6} \equiv 0$, which was to be expected because an axisymmetric body yawing in an ideal fluid can generate no wave. The complex functions $\varphi_{pj}(j = 1, 5)$ must satisfy the Laplace equation, and the usual homogeneous boundary conditions on the free surface $z = 0$ and on fixed surfaces such as the sea bed $z = -h$. Moreover, φ_{pj} must satisfy the radiation condition.

For an axisymmetric body it is easy to see that a particular solution of the type

$$\varphi_{pj} = \varphi_{pj}(r, \theta, z) = \varphi_{pj0}(r, z)\Theta_j(\theta)$$ (5.261)

satisfies the Laplace equation, provided $\Theta_j(\theta)$ is a function of the type given by Eq. (4.210). If we choose

$$\Theta_1(\theta) = \Theta_5(\theta) = \cos\theta$$
$$\Theta_2(\theta) = \Theta_4(\theta) = \sin\theta$$
$$\Theta_3(\theta) = \Theta_6(\theta) = 1,$$ (5.262)

then also the inhomogeneous boundary condition (5.164) on S_p can be satisfied in accordance with Eq. (5.259). (Note, however, that $\dot{\varphi}_{p60} \equiv 0$.)

In the far-field region we have asymptotically, according to Eqs. (4.221) and (4.224),

$$\varphi_{pj} \sim (\text{constant}) \times \Theta_j(\theta)e(kz)H_n^{(2)}(kr)$$ (5.263)

with $n = 0$ for $j = 3$ and $n = 1$ for $j = 1, 2, 4$ and 5.

Because the solution for $\varphi_{pj}(r, \theta, z)$ is of the type of Eq. (5.261) also in the far-field region, we may write the Kochin functions and the far-field coefficients as follows:

$$h_{pj}(\theta) = h_{pj0}\Theta_j(\theta)$$
$$a_{pj}(\theta) = a_{pj0}\Theta_j(\theta).$$ (5.264)

By comparing Eqs. (5.261) and (4.246) we see that in the far-field region all the functions $\varphi_{pj0}(r, z)$ vary in the same way with r and z, because asymptotically we have

$$\varphi_{pj0}(r, z) \sim a_{pj0}\, e(kz)(kr)^{-\frac{1}{2}} e^{-ikr} \tag{5.265}$$

for $kr \to \infty$. Note that a_{pj0} is independent of r, θ and z.

Similarly, we may, according to Haskind relation (5.202), write the excitation-force coefficients as

$$f_{pj}(\beta) = f_{pj0}\Theta_j(\beta \pm \pi), \tag{5.266}$$

where

$$f_{pj0} = \frac{\rho g D}{k} h_{pj0} = \frac{\rho g D}{k}\sqrt{2\pi}\, a_{pj0} e^{i\pi/4}. \tag{5.267}$$

Note from Eq. (5.262) that $\Theta_j(\beta \pm \pi) = -\Theta_j(\beta)$ for $j = 1, 2, 4, 5$.

5.7.1 The Radiation Impedance

Radiation-impedance matrix $\mathbf{Z}$ is, according to Eq. (5.156), partitioned into 6×6 matrices $\mathbf{Z}_{pp'}$, of which each element is of the type [see Eqs. (5.167) and (5.164)]

$$Z_{pj,p'j'} = -i\omega\rho \iint_{S_p} \varphi_{p'j'} \frac{\partial \varphi_{pj}^*}{\partial n}\, dS. \tag{5.268}$$

Let us consider the wet body surface S_p as a surface of revolution generated by curve l_p as shown in Figure 5.20. Then

$$
\begin{aligned}
Z_{pj,p'j'} &= -i\omega\rho \int_0^{2\pi} d\theta \int_{l_p} dl\, R(z)\varphi_{p'j'} \frac{\partial \varphi_{pj}^*}{\partial n} \\
&= -i\omega\rho \int_0^{2\pi} \Theta_{j'}(\theta)\Theta_j(\theta)d\theta \int_{l_p} \varphi_{p'j'0} \frac{\partial \varphi_{pj0}^*}{\partial n} R(z)dl.
\end{aligned} \tag{5.269}
$$

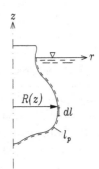

Figure 5.20: Wet surface S_p of an axisymmetric body is a surface of revolution generated by curve l_p.

Note that $\varphi_{pj0} = \varphi_{pj0}(r, z)$. Now because

$$\int_0^{2\pi} (\cos\theta, \sin\theta, \sin\theta\cos\theta)\, d\theta = (0, 0, 0),$$ (5.270)

$$\int_0^{2\pi} (\cos^2\theta, \sin^2\theta, 1^2)\, d\theta = (\pi, \pi, 2\pi),$$ (5.271)

radiation-impedance matrix $\mathbf{Z}_{pp'}$ is of the form

$$\mathbf{Z}_{pp'} = \begin{bmatrix} Z_{p1,p'1} & 0 & 0 & 0 & Z_{p1,p'5} & 0 \\ 0 & Z_{p1,p'1} & 0 & Z_{p1,p'5} & 0 & 0 \\ 0 & 0 & Z_{p3,p'3} & 0 & 0 & 0 \\ 0 & Z_{p5,p'1} & 0 & Z_{p5,p'5} & 0 & 0 \\ Z_{p5,p'1} & 0 & 0 & 0 & Z_{p5,p'5} & 0 \\ 0 & 0 & 0 & 0 & 0 & 0 \end{bmatrix},$$ (5.272)

where we also utilised the fact that for the axisymmetric system we have $Z_{p2,p'2} = Z_{p1,p'1}$, $Z_{p4,p'4} = Z_{p5,p'5}$, $Z_{p2,p'4} = Z_{p1,p'5}$ and $\varphi_{p6} = \varphi_{p60} \equiv 0$. Note matrix $\mathbf{Z}_{pp'}$ is singular. Its rank is at most 5. [An $n \times n$ matrix is of rank k ($k \leq n$) if in the matrix there exists a non-vanishing minor of size $k \times k$, but no larger one.]

The *non-vanishing* elements of $\mathbf{Z}$ may be expressed as

$$Z_{pj,p'j'} = -i\omega\rho(1 + \delta_{3j}\delta_{3j'})\pi \int_{l_p} \varphi_{p'j'0} \frac{\partial\varphi_{pj0}^*}{\partial n} R(z)\,dl,$$ (5.273)

where $\delta_{3j} = 1$ if $j = 3$ and $\delta_{3j} = 0$ otherwise. Thus we have for the radiation-impedance matrix

$$Z_{pj,p'j'} = -i\omega\rho\sigma_{jj'}\pi \int_{l_p} \varphi_{p'j'0} \frac{\partial\varphi_{pj0}^*}{\partial n} R(z)\,dl,$$ (5.274)

where 27 of the 36 values of the numbers $\sigma_{jj'}$ vanish. There are just nine non-vanishing values, namely

$$\sigma_{33} = 2$$ (5.275)

and

$$\sigma_{11} = \sigma_{22} = \sigma_{44} = \sigma_{55} = \sigma_{15} = \sigma_{51} = \sigma_{24} = \sigma_{42} = 1.$$ (5.276)

5.7.2 Radiation Resistance and Excitation Force

Radiation-resistance matrix $\mathbf{R}_{pp'}$ is obtained by taking the real part of Eq. (5.272), which shows that at least 27 of the 36 elements vanish. To calculate the non-vanishing elements $R_{pj,p'j'}$, we take the integral along a generatrix $r = $ constant ($r \to \infty$) of the control cylinder S_∞ in the far-field (Figure 5.13) instead of along the curve l_p; that is, $r = R(z)$ as in Eq. (5.274). Compare Eqs. (5.156), (5.164), (5.169) and (5.175).

Thus the radiation-resistance matrix is

$$R_{pj,p'j'} = -i\omega\rho\sigma_{jj'}\pi \lim_{r\to\infty} \int_{-h}^{0} \varphi_{p'j'0} \frac{\partial \varphi_{pj0}^*}{\partial r} r\,dz. \tag{5.277}$$

From expression (4.224) – also see Eqs. (4.282), (5.261) and (5.264) – we have the asymptotic expression

$$\varphi_{pj0}(r, z) \sim h_{pj0}e(kz)(2\pi kr)^{-1/2}e^{-i(kr+\pi/4)}. \tag{5.278}$$

Thus we have

$$
\begin{aligned}
R_{pj,p'j'} &= -i\omega\rho\sigma_{jj'}\pi h_{p'j'0}h_{pj0}^* \frac{ik}{2\pi k} \int_{-h}^{0} e^2(kz)dz \\
&= \frac{1}{2}\omega\rho\sigma_{jj'}h_{p'j'0}h_{pj0}^* \frac{D(kh)}{2k} = \frac{\omega\rho D(kh)}{4k}\sigma_{jj'}h_{p'j'0}h_{pj0}^* \\
&= \frac{\omega k}{4\rho g^2 D(kh)}\sigma_{jj'}f_{p'j'0}f_{pj0}^* = \frac{\omega k}{4\rho g^2 D(kh)}\sigma_{jj'}f_{p'j'0}^*f_{pj0}, \tag{5.279}
\end{aligned}
$$

where we have used Eqs. (4.112) and (5.267) and also utilised the fact that $R_{pj,p'j'}$ is real. We note from Eq. (5.279) that the diagonal elements $R_{pj,pj}$ of the radiation-resistance matrix are necessarily non-negative. This is a result which was to be expected, because otherwise the principle of conservation of energy would be violated. If the only oscillation is mode j of body p, the radiated power is, in accordance with Eq. (5.177), $\frac{1}{2}R_{pj,pj}|\hat{u}_j|^2$, which cannot, of course, be negative.

Because the radiation-resistance matrix is real, it follows from Eq. (5.279) that for those combinations of j and j' for which $\sigma_{jj'}$ *does not vanish*, we have that

$$f_{pj0}f_{p'j'0}^* = f_{pj0}^*f_{p'j'0} \tag{5.280}$$

is real. This is also in agreement with reciprocity relation (5.173), from which it follows that

$$R_{p'j',pj} = R_{pj,p'j'}. \tag{5.281}$$

Hence,

$$\frac{f_{p'j'0}}{f_{p'j'0}^*} = \frac{f_{pj0}}{f_{pj0}^*} \quad \text{or} \quad \frac{f_{p'j'0}^2}{|f_{p'j'0}|^2} = \frac{f_{pj0}^2}{|f_{pj0}|^2}. \tag{5.282}$$

Taking the square root gives

$$\frac{f_{p'j'0}}{|f_{p'j'0}|} = \pm\frac{f_{pj0}}{|f_{pj0}|}, \tag{5.283}$$

which shows that the excitation-force coefficients $f_{p'j'0}$ and f_{pj0} have equal or opposite phases. Note that this applies only if $\sigma_{jj'} \neq 0$. Moreover, note from

Eq. (5.279) that

$$R_{pj,p'j'} = \pm\sqrt{R_{pj,pj}\,R_{p'j',p'j'}} \tag{5.284}$$

is negative if, and only if, f_{pj0} and $f_{p'j'0}$ have opposite phases.

Considering just body p, we find that the excitation surge force and pitch moment are in equal or opposite phases, and the same can be said about the excitation sway force and roll moment. The only non-vanishing off-diagonal elements of $\mathbf{R}_{pp}$ are

$$R_{p1,p5} = R_{p5,p1} = \pm\sqrt{R_{p1,p1}\,R_{p5,p5}}, \tag{5.285}$$

$$R_{p2,p4} = R_{p4,p2} = \pm\sqrt{R_{p2,p2}\,R_{p4,p4}}, \tag{5.286}$$

which are equal as a result of the axial symmetry. From this it follows that $\mathbf{R}_{pp}$ is a more singular matrix than $\mathbf{Z}_{pp}$. Their ranks are at most 3 and 5, respectively (see Problem 5.13).

Let us now consider a system of two bodies. Then, because $\sigma_{33} = 2 \neq 0$, Eq. (5.283) shows that the heave excitation forces for the two bodies have either equal or opposite phases. If both bodies are floating (as in the upper part of Figure 5.18), we know that they also have the same phase if the frequency is sufficiently low for the small-body approximation (5.248) to be applicable. Then also $R_{p3,p'3}$ is positive. However, if one body is floating while the other is submerged (as if the largest, annular, body were removed from Figure 5.18), then because the latter has a vanishing water plane area, the heave excitation forces are in opposite phases for low frequencies where the small-body approximation is applicable. In this case $R_{p3,p'3}$ is negative. Moreover, for this system the excitation surge forces and pitch moments for the two bodies have equal or opposite phases. The same may be said about the excitation sway forces and roll moments for the two bodies.

For the two-body system the radiation-impedance matrix $\mathbf{Z}$, a 12×12 matrix, has the structure

$$\mathbf{Z} = \begin{bmatrix} \mathbf{Z}_{pp} & \mathbf{Z}_{pp'} \\ \mathbf{Z}_{p'p} & \mathbf{Z}_{p'p'} \end{bmatrix}. \tag{5.287}$$

We see from Eq. (5.272) that at least $4 \times 27 = 108$ of the $4 \times 36 = 144$ matrix elements vanish. Among the $4 \times 9 = 36$ remaining elements not more than 23 different values are apparently possible, because of the symmetry relation $\mathbf{Z}^T = \mathbf{Z}$; see Eq. (5.171). However, there are, in fact, only 13 different non-zero values if we do not forget that even though the matrix (5.272) has only 9 non-vanishing elements, these 9 ones share only 4 or 5 different values (for $p' = p$ or $p' \neq p$, respectively), The matrix $\mathbf{Z}$ is singular, and its rank is at most 10. However, as a result of the relation (5.279), the rank of the radiation resistance matrix $\mathbf{R} = \mathrm{Re}(\mathbf{Z})$ cannot be more than 3. A physical explanation of this fact will be given in the next chapter (see Subsection 6.4.2). Let us here just mention that the maximum rank of 3 is related to the fact that there is only three different linearly independent functions $\Theta_j(\theta)$, as given in Eqs. (5.262).

If a third body is included in our axisymmetric system, the rank of the 18×18 matrix $\mathbf{Z}$ cannot be larger than 15, but the rank of the 18×18 matrix $\mathbf{R}$ is still not more than 3.

If the radiation-resistance matrix is singular, it is possible to force the bodies to oscillate on still water, without radiating a wave in the far-field region. As an example, consider surge and pitch oscillations of a single body. Because $\Theta_1(\theta) = \Theta_5(\theta) = \cos\theta$ according to Eq. (5.262), it is possible to choose a combination of $\hat{u}_1$ and $\hat{u}_5$ such that the far-field waves radiated by the surge mode and the pitch mode cancel each other. Such a cancellation is possible in the far-field region [see asymptotic approximation (5.265)], but not in general in the near-field region, where fluid motion is associated with the added-mass matrix [see Eqs. (5.187)–(5.188)].

5.7.3 Numerical 2-Body Example

We consider two bodies with a common vertical axis of symmetry. One body (number 1) is floating whereas the other (number 2) is submerged, and we take only the heave mode into account. Then, according to the numbering scheme of Eq. (5.152), the force-excitation coefficients of interest are $f_3(\omega)$ and $f_9(\omega)$, both of which are independent of angle β of wave incidence. Correspondingly, radiation-impedance matrix $\mathbf{Z}(\omega) = \mathbf{R}(\omega) + i\omega\mathbf{m}(\omega)$ has only the following elements of interest: Z_{33}, Z_{99} and $Z_{39} = Z_{93}$. It follows from Eqs. (5.279) and (5.275) that

$$\frac{f_i f_{i'}^*}{8R_{ii'}} = \frac{J}{k|A|^2} = \frac{J\lambda}{2\pi|A|^2} = v_g \rho g \lambda/(4\pi), \qquad (5.288)$$

where the subscripts i and i' are 3 or 9. Furthermore, $\lambda = 2\pi/k$ is the wavelength, and v_g is the group velocity. (See also the last remark in Subsection 5.4.3.)

As an example, let us present some numerical results for a case in which the geometry is specified as follows. The upper body is a floating cylinder with a hemisphere at the lower end. The radius is a and the draft is $2.21a$ at equilibrium. The body displaces a water volume of $\pi a^3(1.21 + 2/3) \approx 5.89a^3$. The lower body is completely submerged, and it is shaped as an ellipsoid, with a radius of $2.27a$ and vertical extension $0.182a$. Its volume is $\pi a^3(2.27^2 \cdot 0.182)2/3 = 1.96a^3$. The centre of the ellipsoid is at a submergence $6.06a$, and the water depth is $h = 15a$.

For this geometry, hydrodynamical parameters have been computed by using computer code ©AQUADYN-2.1. The wet surfaces of the upper and lower bodies are approximated by 660 and 540 plane panels, respectively. Numerical results for the computed hydrodynamical parameters are expected to be correct within an accuracy of 2–3%.[60] Excitation-force coefficients $f_3(\omega)$ and $f_9(\omega)$ are given by the curves in Figures 5.21 and 5.22. Observe that there is a frequency range, corresponding to $0.2 < ka < 0.4$ where the heave excitation force is approximately equal but oppositely directed for the two bodies. We may refer to this as "force compensation", and this phenomenon[61,62] has been utilised with some drilling platforms (semisubmersible floating vessels). Radiation resistances $R_{33}(\omega)$ and $R_{99}(\omega)$ are given by the curves in Figure 5.23. From Eq. (5.288) it follows that

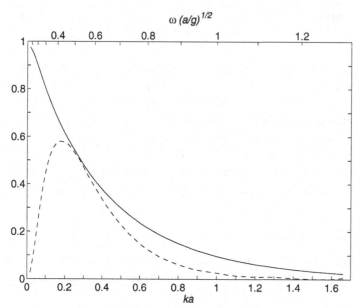

Figure 5.21: Modulus of excitation-force coefficients $|f_i(\omega)|/(\rho g \pi a^2)$ with $i = 3$ for the floating upper body (solid curve) and with $i = 9$ for the submerged lower body (dashed curve).

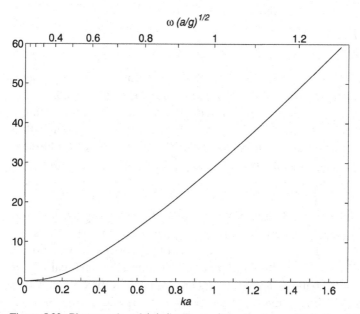

Figure 5.22: Phase angle $\angle f_3(\omega)$ (in degrees) for the heave excitation-force coefficient of the floating upper body. It differs by π (180°) from the corresponding phase angle $\angle f_9(\omega)$ for the submerged lower body.

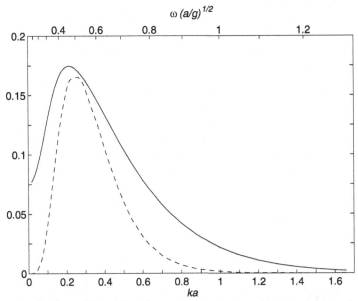

Figure 5.23: Normalised radiation resistances $R_{i,i'}(\omega)/(\omega\rho a^3 2\pi/3)$ for $(i,i') = (3,3)$ (solid curve) and for $(i,i') = (9,9)$ (dashed curve), versus normalised angular repetency ka (lower scale) and normalised frequency $\omega(a/g)^{1/2}$ (upper scale).

$R_{39}^2 = R_{93}^2 = R_{33}R_{99}$, and from the observation that $f_3/f_9 = -|f_3/f_9|$ we have that $R_{39} = R_{93}$ is negative; that is,

$$R_{39} = -(R_{33}R_{99})^{1/2}. \tag{5.289}$$

From the numerical computation it was found[60] that at infinite frequency the elements of the added-mass matrix are in normalised quantity $m_{ii'}(\infty)/(\rho a^3 2\pi/3)$ given by $0.58, 15.0$ and -0.156 for (i,i') equal to $(3,3), (9,9)$ and $(3,9)$, respectively. These numerical values are needed if the dynamics of the two-body system are to be investigated in the time domain [cf. Subsection 5.3.1 and also Eq. (5.337)]. For such an investigation also the causal impulse-response functions $k_{33}(t)$, $k_{99}(t)$ and $k_{39}(t) = k_{93}(t)$ are needed. For $t > 0$ they are, according to Eq. (5.115), equal to twice the inverse Fourier transform of $R_{33}(\omega)$, $R_{99}(\omega)$ and $R_{39}(\omega) = R_{93}(\omega)$, respectively. The curves shown in Figure 5.24 represent computed numerical results for these impulse-response functions. Observe that they, together with the numerical values of $m_{ii'}(\infty)$ stated above, represent, in principle, the same quantitative information as the complex radiation impedances $Z_{33}(\omega)$, $Z_{99}(\omega)$ and $Z_{39}(\omega)$.

Impulse-response functions $f_{t,3}(t)$ and $f_{t,9}(t)$ are inverse Fourier transforms of excitation-force coefficients $f_3(\omega)$ and $f_9(\omega)$, respectively. The curves shown in Figure 5.25 represent computed numerical values for the axisymmetric two-body example. We observe that these impulse-response functions do not vanish for $t < 0$, and hence they are non-causal as explained in Subsection 5.3.2.

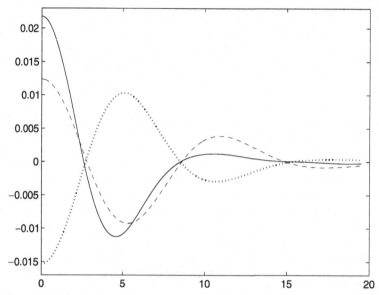

Figure 5.24: Normalised radiation-problem impulse-response functions $k_{ii'}/(\pi\rho ga^2)$ versus normalised time $t(g/a)^{1/2}$ for $(i, i') = (3, 3)$ (solid curve), for $(i, i') = (9, 9)$ (dashed curve) and for $(i, i') = (3, 9)$ (dotted curve).

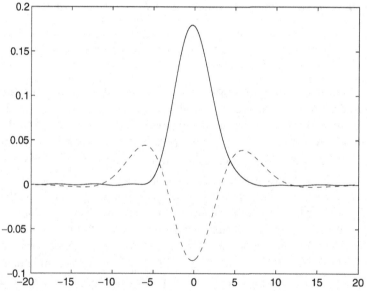

Figure 5.25: Normalised excitation-problem impulse-response function $f_{t,i}(t)(a/g)^{1/2}/(\pi\rho ga^2)$ versus normalised time $t(g/a)^{1/2}$ for $i = 3$ (solid curve) and for $i = 9$ (dashed curve).

5.8 Two-Dimensional System

For a two-dimensional system there is no variation in the y direction, and wave propagation is in the x direction. Each body has three degrees of freedom (or modes of motion), namely surge, heave and pitch, corresponding to $j = 1$, $j = 3$ and $j = 5$, respectively.

Let us use a prime symbol on variables to denote physical quantities per unit width. Thus F', m', Z' and R' denote force, mass, impedance and resistance, respectively, per unit width of the two-dimensional system. The total wave force on body p is per unit width given by

$$\hat{\mathbf{F}}'_{pt} = \hat{\mathbf{F}}'_{pe} + \hat{\mathbf{F}}'_{pr} = \hat{\mathbf{F}}'_{pe} - \sum_{p'} \mathbf{Z}'_{pp'} \hat{\mathbf{u}}_{p'}, \tag{5.290}$$

which is a three-dimensional column vector. The two terms result from the incident wave and from the bodies' radiation, respectively. The 3×3 coefficient matrix $\mathbf{Z}'_{pp'}$ represents the radiative interaction from body p' on body p.

In analogy with Eq. (5.196), we may define $3N$-dimensional force vectors $\mathbf{F}'_t$ and $\mathbf{F}'_e$ and correspondingly $3N$-dimensional velocity vector $\mathbf{u}$, which are related by

$$\hat{\mathbf{F}}'_t = \hat{\mathbf{F}}'_e - \mathbf{Z}' \hat{\mathbf{u}}, \tag{5.291}$$

where the $3N \times 3N$ matrix $\mathbf{Z}'$ represents the radiation impedance per unit width.

The radiated wave has a velocity potential given by

$$\hat{\phi}_r = \sum_{p=1}^{N} \boldsymbol{\varphi}_p^T \hat{\mathbf{u}}_p = \boldsymbol{\varphi}^T \hat{\mathbf{u}}, \tag{5.292}$$

where $\boldsymbol{\varphi}_p$ and $\boldsymbol{\varphi}$ are coefficient column vectors of dimension 3 and $3N$, respectively. In accordance with Eq. (4.256), the far-field (asymptotic) approximation may be written as

$$\boldsymbol{\varphi} \sim -\frac{g}{i\omega} \mathbf{a}^{\pm} e(kz) e^{\mp ikx} \quad \text{as } x \to \pm\infty. \tag{5.293}$$

Moreover, taking the complex conjugate of Eq. (4.258), we find that

$$\mathbf{I}'(\boldsymbol{\varphi}^*, \boldsymbol{\varphi}^T) = -i\left(\frac{g}{\omega}\right)^2 D(kh)(\mathbf{a}^{+*}\mathbf{a}^{+T} + \mathbf{a}^{-*}\mathbf{a}^{-T}) \tag{5.294}$$

and that Eq. (4.259) gives

$$\mathbf{I}'(\boldsymbol{\varphi}, \boldsymbol{\varphi}^T) = 0. \tag{5.295}$$

The two-dimensional Kochin function for radiation is given by

$$\mathbf{h}'(0) = -\frac{k}{D(kh)} \mathbf{I}'\{e(kz)e^{ikx}, \boldsymbol{\varphi}\} = -\frac{gk}{\omega} \mathbf{a}^{+} \tag{5.296}$$

$$\mathbf{h}'(\pi) = -\frac{k}{D(kh)} \mathbf{I}'\{e(kz)e^{-ikx}, \boldsymbol{\varphi}\} = -\frac{gk}{\omega} \mathbf{a}^{-}, \tag{5.297}$$

which is in agreement with Eqs. (4.278), (4.298) and (4.299). The Kochin function $H'_d(\beta)$ ($\beta = 0, \pi$) for the diffracted wave is given by analogous expressions, provided $\boldsymbol{\varphi}, \mathbf{a}^{\pm}$ and $\mathbf{h}'$ are replaced by $\hat{\phi}_d$, $A_d^{\pm}$ and H'_d, respectively, in Eqs. (5.293)–(5.297). For the diffraction problem (i.e., when the bodies are not moving) we may define the reflection coefficient

$$\Gamma_d = A_d^{-}/A \tag{5.298}$$

[see Eq. (4.88)] and the transmission coefficient

$$T_d = 1 + A_d^{+}/A. \tag{5.299}$$

These coefficients quantitatively describe how the incident wave of amplitude $|A|$ is divided into a reflected wave and a transmitted wave, which is propagating beyond the bodies. From energy conservation arguments it is to be expected that

$$|\Gamma_d|^2 + |T_d|^2 = 1. \tag{5.300}$$

This can indeed be proved mathematically (see Problem 5.14).

Now let us consider some reciprocity relations involving the radiation resistance and the excitation force for a two-dimensional system. The radiation resistance per unit width is given by the $3N \times 3N$ matrix

$$
\begin{aligned}
\mathbf{R}' &= \frac{1}{2} i\omega\rho \mathbf{I}'(\boldsymbol{\varphi}^*, \boldsymbol{\varphi}^T) = \frac{\rho g^2\, D(kh)}{2\omega} (\mathbf{a}^{+*}\mathbf{a}^{+T} + \mathbf{a}^{-*}\mathbf{a}^{-T}) \\
&= \frac{\omega\rho\, D(kh)}{2k^2} \{\mathbf{h}'^*(0)[\mathbf{h}'(0)]^T + \mathbf{h}'^*(\pi)[\mathbf{h}'(\pi)]^T\} \\
&= \frac{\omega\rho\, D(kh)}{2k^2} \{\mathbf{h}'(0)[\mathbf{h}'(0)]^{\dagger} + \mathbf{h}'(\pi)[\mathbf{h}'(\pi)]^{\dagger}\}.
\end{aligned}
\tag{5.301}
$$

Compare Eqs. (4.304), (5.174), (5.296) and (5.297). Observe that because $\mathbf{R}'$ is real we may, if we wish, take the complex conjugate of Eq. (5.301) without changing the result. The excitation force per unit width is

$$\mathbf{F}'_e = \mathbf{f}'(\beta)A = i\omega\rho \mathbf{I}'(\hat{\phi}_0, \boldsymbol{\varphi}), \tag{5.302}$$

where, for $\beta = 0$ or $\beta = \pi$,

$$\mathbf{f}'(\beta) = \frac{\rho g\, D(kh)}{k} \mathbf{h}'(\beta \pm \pi). \tag{5.303}$$

That is,

$$\mathbf{f}'(0) = \frac{\rho g\, D(kh)}{k} \mathbf{h}'(\pi) = -\frac{\rho g^2\, D(kh)}{\omega} \mathbf{a}^{-} \tag{5.304}$$

$$\mathbf{f}'(\pi) = \frac{\rho g\, D(kh)}{k} \mathbf{h}'(0) = -\frac{\rho g^2\, D(kh)}{\omega} \mathbf{a}^{+}. \tag{5.305}$$

Compare Eqs. (5.198), (5.296) and (5.297). Also, the two-dimensional version of Eq. (4.281) has been used with $A_j = 0$, $A_i = A$, $\phi_j = \boldsymbol{\varphi}$ and $H'_j = \mathbf{h}'$. Note that the excitation force on a surging piston in a wave channel, derived previously [see Eq. (5.86)], agrees with the general equation (5.302) combined with Eq. (5.304).

For a single symmetric body oscillating in heave the wave radiation is symmetric, and hence

$$
\begin{bmatrix} a_3^- \\ h_3'(\pi) \\ f_3'(0) \end{bmatrix} = \begin{bmatrix} a_3^+ \\ h_3'(0) \\ f_3'(\pi) \end{bmatrix}.
\tag{5.306}
$$

If it oscillates in surge and/or pitch, the radiation is antisymmetric and hence

$$
\begin{bmatrix} a_j^- \\ h_j'(\pi) \\ f_j'(0) \end{bmatrix} = - \begin{bmatrix} a_j^+ \\ h_j'(0) \\ f_j'(\pi) \end{bmatrix} \quad \text{for } j = 1 \text{ and } j = 5.
\tag{5.307}
$$

Thus, according to Eq. (5.301), the radiation-resistance matrix for the two-dimensional symmetric body is

$$
\mathbf{R}' = \frac{\rho g^2\, D(kh)}{\omega} \begin{bmatrix} |a_1^+|^2 & 0 & a_5^+ a_1^{+*} \\ 0 & |a_3^+|^2 & 0 \\ a_1^+ a_5^{+*} & 0 & |a_5^+|^2 \end{bmatrix}.
\tag{5.308}
$$

In this matrix it is possible to find a non-vanishing minor of size 2×2, but not of size 3×3. Hence $\mathbf{R}'$ is a singular matrix of rank not more than 2.

Because $\mathbf{R}'$ is a symmetrical and real matrix, we may use Eqs. (5.304) and (5.305) and then argue as we did in connection with Eqs. (5.280)–(5.286). We then conclude that the off-diagonal element of $\mathbf{R}'$ is positive or negative when the excitation-force coefficients for surge and pitch for the symmetric two-dimensional body are in the same phase or in opposite phases, respectively.

It can be shown, in general, for a two-dimensional system of oscillating bodies that the system's radiation-resistance matrix has a rank which is at most 2. Let $\mathbf{x}$ and $\mathbf{y}$ be two arbitrary column vectors of order $3N$. Then the matrix $\mathbf{xy}^T$ is of rank not more than 1 (see Pease,[12] p. 239). Matrix $\mathbf{R}'$ in Eq. (5.301) is a sum of two such matrices. Hence $\mathbf{R}'$ is a matrix of rank 2 or smaller. (The rank is smaller than 2 if column vectors $\mathbf{a}^+$ and $\mathbf{a}^-$ are linearly dependent.) A physical explanation of why the rank is bounded will be given in the next chapter (see Subsection 6.4.3.)

5.9 Motion Response and Control of Motion

When an immersed body is not moving, it is subjected to an excitation force $\mathbf{F}_e$ if an incident wave exists. If the body is oscillating, it is subjected to two other forces, namely the radiation force $\mathbf{F}_r$, which we have already considered [see Eq. (5.37) or (5.110)], and a hydrostatic buoyancy force $\mathbf{F}_b$. The latter force originates from the static-pressure term $-\rho g z$ in Eq. (4.14), because the body's wet surface experiences varying hydrostatic pressure as a result of its oscillation. However, oscillation in surge, sway or yaw modes of motion does not produce any hydrostatic force. Below let us limit ourselves to the study of heave motion in more detail. For a more general treatment of hydrostatic buoyancy forces and moments, refer to textbooks by Newman[24] (Chap. 6) or by Mei[1] (Chap. 7). In linear theory we shall assume that the hydrostatic buoyancy force $\mathbf{F}_b$ is proportional to the excursion $\mathbf{s}$ of the body from its equilibrium position. We write

$$\mathbf{F}_b = -\mathbf{S}_b \mathbf{s}, \tag{5.309}$$

where the elements of the "buoyancy stiffness" matrix $\mathbf{S}_b$ are coefficients of proportionality. Many of the 36 elements, for instance, all elements associated with surge or sway, vanish. Moreover, we note that the velocity is

$$\mathbf{u} = \frac{d}{dt} \mathbf{s} \tag{5.310}$$

There may be additional forces acting on the oscillating body, such as a viscous force $\mathbf{F}_v$ and a control or load force $\mathbf{F}_u$ from a control mechanism or from a mechanism which delivers or absorbs energy. There may also be a friction force $\mathbf{F}_f$ which is due to this mechanism or to other means for transmission of forces. Moreover, there may be mooring forces, which we shall not consider explicitly. We may assume that mooring forces are included in $\mathbf{F}_u$ (or in $\mathbf{F}_b$, if the mooring is flexible and linear theory applicable).

From Newton's law, the dynamic equation for the oscillating body may be written as

$$\mathbf{m}_m \frac{d^2}{dt^2} \mathbf{s} = \mathbf{F}_e + \mathbf{F}_r + \mathbf{F}_b + \mathbf{F}_v + \mathbf{F}_f + \mathbf{F}_u. \tag{5.311}$$

Matrix $\mathbf{m}_m$ represents the inertia of the oscillating body. The three first diagonal elements are $m_{m11} = m_{m22} = m_{m33} = m_m$, where m_m is the mass of the body. Some of the off-diagonal elements of $\mathbf{m}_m$ involving rotary modes of motion may be different from zero (see Newman,[24] Chap. 6 or Mei,[1] Chap. 7). For j and j' equal to 4, 5 or 6, $m_{mjj'}$ is of the dimension of an inertia moment.

Although sum $\mathbf{F}_e + \mathbf{F}_r$ represents the total wave force, the above-mentioned additional forces are represented by the three last terms in Eq. (5.311). The last term $\mathbf{F}_u$ is due to some purpose. It may be a control force intended, for instance, to reduce the oscillation of a ship or floating platform in rough seas, or it may represent a load force necessary for conversion of ocean-wave energy.

The third and second last terms $\mathbf{F}_v$ and $\mathbf{F}_f$ represent unavoidable viscous and friction effects. Thus, although we in Section 4.1 assumed an ideal fluid, we may, in this time-domain formulation, make a practical correction by introducing non-zero loss forces in Eq. (5.311). In general such forces may depend explicitly on time as well as on $\mathbf{s}$ and $\mathbf{u}$. For mathematical convenience, we shall, however, assume that $\mathbf{F}_f$ may be written as

$$\mathbf{F}_f = -\mathbf{R}_f \mathbf{u} = -\mathbf{R}_f \frac{d\mathbf{s}}{dt}, \tag{5.312}$$

where $\mathbf{R}_f$ is a constant friction resistance matrix. A viscosity resistance matrix $\mathbf{R}_v$ may be defined in an analogous manner.

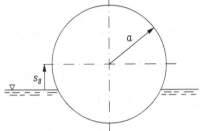

Figure 5.26: Floating sphere displaced a distance s_3 above its equilibrium semisubmerged position.

5.9.1 Dynamics of a Floating Body in Heave

In this subsection let us assume that the body is restricted to oscillation in the heave mode only. Then we may consider the matrices in Eqs. (5.309), (5.311) and (5.312) as scalars S_b, m_m and R_f.

Let us at first determine the buoyancy stiffness S_b. As an example, consider a sphere of radius a and mass $m_m = \rho 2\pi a^3/3$. Thus in equilibrium it is semisubmerged. When the sphere is displaced upward a distance s_3 as shown in Figure 5.26, the volume of displaced water is $\pi(2a^3 - 3a^2 s_3 + s_3^3)/3$, and hence there is a restoring force $F_b = -\rho g \pi (a^2 s_3 - s_3^3/3)$, from which we see that the buoyancy stiffness is

$$S_b = \pi \rho g a^2 \left(1 - s_3^2/3a^2\right) \tag{5.313}$$

for $|s_3| < a$. For small excursions, such that $|s_3| \ll a$, we have

$$S_b \approx \pi \rho g a^2, \tag{5.314}$$

which is independent of s_3. We note that it is ρg multiplied by the water plane area.

For a floating body, in general, the buoyancy stiffness is (for small heave excursion s_3)

$$S_b = \rho g S_w, \tag{5.315}$$

where S_w is the (equilibrium) water plane area of the body. For a floating vertical cylinder of radius a we have $S_w = \pi a^2$. (For such a cylinder the water plane area is independent of heave position s_3. Then hydrostatic stiffness S_b is constant even if s_3 is not very small.) For a freely floating body the weight equals $\rho g V$, where V is the volume of displaced water at equilibrium.

For a body restricted to oscillate in the heave mode only, dynamic equation (5.311) becomes

$$m_m \ddot{s}_3 = F_{e3} + F_{r3} + F_{b3} + F_{v3} + F_{f3} + F_{u3}. \tag{5.316}$$

In terms of complex amplitudes this may be written as

$$\{-\omega^2 [m_m + m_{33}(\omega)] + i\omega [R_v + R_f + R_{33}(\omega)] + S_b\} \hat{s}_3 = \hat{F}_{e3} + \hat{F}_{u3}, \tag{5.317}$$

when use has also been made of Eqs. (5.38), (5.44), (5.309), (5.310) and (5.312). Note that Eq. (5.317) corresponds to the Fourier transform of the time-domain equation (5.316) [cf. Eqs. (2.161) and (2.169)]. Introducing the "intrinsic" transfer function

$$G_i(\omega) = S_b + i\omega R_l - \omega^2 m_m + i\omega Z_{33}(\omega) \tag{5.318}$$

where $R_l = R_v + R_f$ is the loss resistance, we write Eq. (5.317) simply as

$$G_i(\omega)\hat{s}_3 = \hat{F}_{e3} + \hat{F}_{u3} \equiv \hat{F}_{ext}. \tag{5.319}$$

We have adopted the adjective "intrinsic" because all terms in G_i relate to the properties of the oscillating system, and Eq. (5.319) describes quantitatively how the system responds to an external force $F_{ext} = F_{e3} + F_{u3}$, of which the first term F_{e3} is given by an incident wave and the second term F_{u3} is determined by an operator or by a control device assisting the operation of the oscillating body. In particular cases, F_{e3} or F_{u3} may vanish. In the pure radiation problem, for instance, there is no incident wave, and $F_{e3} = 0$ ($F_{ext} = F_{u3}$).

We also introduce the "intrinsic mechanical impedance"

$$Z_i(\omega) = Z_{33}(\omega) + R_l + i\omega m_m + S_b[1/i\omega + \pi\delta(\omega)] \tag{5.320}$$

such that

$$G_i(\omega) = i\omega Z_i(\omega) \tag{5.321}$$

and

$$Z_i(\omega)\hat{u}_3 = \hat{F}_{e3} + \hat{F}_{u3} = \hat{F}_{ext}. \tag{5.322}$$

The last term with $\delta(\omega)$ in Eq. (5.320) secures that the inverse Fourier transform of $Z_i(\omega)$ is causal (see Problem 2.15). This term, which is of no significance if $\omega \neq 0$, may be of importance if the input and output functions in Eq. (5.322) are Fourier transforms rather than complex amplitudes. Also note that this term does not explicitly show up in expression (5.318) for $G_i(\omega)$ because $i\omega\delta(\omega) = 0$.

With given external force $\hat{F}_{e3} + \hat{F}_{u3}$ the solution of Eq. (5.322), for $\omega \neq 0$, is

$$\hat{u}_3 = \frac{\hat{F}_{ext}}{Z_i(\omega)} = \frac{\hat{F}_{ext}}{R_{33}(\omega) + R_l + i[\omega m_{33}(\omega) + \omega m_m - S_b/(i\omega)]}, \tag{5.323}$$

where we have used Eq. (5.44) in Eq. (5.320). When the intrinsic reactance vanishes, we say that the system is in resonance. This happens for a frequency $\omega = \omega_0$ for which

$$\text{Im}\{Z_i(\omega_0)\} = 0 \quad \text{or} \quad \omega_0 = \sqrt{S_b/[m_m + m_{33}(\omega_0)]}. \tag{5.324}$$

Then the heave velocity is in phase with the external force. If $R_{33}(\omega)$ has a relatively slow variation with ω, we also have maximum velocity amplitude response $|\hat{u}_3/\hat{F}_{ext}|$ at a frequency close to ω_0.

Let us now for a moment consider a floating sphere, of radius a, having a semisubmerged equilibrium position. When this sphere is heaving on deep water, the radiation impedance depends on the frequency in accordance with Figure 5.6. The hydrostatic stiffness is $S_b = \rho g \pi a^2$ and the mass is $m_m = \rho(2\pi/3)a^3$. Furthermore, the added mass is $m_{33}(\omega) = m_m \mu_{33}$ where the dimensionless parameter $\mu_{33} = \mu_{33}(ka)$ is in the region $0.38 < \mu_{33} < 0.9$ as appears from Figure 5.6. Note that $\mu_{33} \to 0.5$ as $ka \to \infty$. Insertion into Eq. (5.324) gives the following condition for resonance of the heaving semisubmerged sphere:

$$\omega_0 = \sqrt{\frac{3g}{2a[1 + \mu_{33}(\omega_0)]}}. \tag{5.325}$$

We may solve for ω_0 by using iteration. If, as a first approximation, we assume a value $\mu_{33} \approx 0.5$, we have $\omega_0 \approx \sqrt{g/a}$. Next we may improve the approximation. For $\omega_0 = \sqrt{g/a}$, that is, $k_0 a = \omega_0^2 a/g = 1$, we have $\mu_{33} = 0.43$. Then we have the following corrected value for the angular eigenfrequency:

$$\omega_0 \approx \sqrt{3g/2a(1 + 0.43)} = 1.025 \sqrt{g/a}, \tag{5.326}$$

and correspondingly $k_0 a = \omega_0^2 a/g \approx 1.05$. For a sphere that is 20 m in diameter ($a = 10$ m) the resonance frequency is given by $\omega_0 \approx 1.01$ rad/s, $f_0 = \omega_0/2\pi \approx 0.161$ Hz. The resonance period is $T_0 = 1/f_0 \approx 6.2$ s.

As another example let us consider a floating slender cylinder of radius a and depth of submergence l, as shown in Figure 5.27. We assume that $a \ll l$. The hydrostatic stiffness is $S_b = \rho g \pi a^2$ and the mass is $m_m = \rho \pi a^2 l$. Furthermore, the added mass is $m_{33} = m \mu_{33}$ where the dimensionless added-mass coefficient μ_{33} depends on $ka, l/a$ and kh. From the book by McCormick[63] (p. 49), we take the value

$$m_{33} = 0.167\rho(2a)^3 = 0.64(2\pi/3)a^3\rho, \tag{5.327}$$

which is the high-frequency limit ($ka \gg 1$) on deep water ($kh \gg 1$) when the cylinder is relatively high ($l/a \gg 1$). If $l \gg a$, we have $m_m \gg m_{33}$ and the angular eigenfrequency (natural frequency) is

$$\omega_0 = \sqrt{S_b/(m_m + m_{33})} \approx \sqrt{S_b/m_m} = \sqrt{g/l}. \tag{5.328}$$

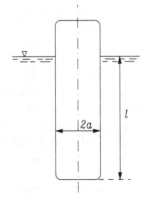

Figure 5.27: Floating vertical cylinder. The diameter is $2a$ and the depth of submergence is l.

Correspondingly, the resonance period is $T_0 \approx 2\pi \sqrt{l/g}$, which amounts to $T_0 = 10$ s for a cylinder of draft $l = 25$ m.

Looking at Eqs. (5.319) and (5.322), we interpret $G_i(\omega)$ and $Z_i(\omega)$ as transfer functions of linear systems where the input functions are the heave excursion and heave velocity, respectively. For both systems the output is the force, which we called external force. These linear systems "produce" radiation force, friction force, restoring force and inertial force which are balanced by the external force.

However, if we consider the heave oscillation to be caused by the external force, it would seem more natural to consider linear systems in which input and output roles have been interchanged. Mathematically these systems may be represented by

$$s_3(\omega) = H_i(\omega) F_{ext}(\omega), \tag{5.329}$$
$$u_3(\omega) = Y_i(\omega) F_{ext}(\omega), \tag{5.330}$$

where we have written Fourier transforms rather than complex amplitudes for the input and output quantities [cf. Eqs. (2.161) and (2.169)]. The transfer functions for these linear systems are

$$H_i(\omega) = 1/G_i(\omega), \tag{5.331}$$
$$Y_i(\omega) = H_i(\omega)/i\omega. \tag{5.332}$$

[Note that we could not define Y_i as the inverse of Z_i, because the inverse of $\delta(\omega)$ in Eq. (5.320) does not exist; also see Problem 2.15.]

By taking inverse Fourier transforms, we have the following time-domain representation of the four considered linear systems:

$$g_i(t) * s_3(t) = z_i(t) * u_3(t) = F_{ext}(t), \tag{5.333}$$
$$h_i(t) * F_{ext}(t) = s_3(t), \quad y_i(t) * F_{ext}(t) = u_3(t). \tag{5.334}$$

The impulse response functions $g_i(t)$, $z_i(t)$ $h_i(t)$ and $y_i(t)$ entering into these convolutions are inverse Fourier transforms of $G_i(\omega)$, $Z_i(\omega)$, $H_i(\omega)$ and $Y_i(\omega)$, respectively. Using Table 2.2 and Eqs. (5.109) and (5.112) with Eqs. (5.320) and (5.321), we find

$$z_i(t) = k_{33}(t) + R_l \delta(t) + [m_{33}(\infty) + m_m]\dot{\delta}(t) + S_b U(t), \tag{5.335}$$
$$g_i(t) = \dot{z}_i(t) = \dot{k}_{33}(t) + R_l \dot{\delta}(t) + [m_{33}(\infty) + m_m]\ddot{\delta}(t) + S_b \delta(t). \tag{5.336}$$

As explained in Section 5.3, we know that impulse response function $k_{33}(t)$ for radiation is causal. Then it is obvious that also $z_i(t)$ and $g_i(t)$ are causal. Note that without the term with $\delta(\omega)$ in Eq. (5.320), we would have had $\frac{1}{2}$ sgn(t) instead of the causal unit step function $U(t)$ in Eq. (5.335), but Eq. (5.336) would have remained unchanged. Combining Eqs. (5.333) and (5.335), we may write the time-domain dynamic equation as

$$[m_{33}(\infty) + m_m]\ddot{s}_3(t) + R_l \dot{s}_3(t) + k_{33}(t) * \dot{s}_3(t) + S_b s_3(t) = F_{ext}(t) \tag{5.337}$$

when we observe that $\dot{k}_{33}(t) * s_3(t) = k_{33}(t) * \dot{s}_3(t)$, because $k_{33}(t) \to 0$ as $t \to \infty$. [This may be easily shown by partial integration. See also Eq. (2.148).] Similarly, $\delta(t) * s_3(t) = \delta(t) * \dot{s}_3(t) = \dot{s}_3(t)$ and $\ddot{\delta}(t) * s_3(t) = \dot{\delta}(t) * \dot{s}_3(t) = \ddot{s}_3(t)$.

It is reasonable to assume that also impulse response functions $h_i(t)$ and $y_i(t)$ are causal; this view may obtain support by studying Problem 2.15. A rigorous proof has, however, been delivered by Wehausen.[64]

Problems

Problem 5.1: Surge and Pitch of Hinged Plate

A rectangular plate is placed in a vertical position with its upper edge above a free water surface, while its lower edge is hinged below the water surface, at depth c. We choose a coordinate system such that the plate is in the vertical plane $x = 0$, and the hinge at the horizontal line $x = 0$, $z = -c$. Further, for this immersed plate we choose a reference point $(x, y, z) = (0, 0, z_0)$; see Figure 5.3. The plate performs oscillatory sinusoidal rotary motion with respect to the hinge. The angular amplitude is β_5 (assumed to be small), and the angular frequency is ω. Determine, for this case, the six-dimensional vectors $\mathbf{u}$ and $\mathbf{n}$ defined in Subsection 5.1.1 for the three cases

(a) $z_0 = -c$ (reference point at the hinge),
(b) $z_0 = 0$ (reference point at the free water surface),
(c) $z_0 = -c/2$ (reference point midway between hinge and $z = 0$).

Problem 5.2: Excitation Force on a Surging Piston in Wave Channel

A wave channel is assumed to extend from $x = 0$ to $x \to -\infty$. The water depth is h and the channel width is d. Choose a coordinate system such that the wet surface of the end wall is $x = 0$, $0 < y < d$, $0 > z > -h$.

In the vertical end wall there is a rectangular piston of width d_1 and height $a_2 - a_1$, corresponding to the surface

$$x = 0, \quad b < y < c, \quad -a_2 < z < -a_1,$$

where $c - b = d_1$ $(0 < b < c < d)$. In a situation in which the piston is not oscillating, the velocity potential is as given by Eq. (5.31). Derive an expression for the surge excitation force on the piston, in terms of $A, \rho, g, k, h, a_1, a_2$ and d_1.

Problem 5.3: Radiation Resistance for a Vertical Plate

A wave channel of width d and water depth h has in one end $(x = 0)$ a wave generator in the form of a stiff vertical plate (rectangular piston) which oscillates harmonically with velocity amplitude $\hat{u}_1$ (pure surge motion) and with angular frequency ω. In the opposite end of the wave channel there is an ideal absorber.

That is, we could consider the channel to be of infinite length $(0 < x < \infty)$.

(a) Find an exact expression for the velocity potential of the generated wave, in terms of an infinite series, where some parameters are defined implicitly through a transcendental equation.

(b) By means of elementary functions, express the radiation resistance R_{11} in terms of the angular repetency k, the fluid density ρ and the lengths h and d. Draw a curve for $(R_{11}\omega^3/\rho g^2 d)$ versus kh in a diagram.

(c) Also solve the same problem for the case in which the plate is hinged at its lower edge (at $z = -h$), and let now $\hat{u}_1$ represent the horizontal component of the velocity amplitude at the mean water level.

Problem 5.4: Flap as a Wave Generator

An incident wave $\hat{\eta} = Ae^{-ikx}$ hits a vertical plate hinged with its lower end at water depth $z = -c$ at a horizontal position $x = 0$, where $0 < c < h$. At $x = 0$, $-h < z < -c$ (below the hinge) there is a fixed vertical wall. Consider the system as a wave channel of width $\Delta y = d$.

(a) Determine the excitation pitch moment $\hat{M}_y = \hat{F}_5$ expressed by ρ, g, k, A, h and d. Give simpler expressions for the two special cases $h \to \infty$ and $c = h$.

(b) Solve the radiation problem by obtaining an expression for $\varphi_5 = \hat{\phi}_r/\hat{u}_5$. Further derive a formula for the radiation impedance $Z_{55} = R_{55} + \omega m_{55}$.

(c) Check that $R_{55} = |\hat{F}_5/A|^2\omega/(2\rho g^2 Dd)$. Check also that the excitation pitch moment $\hat{F}_5$ is recovered through application of the Haskind relation.

Problem 5.5: Pivoting Vertical Plate as a Wave Generator

This is the same as Problem 5.4, but with the plate extending down to $z = -h$, while the hinge is still at $z = -c$. Show for this case that with proper choice of c/h $(0 < c < h/2)$ it is possible to find a frequency where the radiated wave vanishes, that is, $R_{55} = 0$. [Hint: to find the repetency (wave number) corresponding to this frequency a transcendent equation involving hyperbolic functions has to be solved. The solution may be graphically represented as the intersection of a straight line with the graph for $\tanh(kh/2)$.]

Problem 5.6: Circular Wave Generator

A vertical cylinder of radius a and height larger than the water depth stands on the sea bed, $z = -h$. The cylinder wall is pulsating in the radial direction with a complex velocity amplitude $\hat{u}_r(z) = \hat{u}_0 c(z)$, where the constant $\hat{u}_0 = \hat{u}_r(0)$ represents the pulsating velocity at the mean water level $z = 0$. Note that $c(0) = 1$. Assume that the system is axisymmetric; that is, there is no variation with the coordinate angle θ.

(a) Express the velocity potential $\hat{\phi} = \varphi \hat{u}_0$ for the generated wave as an infinite series in terms of the normalised vertical eigenfunctions $\{Z_n(z)\}$ and the corresponding eigenvalues $\lambda_0 = k^2$, $\lambda_1 = -m_1^2$, $\lambda_2 = -m_2^2, \ldots$.

(b) Express the radiation impedance Z_{00} as an infinite series, and find a simple expression for the radiation resistance R_{00}.

[Hint: for Bessel functions we have the relations $(d/dx)J_0(x) = -J_1(x)$ and correspondingly for $N_0(x)$, $H_0^{(2)}(x)$ and $K_0(x)$, whereas $(d/dx)I_0(x) = I_1(x)$. Here I_n and K_n are modified Bessel functions of order n and of first kind and second kind, respectively. Asymptotic expressions are $I_0(x) \to e^x/\sqrt{2\pi x}$ and $K_0(x) \to \pi e^{-x}/\sqrt{2\pi x}$ as $x \to +\infty$. Moreover, we have $J_1(x)N_0(x) - J_0(x)N_1(x) = 2/\pi x$.]

Problem 5.7: Radiation Resistance in Terms of Far-Field Coefficients

The time-average mechanical power radiated from an oscillating body may be expressed as

$$P_r = \sum_{j=1}^{6} \sum_{j'=1}^{6} \frac{1}{2} R_{jj'} \hat{u}_j^* \hat{u}_{j'}.$$

In an ideal fluid this time-average power may be recovered in the radiated wave-power transport, for instance, in the far-field region where the formula for wave-power transport with a plane wave is applicable. In the far-field region the radiated wave is

$$\hat{\phi}_r = \hat{\phi}_l(r, \theta, z) + A_r(\theta) e(kz)(kr)^{-1/2} e^{-ikr},$$

where the near-field part $\hat{\phi}_l$ is negligible, because $\hat{\phi}_l = \mathcal{O}\{r^{-1}\}$ as $r \to \infty$. Use the decomposition

$$A_r(\theta) = \sum_{j=1}^{6} a_j(\theta) \hat{u}_j$$

in deriving an expression for the radiation-resistance matrix $R_{jj'}$ in terms of the far-field coefficient vector $a_j(\theta)$.

Problem 5.8: Excitation-Force Impulse Response

Consider the surge excitation force per unit width of a horizontal strip of a vertical wall normal to the wave incidence. The strip of height Δz covers the interval $(z - \Delta z, z)$ where $z < 0$. Show that for the excitation-force impulse-response function we have

$$\Delta f'_{t,1}(t) = \rho g(g/\pi|z|)^{1/2} e^{+gt^2/4z} \Delta z \to 2\rho g\delta(t)\Delta z$$

as $z \to 0$. Compare Subsections 4.9.2 and 5.3.2.

Problem 5.9: Froude-Krylov Force for an Axisymmetric Body

The radius of the wet surface of an axisymmetric body with vertical axis ($x = y = 0$) is given as the function $a = a(z)$. Show that the heave component of the Froude-Krylov force which is due to a plane incident wave $\hat{\eta} = Ae^{-ikx}$ is given by

$$f_{\text{FK},3} = F_{\text{FK},3}/A = \pi\rho g \int_{-b}^{-c} J_0(ka)\frac{da^2(z)}{dz}e(kz)\,dz,$$

where J_0 is the zero-order Bessel function of first kind. Furthermore, b and c are the depth of submergence of the wet surface's bottom and top, respectively. If the body is floating, then $c = 0$. The lower end of the body is at $z = -b$.

For a floating vertical cylinder with hemispherical bottom, we have

$$a^2(z) = \begin{cases} a_0^2 & \text{for } -l < z < 0 \\ a_0^2 - (z+l)^2 & \text{for } -l - a_0 < z < -l \end{cases},$$

where $l = b - a_0$ is the length of the cylindrical part of the wet surface. We may expand the above exact formula for $f_{\text{FK},3}$ as a power series in (ka_0) and (kl). Assume deep water ($kh \gg 1$), and find the first few terms of this expansion. Compare with the Froude-Krylov force as obtained from the small-body approximation.

[Hint: assume as known the following.

$$\frac{1}{2\pi}\int_0^{2\pi} e^{-ix\cos\theta}\,d\theta = J_0(x) = 1 - \frac{(x/2)^2}{(1!)^2} + \frac{(x/2)^4}{(2!)^2} - +\cdots.]$$

Problem 5.10: Radiation-Resistance Limit for Zero Frequency

For a floating axisymmetric body we write the heave-mode radiation resistance as

$$R_{33} = \epsilon_{33}\omega\rho a^3 2\pi/3,$$

where a is the radius of the water-plane area. Show that if the water depth is infinite, then

$$\epsilon_{33} \to ka3\pi/4 \quad \text{as } ka \to 0,$$

whereas for finite water depth h

$$\epsilon_{33} \to 3\pi a/(8h) \quad \text{as } ka \to 0.$$

[Hint: make use of Eqs. (5.253) and (5.279).]

Problem 5.11: Heave Excitation Force for a Semisubmerged Sphere

For a semisubmerged floating sphere of radius a we write the heave excitation force as $F_3 = \kappa SA = \kappa\rho g\pi a^2 A$, where S is the buoyancy stiffness and A is the complex amplitude of the undisturbed incident wave at the position of the centre of the sphere. Furthermore, κ is the non-dimensionalised excitation force. The radiation impedance is

$$Z_{33} = R_{33} + i\omega m_{33} = \omega\rho\tfrac{2}{3}\pi a^3(\epsilon + i\mu).$$

For deep water $(kh \gg 1)$, values for $\epsilon = \epsilon(ka)$ and $\mu = \mu(ka)$, computed by Hulme,[43] are given by the table

$ka = 0$	0.05	0.1	0.2	0.3	0.4
$\mu = 0.8310$	0.8764	0.8627	0.7938	0.7157	0.6452
$\epsilon = 0$	0.1036	0.1816	0.2793	0.3254	0.3410
$ka = 0.5$	0.6	0.7	0.8	0.9	1
$\mu = 0.5861$	0.5381	0.4999	0.4698	0.4464	0.4284
$\epsilon = 0.3391$	0.3271	0.3098	0.2899	0.2691	0.2484

Compute numerical values of $|\kappa|$ from the values of ϵ by using the exact reciprocity relation between the radiation resistance and the excitation force. Also compute approximate values of κ from the values of μ by using the small-body approximation for the excitation force. Compare the values by drawing a curve for the exact $|\kappa|$ and the approximate κ in the same diagram, versus ka $(0 < ka < 1)$.

Problem 5.12: Radiation Resistance for a Cylindric Body

A vertical cylinder of diameter $2a$ and of height $2l$ is (in order to reduce viscous losses) extended with a hemisphere in its lower end. When the cylinder is floating in equilibrium position, it has a draught of $l + a$ (and the cylinder's top surface is a distance of l above the free water surface). We shall assume deep water in the present problem.

If we assume, for simplicity, that the heave excitation force is

$$\hat{F}_3 \equiv F(l) \approx e^{-kl}F(0), \tag{1}$$

where $F(0)$ is the heave excitation force for a semisubmerged sphere of radius a, then it follows from the reciprocity relationship between the radiation resistance and the excitation force [cf. Eq. (4) below] that the radiation resistance for heave is

$$R_{33} \equiv R_0 \approx R_H e^{-2kl}, \tag{2}$$

where R_H is the radiation resistance for the semisubmerged sphere,

$$R_H = \omega\rho(2\pi/3)a^3\epsilon, \tag{3}$$

where $\epsilon = \epsilon(ka)$ is Havelock's dimensionless damping coefficient as computed by Hulme.[43] Compare the table below or the table given in Problem 5.11. [It may be observed from the solution of Problem 5.9 that a relationship such as (1) is valid exactly for the Froude-Krylov force. In the small-body case approximation (1) is valid for the diffraction force if the variation of the added mass with l is neglected. Note, however, that this assumption is not exactly true, because a modification of the resistance according to approximation (2) is associated with some modification of the added mass, according to the Kramers-Kronig relations. See Subsection 5.3.1.]

For an axisymmetric body on deep water we have the reciprocity relation

$$R_0 = R_{33} = (\omega k/2\rho g^2) |f_3|^2, \tag{4}$$

where $f_3 = \hat{F}_3/A$ is the heave excitation force coefficient and A is the incident wave-elevation amplitude (complex amplitude at the origin). Express the dimensionless parameters $|f_3|/S$ and $\omega R_{33}/S$ in terms of ϵ, ka and kl, where $S = \rho g \pi a^2$ is the hydrostatic stiffness of the body. Compute numerical values for both these parameters with $l/a = 0, 4/3$ and $8/3$ for each of the ka values from the following table:

$ka =$	0.01	0.05	0.1	0.15	0.2	0.3
$\epsilon =$	0.023	0.1036	0.1816	0.24	0.2793	0.3254

Problem 5.13: Eigenvalues of a Radiation-Resistance Matrix

Given that the diagonal matrix elements are non-negative, show that the 6×6 radiation resistance matrix for an axisymmetric body has at least three vanishing eigenvalues (or, expressed differently, one triple eigenvalue equal to zero), and hence that its rank is at most $6 - 3 = 3$. Also show that the three remaining eigenvalues are positive (or zero in exceptional cases) and that two of them have to be equal (or, expressed differently, there is a double eigenvalue which is not necessarily zero).

[Hint: consider a real matrix of the form

$$\mathbf{r} = \begin{pmatrix} r_{11} & 0 & 0 & 0 & r_{15} & 0 \\ 0 & r_{22} & 0 & r_{24} & 0 & 0 \\ 0 & 0 & r_{33} & 0 & 0 & 0 \\ 0 & r_{24} & 0 & r_{44} & 0 & 0 \\ r_{15} & 0 & 0 & 0 & r_{55} & 0 \\ 0 & 0 & 0 & 0 & 0 & 0 \end{pmatrix},$$

where $r_{33} \geq 0, r_{22} = r_{11} \geq 0, r_{44} = r_{55} \geq 0, r_{24} = r_{15}$ and $r_{15}^2 = r_{11}r_{55}$; cf. Eqs. (5.272), (5.279) and (5.286). The eigenvalues λ are the values which make the determinant of $\mathbf{r} - \lambda \mathbf{I}$ equal to zero, where $\mathbf{I}$ is the identity matrix. Factorising the determinant, which is a sixth-degree polynomial in λ, gives the eigenvalues in a straightforward manner.]

Problem 5.14: Reciprocity Relations Between Diffracted Waves

In agreement with Eq. (5.166), let S be the totality of wet surfaces of oscillating bodies as indicated in Figure 5.13. It follows from boundary condition (5.163) and definition (4.240) that

$$I(\hat{\phi}_i, \hat{\phi}_j) = 0, \quad I(\hat{\phi}_i^*, \hat{\phi}_j^*) = 0$$

if $\hat{\phi}_i$ and $\hat{\phi}_j$ are sums of an incident wave and a diffracted wave, namely

$$\hat{\phi}_j = (-g/i\omega) A_j \, e(kz) \exp\{-ikr(\beta_j)\} + \hat{\phi}_{d,j},$$

and similarly for $\hat{\phi}_i$ with subscript j replaced by i. (It is assumed that the bodies indicated in Figure 5.13 are held fixed.) Let us consider two-dimensional cases with two waves either from opposite directions, $\beta_i = \beta_1 = 0$ and $\beta_j = \beta_2 = \pi$, that is, $r(\beta_i) = -x$ and $r(\beta_j) = x$, respectively, or from the same direction ($\beta_i = \beta_j$).

(a) With Kochin functions $H'(\beta) = H_d'(\beta)$ for the diffracted waves, show that

$$A_i \, H_j'^*(\beta_i) + A_j^* \, H_i'(\beta_j) = \frac{\omega}{gk} [H_i'(0) H_j'^*(0) + H_i'(\pi) H_j'^*(\pi)]$$

[Hint: use Eq. (4.304) and the two-dimensional version of Eq. (4.308).]

(b) By considering the diffracted plane waves in the far-field region, let us define transmission coefficients T and reflection coefficients Γ as follows:

$$T_1 = 1 + A_1^+/A_1, \quad \Gamma_1 = A_1^-/A_1,$$

$$T_2 = 1 + A_2^-/A_2, \quad \Gamma_2 = A_2^+/A_2,$$

where the relationship between the far-field coefficients $A_j^\pm$ and the Kochin functions H_j' are given by Eqs. (4.298) and (4.299).

Derive the following three relations:

$$\Gamma_j \Gamma_j^* + T_j T_j^* = 1 \quad \text{for } j = 1, 2, \quad T_1 \Gamma_2^* + T_2^* \Gamma_1 = 0.$$

(The former of these relations represents energy conservation. For a geometrically symmetrical case the latter relation reduces to a result derived in Problem 4.13.)

[Hint: Consider the three cases when (i, j) equals $(1, 1)$, $(2, 2)$ and $(1, 2)$.]

(c) Use the two-dimensional version of Eq. (4.281) to show that $T_2 = T_1$, as derived by Newman.[30]

Problem 5.15: Diffracted and Radiated Two-Dimensional Waves

According to boundary condition (5.164), $\partial\varphi/\partial n$ is real on S, the totality of wet surfaces of wave-generating bodies. Hence, we may replace φ by φ^* in Eq. (5.198) as well as in Eqs. (5.141)–(5.143). By combining Eqs. (5.198) and (5.200) this gives

$$I'(\hat{\phi}_0 + \hat{\phi}_d, \varphi^*) = \frac{g D(kh)}{i\omega k} h'(\beta \pm \pi) A$$

for the two-dimensional case, where $\hat{\phi}_0 + \hat{\phi}_d = \hat{\phi}$ is as given in Problem 5.14 (when the subscript j is omitted).

(a) Use Eq. (4.304) and the two-dimensional version of Eq. (4.308) to show that (for $\beta = 0$)

$$\Gamma_d \mathbf{h}'^*(\pi) + T_d \mathbf{h}'^*(0) = \mathbf{h}'(\pi),$$

where $\Gamma_d = \Gamma_1$ and $T_d = T_1$. (Here reflection coefficient Γ_1 and transmission coefficient T_1 are defined in Problem 5.14.)
[Hint: In Eqs. (4.304) and (4.308) set $H_i' = H_d'$, $H_j' = \mathbf{h}'$, $A_i = A$ and $A_j = 0$; cf. Newman.[30]]

(b) Consider now a single (two-dimensional) body which has a symmetry plane $x = 0$ and which oscillates in the heave mode only. Show that in the far-field region beyond the body ($x \to \infty$) there is no propagating wave, provided the heave velocity is given

$$\hat{u}_3 = T_d \, (gk/\omega h_3') \, A,$$

because the radiated wave then cancels the transmitted wave for the diffraction problem.

(c) Show that for maximum absorbed wave power, the heave velocity is given by

$$\hat{u}_3 = \hat{u}_{3,\text{OPT}} = (gk/2\omega h_3'^*) \, A$$

[Hint: use Eqs. (6.9) and (5.301)–(5.304).]
What is the absorbed power if $\hat{u}_3 = 2\hat{u}_{3,\text{OPT}}$?

(d) Find a possible value of $\hat{u}_3$ for which the resulting far-field wave propagating in the negative x direction ($x \to -\infty$) is as large as the incident wave. Note that this value of $\hat{u}_3$ is not unique, because the phase of the wave propagating in the negative x direction has not been specified.

(e) Make a comparative discussion of the various results for $\hat{u}_3$ obtained above, when we assume that the body is so small that $\Gamma_d \to 0$. Can you say something about the above-mentioned unspecified phase for the two "opposite" cases $\Gamma_d \to 0$ and $\Gamma_d \to 1$? (In the latter case the body has to be large.)

Problem 5.16: Rectangular Piston in the End Wall of a Wave Channel

A wave channel is assumed to extend from $x = 0$ to $x \to -\infty$. The water depth is h and the channel width is d. Choose a coordinate system such that the wet surface of the end wall is $0 > z > -h$, $0 < y < d$.

In the vertical end wall there is a rectangular piston of width d_1 and height $a_2 - a_1$, corresponding to the surface

$$x = 0, \quad b < y < c, \quad -a_2 < z < -a_1,$$

where $c - b = d_1$. The piston has a surging velocity of complex amplitude $\hat{u}_1$.
The velocity potential has a complex amplitude $\hat{\phi} = \varphi_1 \hat{u}_1$ where

$$\varphi_1 = \sum_{n=0}^{\infty} \sum_{q=0}^{\infty} b_{nq} \exp\{\gamma_{nq} x\} \cos\left(\frac{q\pi y}{d}\right) Z(m_n z)$$

with $q = 0, 1, 2, 3, \ldots$. Moreover,

$$\gamma_{nq} = \{m_n^2 + (q\pi/d)^2\}^{1/2},$$
$$Z_n(m_n z) = N_n^{-1/2} \cos\{m_n(z + h)\},$$
$$N_n = \tfrac{1}{2} + \sin(2m_n h)/(4m_n h),$$

and m_n is the $(n + 1)$th solution of

$$\omega^2 = -gm\tan(mh)$$

such that, with $m_0 = ik$,

$$\lambda_0 = k^2 = -m_0^2 > \lambda_1 = -m_1^2 > \lambda_2 = -m_2^2 > \cdots > \lambda_n = -m_n^2 > \cdots.$$

Show that $\hat{\phi}$ satisfies the Laplace equation in the fluid, the radiation condition at $x = -\infty$, and the homogeneous boundary conditions on the free surface $z = 0$, on the bottom $z = -h$ and on the walls $y = 0$ and $y = d$ of the wave channel. Further, show that the boundary conditions on the wall $x = 0$ are satisfied if the unknown coefficients are given by

$$b_{nq} = \sigma_q(d_1/d)t_q N_n^{-1/2} s_n/\gamma_{nq},$$

where $\sigma_0 = 1$ and $\sigma_q = 2$ for $q \geq 1$. Furthermore,

$$t_d = (d/q\pi d_1)\{\sin(q\pi c/d) - \sin(q\pi b/d)\},$$
$$s_n = \{\sin[m_n(h - a_1)] - \sin[m_n(h - a_2)]\}/m_n h.$$

Moreover, show that the radiation resistance is

$$R_{11} = \frac{\omega\rho s_0^2 d_1^2 h}{k N_0 d}\left(1 + \sum_{q=1}^{q_0} \frac{2kt_q^2}{\{k^2 - (q\pi/d)^2\}^{1/2}}\right),$$

where q_0 is the integer part of kd/π. Finally, determine the added mass m_{11} expressed as a double sum over n and q.

CHAPTER SIX

Wave-Energy Absorption by Oscillating Bodies

This chapter starts with a consideration of absorption as a destructive interference between an incident wave and a radiated wave. Then absorption by a body oscillating in one mode only is described in some detail. The optimum motion for maximising the absorbed power is a particular subject of interest. The last part of the chapter concerns absorption by a group of bodies oscillating in several degrees of freedom, and an explanation is given of why radiation-resistance matrices may be singular.

The large-scale utilisation of ocean-wave energy is still in a rather premature state of technological development. A review of the present state of the art is given by the ECOR Working Group on Wave Energy Conversion.[7] Some shorter reviews[5,6] described the state approximately a decade ago. The theory of wave-energy absorption was reviewed in 1981 by Evans.[65]

6.1 Absorption Considered as Wave Interference

A body oscillating in water will produce waves. A big body and a small body may produce equally large waves, provided the smaller body oscillates with larger amplitude. This may be utilised for the purpose of wave-energy conversion, for instance by a small floating body heaving in response to an incident wave, particularly if it can be arranged that the body oscillates with a larger amplitude than the wave amplitude.

Generally it can be said that a good wave absorber must be a good wavemaker. Hence, in order to absorb wave energy it is necessary to displace water in an oscillatory manner and with correct phase (timing). Absorbing wave energy for conversion means that energy has to be removed from the waves. Hence there must be a cancellation or reduction of waves which are passing the energy-converting device or are being reflected from it. Such a cancellation or reduction of waves can be realised by the oscillating device, provided it generates waves which oppose (are in counterphase with) the passing and/or reflected waves. In other words, the generated wave has to interfere destructively with the other waves. This explains

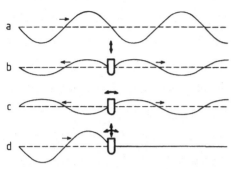

Figure 6.1: To absorb waves means to generate waves. Curve *a* represents an undisturbed incident wave. Curve *b* illustrates symmetric wave generation (on otherwise calm water) by means of a straight array of evenly spaced, small floating bodies oscillating in heave (up and down). Curve *c* illustrates antisymmetric wave generation. Curve *d*, which represents the superposition (sum) of the above three waves, illustrates complete absorption of the incident wave energy.

the paradoxical but general statement that "to destroy a wave means to create a wave". An illustrative example, in which 100% absorption of wave energy is possible, is shown in Figure 6.1. This corresponds to an infinite line (perpendicular to the figure) of oscillating small floating bodies, evenly interspaced a short distance (shorter than one wavelength). Complete absorption of the incident wave energy is also possible with an elongated body, of cross section as shown in Figure 6.1, and aligned perpendicular to the plane of the figure, provided the body oscillates vertically and horizontally in an optimum manner.

It will be shown later in this chapter [see Eq. (6.113)] that only 50% absorption is possible if there is only the symmetrical radiated wave, as shown by curve *b* in Figure 6.1, when the wave is generated by a symmetrical body oscillating in only one mode of motion, the vertical (heave) oscillation. Likewise, if there is only the antisymmetric radiated wave (curve *c*) from the symmetric body, more than 50% absorption is theoretically impossible. However, if a sufficiently non-symmetric body is oscillating in only one mode of motion, it may have the ability to absorb almost all the incident wave energy [see Eq. (6.108)].

Another example is shown in Figure 6.2. Here a heaving point absorber, absorbing wave energy, has to radiate circular waves which interfere destructively with the incident plane wave. A *point absorber*, which may be a heaving body, is (by definition) of very small extension compared with the wavelength. It will be shown later in this chapter [see Eq. (6.77)] that the maximum energy which may be absorbed by a heaving axisymmetric body equals the wave energy transported by the incident wave front of width equal to the wavelength divided by 2π. This width may be termed the "absorption width".

For maximum energy to be obtained from the waves, it is necessary to have optimum oscillation of the wave-energy converter (WEC). For a sinusoidal incident wave there is an optimum phase and an optimum amplitude for the oscillation.

To illustrate this, let us once more refer to Figure 6.1. In this case the amplitudes of the radiated waves (curves *b* and *c*) have to be exactly half of the amplitude of the

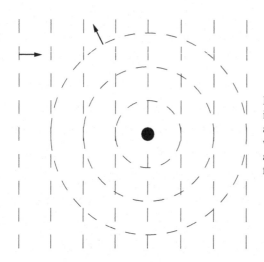

Figure 6.2: Wave pattern of two interfering waves seen from above. When a point absorber absorbs energy from an incident wave, it generates a circular wave radiating away from the absorber's immersed surface.

incident wave (curve a). Thus it is required that the amplitudes of the vertical and horizontal oscillations of the WEC have proper values. Note that these optimum amplitudes are proportional to the amplitude of the incident wave.

Moreover, with optimum phase conditions for the two modes of oscillation, the two corresponding waves radiated toward the right have to have the same phase (i.e., coinciding wave crests and coinciding wave troughs). This also means that the symmetric and antisymmetric radiated waves cancel each other toward the left. Furthermore, the phases of the two oscillations have to be correct with respect to the phase of the incident wave, because the crests of the waves radiated toward the right (curves b and c) must coincide with the troughs of the incident wave (curve a).

For a case with only one mode of oscillation, such as heave, the resulting wave corresponds to the superposition (sum) of waves a and b in Figure 6.1. Then the optimum heave (vertical) amplitude and phase are as above, in the case of two modes. The wave radiated toward the left and the resulting wave transmitted toward the right both have an amplitude equal to half of the amplitude of the incident wave. Because wave energy is proportional to the square of the wave amplitude, this means that 25% of the incident wave energy is reflected toward the left, and also 25% of it is transmitted toward the right. The remaining 50% is absorbed by the WEC, and this is the theoretical maximum, as mentioned previously.

A one-mode oscillating system happens to have the optimum phase condition if it is at resonance with the wave. That means that the wave frequency (reciprocal of the period) is the same as the natural frequency of the oscillating system. Then the oscillatory velocity of the system is in phase with the wave's excitation force which acts on the system (see Section 3.5).

6.2 Absorption by a Body Oscillating in One Mode of Motion

For simplicity let us first consider a single body which has only one degree of motion, only heave ($j = 3$) or only pitch ($j = 5$), say. An incident wave produces an excitation force $F_{e,j}$. The body responds to this force by oscillating with a velocity u_j. The total force is given by Eq. (5.153), which in this case simplifies to

$$\hat{F}_{t,j} = \hat{F}_{e,j} - Z_{jj}\hat{u}_j. \tag{6.1}$$

If we multiply through this equation by $\frac{1}{2}\hat{u}_j^*$ and take the real part, we obtain, according to Eq. (2.78), the time-averaged *absorbed power*

$$P = \tfrac{1}{2}\text{Re}\{\hat{F}_{t,j}\hat{u}_j^*\} = P_e - P_r, \tag{6.2}$$

where

$$P_e = \tfrac{1}{2}\text{Re}\{\hat{F}_{e,j}\hat{u}_j^*\} = \tfrac{1}{4}(\hat{F}_{e,j}\hat{u}_j^* + \hat{F}_{e,j}^*\hat{u}_j) \tag{6.3}$$

is the *excitation power* from the incident wave, and

$$P_r = \tfrac{1}{2}\text{Re}\{Z_{jj}\hat{u}_j\hat{u}_j^*\} = \tfrac{1}{2}R_{jj}|\hat{u}_j|^2 \tag{6.4}$$

is the *radiated power* caused by the oscillation. Here $R_{jj} = \text{Re}\{Z_{jj}\}$ is the radiation resistance.

Note that the excitation power

$$P_e = \overline{F_{e,j}(t)u_j(t)} = \tfrac{1}{2}\text{Re}\{\hat{F}_{e,j}\hat{u}_j^*\} = \tfrac{1}{2}|\hat{F}_{e,j}| \cdot |\hat{u}_j|\cos(\gamma_j) \tag{6.5}$$

(where $\gamma_j = \varphi_u - \varphi_F$ is the phase difference between $\hat{u}_j$ and $\hat{F}_{e,j}$) is linear in u_j whereas the radiated power

$$P_r = -\overline{F_{r,j}(t)u_j(t)} = \tfrac{1}{2}R_{jj}|\hat{u}_j|^2 \tag{6.6}$$

is quadratic in u_j because $F_{r,j}$ is linear in u_j. Thus a graph of P versus $|\hat{u}_j|$ is a parabola, as indicated in Figure 6.3.

We observe that in order to absorb wave power – which is possible if $0 < |\hat{u}_j| < (|\hat{F}_{e,j}|/R_{jj})\cos(\gamma_j)$ – a certain fraction of the excitation power P_e arriving at the oscillating body is necessarily returned to the sea as radiated power P_r. Wave interference is a clue to the explanation of this fact. The radiated wave interferes

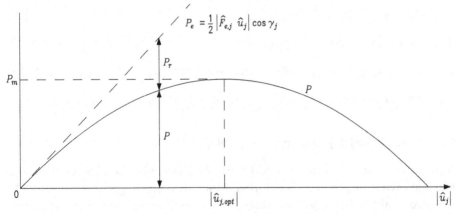

Figure 6.3: Power absorbed (solid curve) versus velocity amplitude. Radiated power P_r is the difference between excitation power P_e (dashed line) and absorbed power P.

destructively with the incident wave. The resulting wave, which propagates beyond the wave absorber, transports less power than the incident wave.

6.2.1 Maximum Absorbed Power

For a given body and a given incident wave, the excitation force $F_{e,j}$ is a given quantity. Let us discuss how the absorbed power P depends on the oscillating velocity u_j.

Observing Eqs. (6.2)–(6.5) and the parabola in Figure 6.3, we note the following. The optimum velocity amplitude is

$$|\hat{u}_j|_{\text{opt}} = (|\hat{F}_{e,j}|/2R_{jj})\cos(\gamma_j), \tag{6.7}$$

giving the maximum absorbed power

$$P = P_{\text{max}} \equiv (|\hat{F}_{e,j}|^2/8R_{jj})\cos^2(\gamma_j) = (P_r)_{\text{opt}} = \tfrac{1}{2}(P_e)_{\text{opt}}. \tag{6.8}$$

With this optimum condition the radiated power is as large as the absorbed power. If the amplitude is twice that of the optimum value given by Eq. (6.7), then $P = 0$, and $P_r = P_e = 4P_{\text{max}}$. This is the situation when a linear array of resonant (i.e., $\gamma_j = 0$) heaving slender bodies is used as a dynamic wave reflector.[66] Note that with resonance $\gamma_j = 0$, in which case we write (as below) $\hat{u}_{j,\text{OPT}}$ instead of $\hat{u}_{j,\text{opt}}$.

It can be shown (see Problem 6.5) that P_{max} as given by Eq. (6.8) is greater than $P_{a,\text{max}}$ as given by Eq. (3.41), except that they are equally large when $\gamma_j = 0$, that is, at resonance (see Problem 6.5). The reason why $P_{a,\text{max}}$ may be smaller than P_{max} is that in the former case the oscillation amplitude is restricted by the simple dynamic equation (3.30), whereas $\hat{u}_j$ in the present section is considered as a quantity to be freely selected. Thus, in order to realise the optimum $\hat{u}_j$, we may have to include a control device in the oscillation system.

Assuming that, in addition to selecting the optimum amplitude according to Eq. (6.7), we also select the optimum phase $\gamma_j = 0$, that is, an optimum complex amplitude

$$\hat{u}_{j,\text{OPT}} = (\hat{F}_{e,j}/2R_{jj}). \tag{6.9}$$

Then the maximum absorbed power is

$$P = P_{\text{MAX}} \equiv |\hat{F}_{e,j}|^2/8R_{jj} = (P_r)_{\text{OPT}} = \tfrac{1}{2}(P_e)_{\text{OPT}}. \tag{6.10}$$

Note that here P_{MAX} agrees with Eq. (3.45) in the section on resonance absorption.

6.2.2 Upper Bound of Power-to-Volume Ratio

Note from Eq. (6.10) that, on one hand, when $|\hat{u}_j| = |\hat{u}_j|_{\text{OPT}}$, then half of the incident excitation power P_e is absorbed while the remaining half is radiated. On the other hand, if $|\hat{u}_j| \ll |\hat{u}_j|_{\text{OPT}}$, then

$$P_r \ll P \approx P_e. \tag{6.11}$$

That is, the excitation power is essentially absorbed and very little power is rera-diated. However, in this case only a small fraction of the potentially available wave power in the ocean is absorbed. Most of the wave power remains in the sea. Note that this situation may, from an economic point of view, be desirable for a wave-energy converter. Because wave power in the ocean is free, whereas the realisation of a large body-velocity amplitude requires economical expenditure, it may probably be advantageous to aim at a velocity amplitude which is substan-tially smaller than $|\hat{u}_j|_{\text{OPT}}$, except during rather rare wave conditions, when the wave amplitude and hence also $|\hat{F}_{e,j}|$ and $|\hat{u}_j|_{\text{OPT}}$ are rather small.

What should be maximised is the ratio between the energy produced and the total cost including investment, maintenance and operation. However, it is very difficult to find the solution of this problem. A much simpler problem is to consider the ratio between the produced power and the volume of the wave-absorbing body. The following paragraphs present an upper bound[67,68] to this ratio P/V.

The converted useful power is at most equal to absorbed power P. For the absorbed power we have from Eqs. (6.2), (6.4) and (6.5) that

$$P < (1/2)|\hat{F}_{e,j}| \cdot |\hat{u}_j| \cos(\gamma_j) < (1/2)|\hat{F}_{e,j}| \cdot |\hat{u}_j|. \tag{6.12}$$

The two inequalities here approach equality if $|\hat{u}_j| \ll |\hat{u}_j|_{\text{OPT}}$, and if the phase is approximately optimum, respectively. Let us now assume that the body, of volume V and water-plane area S_w, is oscillating in heave. Then it is reasonable (at least for a body of cylindrical shape) to assume that the design amplitude $|s_3|_{\max}$ for heave excursion does not exceed $V/2S_w$. Then we have the further inequalities

$$|\hat{u}_3| < \omega V/2S_w, \quad |\hat{F}_{e,3}| < \rho g S_w A, \tag{6.13}$$

where we have made use of the small-body approximation (5.253). The latter inequality approaches equality if the body is so small that diffraction from it is negligible. Then the linear extension of the body is very small compared with the wavelength. Combining inequalities (6.13) with the last of inequalities (6.12) yields the upper bound

$$\frac{P}{V} < \frac{\rho g \omega A}{4} = \frac{\pi \rho g A}{2T} = \frac{\pi \rho g H}{4T}, \tag{6.14}$$

where $H = 2A$ is the wave height of the sinusoidal wave and T is its period.

Taking as typical values $T = 8$ s and $H = 2$ m, we find that this gives an upper bound $P/V < 2.0$ kW/m^3. From the above discussion it becomes evident that for approaching this upper bound it is necessary that the body volume V tends to zero, which is, of course, not very practical for a wave-power converter. In Problem 6.2 a case is discussed of a heaving body of diameter $2a = 6$ m, which is rather small compared with the wavelength, which is 100 m for the 8 s wave. Although the body volume is as small as 283 m^3, we obtain in this case only 0.8 kW/m^3 for P/V, which is significantly less than the upper bound 2.0 kW/m^3. For the case of larger wavelengths, or longer wave periods, which is also discussed in Problem 6.2, a performance somewhat closer to the upper bound is obtained.

6.2.3 Maximum Converted Useful Power

In a wave-power device the absorbed power is divided between converted useful power P_u and a power P_f which is lost in viscous effects, friction and non-ideal energy conversion devices.

For simplicity we consider a power takeoff system with linear characteristics as indicated schematically in Figure 6.4, where resistance R_f represents the lost power [see Eq. (5.312)]. In a similar way we introduce a load resistance R_u such that load force $F_{u,j}$ [see Eqs. (5.311) and (5.316)] is

$$F_{u,j} = -R_u u_j. \tag{6.15}$$

Here we assume that we are able to vary R_u in a controllable manner. Mass m includes the mass of the oscillating body, and stiffness S includes the hydrostatic buoyancy effect. Radiation resistance R_{jj} and added mass m_{jj} are not shown in the schematic diagram of Figure 6.4.

The lost power is

$$P_f = \tfrac{1}{2} R_f |\hat{u}_j|^2, \tag{6.16}$$

and the *useful power* is

$$
\begin{aligned}
P_u &= \tfrac{1}{2} R_u |\hat{u}_j|^2 = P - P_f = P_e - (P_r + P_f) \\
&= \tfrac{1}{2} |\hat{F}_{e,j}||\hat{u}_j| \cos(\gamma_j) - \tfrac{1}{2} (R_{jj} + R_f)|\hat{u}_j|^2.
\end{aligned} \tag{6.17}
$$

For the particular case of $\gamma_j = 0$ a graphical representation of P_u, P, P_e and P_f versus $|u_j|$ are given in the diagram of Figure 6.5. If

$$\hat{u}_j = \hat{F}_{e,j}/2(R_{jj} + R_f), \tag{6.18}$$

that is, if $\gamma_j = 0$ and $|\hat{u}_j| = |\hat{F}_{e,j}|/2(R_{jj} + R_f)$, then we have the maximum useful power

$$P_{u,\text{MAX}} = |\hat{F}_{e,j}|^2/8(R_{jj} + R_f). \tag{6.19}$$

Note that a lower oscillation amplitude is required to obtain the maximum useful power than to obtain the maximum absorbed power. The phase condition $\gamma_j = 0$ for optimum is the same in both cases. The velocity has to be in phase with the excitation force.

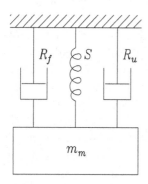

Figure 6.4: Schematic model of a linear power takeoff system of an oscillating system. Mechanical resistances R_u and R_f account for converted useful power and power lost by friction and viscous damping, respectively.

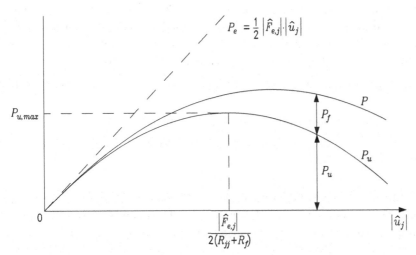

Figure 6.5: Curves showing absorbed power P, useful power P_u and excitation power P_e versus velocity amplitude $|\hat{u}_j|$ at optimum phase.

In the above discussion we have assumed that $\hat{u}_j$ is a variable at our disposal; that is, we may choose $|\hat{u}_j|$ and γ_j as we like. To achieve the desired value(s) we may find it necessary to provide the dynamic system with a certain device for power takeoff and oscillation control. This device supplies a "load" force $F_{u,j}$ in addition to excitation force $F_{e,j}$. In analogy with Eq. (5.322), the oscillation velocity has to obey the equation

$$Z_i(\omega)\hat{u}_j = \hat{F}_{e,j} + \hat{F}_{u,j}, \tag{6.20}$$

where $Z_i(\omega)$ is the intrinsic mechanical impedance for oscillation mode j. It is the mechanical impedance of the oscillating system, and it includes the radiation impedance, but not effects of the mentioned device for control and power takeoff. These effects are represented by the last term $\hat{F}_{u,j}$ in Eq. (6.20). As an extension of Eq. (6.15), let us now assume that the load force may be written as

$$\hat{F}_{u,j} = -Z_u(\omega)\hat{u}_j, \tag{6.21}$$

where $Z_{u,j}(\omega)$ is a load impedance. Introducing this into Eq. (6.20) gives

$$[Z_i(\omega) + Z_u(\omega)]\hat{u}_j = \hat{F}_{e,j}. \tag{6.22}$$

The converted useful power is

$$\begin{aligned}
P_u &= \frac{1}{2}\mathrm{Re}(-\hat{F}_u\hat{u}_j^*) = \frac{1}{2}\mathrm{Re}[Z_u(\omega)]|\hat{u}_j|^2 \\
&= \frac{1}{2}\mathrm{Re}[Z_u(\omega)]\frac{|\hat{F}_{e,j}|^2}{|Z_i(\omega)+Z_u(\omega)|^2} \\
&= \frac{R_u(\omega)|\hat{F}_{e,j}|^2/2}{[R_i(\omega)+R_u(\omega)]^2+[X_i(\omega)+X_u(\omega)]^2},
\end{aligned} \tag{6.23}$$

where we, in the last step, have split impedances $Z(\omega)$ into real parts $R(\omega)$ and imaginary parts $X(\omega)$. Note that the last expression here agrees with Eq. (3.39). Utilising the results of Eqs. (3.42) and (3.44), we find that the optimum values for $R_u(\omega)$ and $X_u(\omega)$ are $R_i(\omega)$ and $-X_i(\omega)$, respectively. Hence, for

$$Z_u(\omega) = Z_i^*(\omega) \equiv Z_{u,\text{OPT}}(\omega) \tag{6.24}$$

we have the maximum useful power [in agreement with Eq. (6.19)]

$$P_u = \frac{|\hat{F}_{e,j}|^2}{8R_i} \equiv P_{u,\text{MAX}}. \tag{6.25}$$

Note that the reactive part of the total impedance is cancelled under the optimum condition. This is automatically fulfilled in the case of resonance. Moreover, under the optimum condition, we have resistance matching [$R_u(\omega) = R_i(\omega)$].

6.3 Optimum Control for Maximising Converted Energy

The purpose of this section is to study wave-energy conversion when the waves are not sinusoidal, and to discuss conditions for maximising the converted energy. Because we, for simplicity, are discussing oscillation in only one mode, for instance heave ($j = 3$) or pitch ($j = 5$), we shall omit the subscript j in the present section. Replacing the complex amplitudes $\hat{u}$, $\hat{F}_e$ and $\hat{F}_u$ by corresponding Fourier transforms $u(\omega)$, $F_e(\omega)$ and $F_u(\omega)$, we may write Eq. (6.20) as

$$Z_i(\omega)u(\omega) = F_e(\omega) + F_u(\omega) = F_{\text{ext}}(\omega). \tag{6.26}$$

We rewrite the corresponding time-domain equation as

$$z_i(t) * u_t(t) = F_{e,t}(t) + F_{u,t}(t), \tag{6.27}$$

which for the heave mode specialises to Eq. (5.333). The causal impulse-response function $z_i(t)$ is the inverse Fourier transform of intrinsic mechanical impedance $Z_i(\omega)$; see Section 2.6.

In analogy with Eq. (6.6), the average useful power is $P_u = -\overline{F_{u,t}(t)u_t(t)}$ and thus the converted useful energy is

$$W_u = -\int_{-\infty}^{\infty} F_{u,t}(t)u_t(t)\,dt. \tag{6.28}$$

(This integral exists if the integrand tends sufficiently fast to zero as $t \to \pm\infty$.) By applying the frequency convolution theorem (2.150) for $\omega = 0$ and utilising the fact that $F_{u,t}(t)$ and $u_t(t)$ are real, we obtain, in analogy with Eqs. (2.208) and (2.212),

$$W_u = \frac{1}{2\pi}\int_0^{\infty} \{-F_u(\omega)u^*(\omega) - F_u^*(\omega)u(\omega)\}\,d\omega. \tag{6.29}$$

By algebraic manipulation of the integrand we may rewrite this as

$$W_u = \frac{1}{2\pi} \int_0^\infty \left\{ \frac{|F_e(\omega)|^2}{2R_i(\omega)} - \left[\frac{F_e(\omega)F_e^*(\omega)}{2R_i(\omega)} + F_u(\omega)u^*(\omega) + F_u^*(\omega)u(\omega) \right] \right\} d\omega$$

$$= \frac{1}{2\pi} \int_0^\infty \left\{ \frac{|F_e(\omega)|^2}{2R_i(\omega)} - \frac{\alpha(\omega)}{2R_i(\omega)} \right\} d\omega, \tag{6.30}$$

where

$$\alpha(\omega) = F_e(\omega)F_e^*(\omega) + 2R_i(\omega)[F_u(\omega)u^*(\omega) + F_u^*(\omega)u(\omega)]. \tag{6.31}$$

Here $F_e(\omega)$ is the Fourier transform of excitation force $F_{e,t}(t)$ and

$$R_i(\omega) = \frac{Z_i(\omega) + Z_i^*(\omega)}{2} \tag{6.32}$$

is the intrinsic mechanical resistance, that is, the real part of the intrinsic mechanical impedance. In the following paragraphs we are going to prove that $\alpha(\omega) \geq 0$ and to examine under which optimum conditions we have $\alpha(\omega) \equiv 0$. Then, because $R_i(\omega) > 0$, Eq. (6.30) means that the maximum converted useful energy is

$$W_{u,\mathrm{MAX}} = \frac{1}{2\pi} \int_0^\infty \frac{|F_e(\omega)|^2}{2R_i(\omega)} d\omega \tag{6.33}$$

under the condition that $\alpha(\omega) = 0$ for all ω. Otherwise we have, in general,

$$W_u = W_{u,\mathrm{MAX}} - W_{u,P}, \tag{6.34}$$

where

$$W_{u,P} = \frac{1}{2\pi} \int_0^\infty \frac{\alpha(\omega)}{2R_i(\omega)} d\omega, \tag{6.35}$$

which is a lost-energy penalty for not operating at optimum. Observe that $W_{u,P} \geq 0$. We wish to minimise $W_{u,P}$ and, if possible, reduce it to zero. It can be shown that Eq. (6.25), which applies for the case of a sinusoidal wave, is in agreement with Eq. (6.33); see Problem 2.14.

In order to show that $\alpha \geq 0$, we shall make use of dynamic equation (6.26) to eliminate either $F_e(\omega)$ or $F_u(\omega)$ from Eq. (6.31).

Omitting for a while the argument ω, we find that Eq. (6.26) gives $F_e = Z_i u - F_u$ and thus

$$F_e F_e^* = F_u F_u^* + Z_i Z_i^* uu^* - F_u Z_i^* u^* - F_u^* Z_i u. \tag{6.36}$$

Furthermore, using Eq. (6.32), we have

$$2R_i(F_u u^* + F_u^* u) = (Z_i + Z_i^*)(F_u u^* + F_u^* u)$$
$$= F_u Z_i u^* + F_u^* Z_i^* u + F_u Z_i^* u^* + F_u^* Z_i u. \tag{6.37}$$

Inserting these two equations into Eq. (6.31) gives

$$\alpha = F_u F_u^* + Z_i Z_i^* uu^* + F_u Z_i u^* + F_u^* Z_i^* u = (F_u + Z_i^* u)(F_u^* + Z_i u^*). \tag{6.38}$$

Hence we have

$$\alpha(\omega) = |F_u(\omega) + Z_i^*(\omega)u(\omega)|^2 \geq 0, \tag{6.39}$$

and from this we see that the optimum condition is

$$F_u(\omega) = -Z_i^*(\omega)u(\omega), \tag{6.40}$$

which for sinusoidal waves and oscillations agrees with Eqs. (6.21) and (6.24).

Eliminating differently, we have $F_u = Z_i u - F_e$ from dynamic equation (6.26), which by insertion into Eq. (6.31) gives

$$\begin{aligned}
\alpha &= F_e F_e^* + 2R_i(Z_i uu^* - F_e u^* + Z_i^* uu^* - F_e^* u) \\
&= F_e F_e^* + (2R_i)^2 uu^* - F_e 2R_i u^* - F_e^* 2R_i u \\
&= (F_e - 2R_i u)(F_e^* - 2R_i u^*)
\end{aligned} \tag{6.41}$$

when Eq. (6.32) has also been observed. Hence we have

$$\alpha(\omega) = |F_e(\omega) - 2R_i(\omega)u(\omega)|^2 \geq 0. \tag{6.42}$$

From this second proof of the relation $\alpha \geq 0$ we obtain the optimum condition written in the alternative way:

$$F_e(\omega) = 2R_i(\omega)u(\omega) \tag{6.43}$$

or

$$u(\omega) = \frac{F_e(\omega)}{2R_i(\omega)}, \tag{6.44}$$

which for sinusoidal waves and oscillations agrees with Eq. (6.18).

The two alternative ways to express the optimum condition, as Eq. (6.40) or as Eq. (6.44), have been used for quite some time. To achieve the optimum condition (6.40) has been called *reactive control*[69] or *complex-conjugate control*[70] and it was already in the mid-1970s tested experimentally in sinusoidal waves.[71] "Reactive control" refers to fact that reactance X_u (the imaginary part of $Z_u = -F_u/u$, assumed to be optimal, i.e., Z_i^*) cancels reactance X_i (the imaginary part of Z_i). "Complex-conjugate control" refers to the fact that optimum load impedance Z_u equals the complex conjugate of intrinsic impedance Z_i. In contrast; to achieve the optimum condition (6.44) has been called *phase control* and *amplitude control*[72] because condition (6.44) means, firstly, that the oscillation velocity u has to be in phase with excitation force F_e and, secondly, that velocity amplitude $|u|$ has to equal $|F_e|/2R_i$. A similar method was proposed and investigated before 1970.[73] Then an attempt was made to achieve optimum oscillation on the basis of wave measurement at an "upstream" position in a wave channel. In this particular case, the object was not to utilise the wave energy, but to attempt achieving complete wave absorption in a wave channel.

It should be noted that in order to obtain the optimum condition $\alpha(\omega) \equiv 0$, then unless Z_i is real, it is necessary that reactive power is involved to achieve

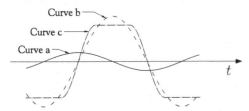

Figure 6.6: Resonance and phase control. The curves indicate incident wave elevation and vertical displacement of (different versions of) a heaving body as functions of time. Curve a: Elevation of the water surface caused by the incident wave (at the position of the body). This would also represent the vertical position of a body with negligible mass. For a body of diameter very small compared with the wavelength, curve a also represents the wave's heave force on the body. Curve b: Vertical displacement of a heaving body whose mass is so large that its natural period is equal to the wave period (resonance). Curve c: Vertical displacement of a body with smaller mass, and hence a shorter natural period. Phase control is then obtained by keeping the body in a fixed vertical position during certain time intervals.

optimum. This means that the load-and-control machinery which supplies load force F_u not only receives energy but also has to return some energy during part of the oscillation cycle. (The final part of Subsection 2.3.1 helps us understand this fact.) Obviously it is very desirable that this machinery has a very high energy-conversion efficiency (preferably close to 1). If, for some reason, we do not want to, or we are not able to, return the necessary amount of energy, then we have a suboptimal control, for which $\alpha(\omega) \neq 0$, and hence, $E_{u,P} > 0$ [see Eq. (6.35)]. An example of such a suboptimal method is phase control by latching, a principle[67] which is illustrated by Figure 6.6 for a heaving body which is so small that the excitation force is in phase with the incident wave elevation (sinusoidal in this case). By comparing curves b and c in Figure 6.6, we find it obvious that if the resonance (curve b) represents the optimum in agreement with condition (6.44), then the latching phase control (curve c) is necessarily suboptimal. With latching-phase-control experiments made so far,[74,75] means have not been provided to enable for the power takeoff machinery to return any energy during parts of the oscillation cycle.

An overview of the problem of optimum control for maximising the useful energy output from a physical dynamic system, represented by Eq. (6.26), is illustrated by the block diagram in Figure 6.7. The oscillation velocity u is the system's response to the external force input $F_{\text{ext}} = F_e + F_u$. A switch in the block diagram indicates the possibility to select between two alternatives to determine the optimum load force F_u. One alternative is the reactive-control (or complex-conjugate-control) method corresponding to optimum condition (6.40). The other alternative is the amplitude-phase-control method corresponding to optimum condition (6.44), for which knowledge on the wave excitation force F_e is used to determine the optimum oscillation velocity $u_{\text{opt}} = F_e/2R_i$, which is compared with the measured actual velocity u. The deviation between u and u_{opt} is used as input to an indicated (but here not specified) controller, which provides the optimum load force F_u. This alternative control method, contrary to the reactive-control method, requires knowledge on excitation force F_e.

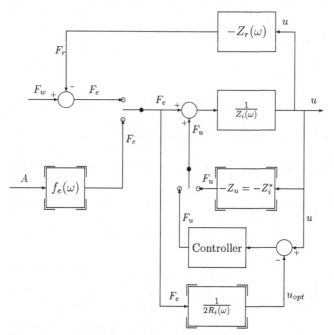

Figure 6.7: Block diagram overview of the control problem for maximising the converted energy, by a system represented by the transfer function $1/Z_i(\omega)$ where the external force $F_{\text{ext}} = F_e + F_u$ is the input and the velocity u the response. Two possibilities are indicated to determine both the excitation force $F_e = f_e A = F_w - F_r$ and the optimum load force. Three of the blocks, which are marked by a hook on each corner, represent non-causal systems.

The other switch in the block diagram of Figure 6.7 indicates the possibility to select between two alternatives to determine the excitation force from measurement either of the total wave force $F_w = F_e + F_r = F_e - Z_r u$ acting on the dynamic system (Figure 6.8) or of the incident wave elevation A at a reference position some distance away from the system (Figure 6.9). Even if this position is upstream (referred to the wave-propagation direction), the excitation-force coefficient $f_e(\omega)$ may correspond to a non-causal impulse-response function.[31] (See also Subsection 5.3.2.)

The two switches for choice of alternatives indicated in the block diagram of Figure 6.7 represent a redundancy, if equipment is installed so that two methods are available both to determine F_e and to determine F_u. Obtained alternative

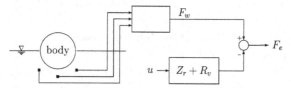

Figure 6.8: Pressure transducers, placed on or near the wet surface of the immersed body, may be used for the measurement of the hydrodynamic pressure in order to estimate the total wave force F_w on the body. Deducting from F_w a wave force $(Z_r + R_v) u$ which is due to the body oscillation, one obtains the wave excitation force F_e.

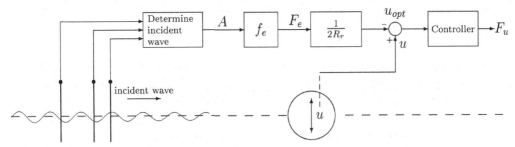

Figure 6.9: Measurement of incident wave elevation A, at some distance from the immersed body, for determination of excitation force $F_e = f_e A$ and of optimum oscillation velocity $u_{\mathrm{opt}} = F_e/2R_i$, whose deviation from the measured actual velocity u is utilised as a signal to a controller for supplying the correct load force F_u to the immersed body.

results may then be compared. Such a redundancy may be useful if it is difficult to achieve a satisfactory optimum control in practice.

Figure 6.9 illustrates the situation if a measurement of wave elevation is chosen to be the method used for determining excitation force F_e, and if the amplitude-and-phase-control method is chosen in order to maximise the useful power.

A method for a more direct measurement of total wave force F_w is indicated in Figure 6.8. An array of pressure transducers is placed on the wet body surface or in fixed positions near this surface. From measurement of the pressure in the fluid, force F_w may be estimated by application of Eq. (5.22). If the body's velocity u is also measured, the wave excitation force F_e may be estimated from

$$F_e = F_w - F_r - F_v + [-F_b] = F_w + Z_r u + R_v u + [S_b s], \qquad (6.45)$$

where Z_r is the radiation impedance (as in the uppermost block of Figure 6.7), and R_v is the viscous resistance (as in Section 5.9). Moreover, $s(\omega) = u(\omega)/i\omega$ is the Fourier transform of the oscillation excursion. Observe that, compared with the uppermost block in Figure 6.7, we have in Figure 6.8 made the generalisation of replacing Z_r by $Z_r + R_v$. The terms within brackets in Eq. (6.45) should be included if the pressure transducers indicated in Figure 6.8 are placed on the body's wet surface, and they should be excluded if the transducers are placed in fixed positions in the water.

Let us next discuss the significance of optimum conditions (6.40) and (6.43) or (6.44) in the time domain. Obviously $z_i(t)$, the inverse Fourier transform of $Z_i(\omega)$, is a causal function (cf. Subsection 5.9.1). Thus, in agreement with Eqs. (2.181)–(2.185), we have the following. Let

$$\mathcal{F}^{-1}\{Z_i(\omega)\} = z_i(t) = r_i(t) + x_i(t) = \begin{cases} 2r_i(t) & \text{for } t > 0 \\ r_i(0) & \text{for } t = 0 , \\ 0 & \text{for } t < 0 \end{cases} \qquad (6.46)$$

where the even and odd parts of $z_i(t)$, namely, $r_i(t)$ and $x_i(t)$, are the inverse Fourier transforms of $R_i(\omega) = \mathrm{Re}\{Z_i(\omega)\}$ and $X_i(\omega) = \mathrm{Im}\{Z_i(\omega)\}$, respectively.

Consequently, the inverse Fourier transform of $Z_i^*(\omega)$ is

$$\mathcal{F}^{-1}\{Z_i^*(\omega)\} = r_i(t) - x_i(t) = r_i(-t) + x_i(-t) = z_i(-t). \tag{6.47}$$

Thus, whereas $Z_i(\omega)$ corresponds to a causal impulse-response function $z_i(t)$, its conjugate $Z_i^*(\omega)$ corresponds to a non-causal (and in particular an "anticausal") impulse-response function $z_i(-t)$, which vanishes for $t > 0$. Thus, apart from exceptional cases (such as, e.g., when the incident wave is purely sinusoidal), the optimum condition (6.40) cannot be exactly satisfied in practice. Also the optimum condition (6.43) or (6.44) cannot, in general, be realised exactly, because the transfer function $R_i(\omega)$ or $1/R_i(\omega)$ is even in ω and consequently corresponds to an even (and hence non-causal) impulse-response function of time.

Moreover, also the transfer function $f_e(\omega)$ corresponds to a non-causal impulse-response function $f_t(t)$, as explained in Subsection 5.3.2. Thus, three of the blocks shown in Figure 6.7 represent non-causal impulse-response functions, and consequently they cannot be realised exactly. As an example, let us now discuss in some detail the relation $F_e(\omega) = f_e(\omega)A(\omega)$ in the time domain. Using Eq. (5.116), we have for the excitation force

$$F_{e,t}(t) = \int_{-\infty}^{\infty} f_t(t - \tau)a(\tau)\,dt, \tag{6.48}$$

where $f_t(t)$ and $a(t)$ are the inverse Fourier transforms of $f_e(\omega)$ and $A(\omega)$, respectively. Now, because f_t is non-causal, that is, $f_t(t - \tau) \neq 0$ also for $\tau > t$, there is a finite contribution to the integral from the interval $t < \tau < \infty$. This means that future values of the incident wave elevation $a(t)$ are required to compute the present-time excitation force $F_{e,t}(t)$. Similarly, because $-Z_i^*(\omega)$ and $1/R_i(\omega)$ represent noncausal impulse-response functions, future values of the actual body velocity $u(t)$ and of the actual excitation force $F_{e,t}(t)$ are needed to compute the present-time values of the optimum load force $F_{u,t}(t)$ and the optimum body velocity $u_t(t)$, respectively.

Two different strategies may be attempted to solve these non-realisable (non-causal) optimum-control problems in an approximate manner. We may refer to these approximate strategies as suboptimal control methods. One strategy is to predict the relevant input quantities a certain time distance into the future, which is straightforward for purely sinusoidal waves and oscillations, and more difficult for a broad-band than for a narrow-band wave spectrum associated with a real-sea situation. The other strategy is to replace the transfer functions $f_e(\omega)$, $-Z_i^*(\omega)$, and $1/2R_i(\omega)$ by practically realisable transfer functions, which represent causal impulse-response functions, and which approximate the ideal, but non-realisable, transfer functions. The aim is to make the approximation good at the frequencies which are important, that is, the interval of frequencies which contains most of the wave-energy. Outside this interval the chosen realisable transfer function may deviate very much from the corresponding ideal but non-realisable transfer function. Also with this version of a suboptimal strategy, better performance may be expected for a narrow-band than for a broad-band spectrum.

Several investigations have been carried out for future prediction of the wave elevation or hydrodynamic pressure at a certain point,[35,36] and also to utilise future predictions for optimum wave-energy conversion.[74,75] The method of replacing a non-realisable ideal transfer function by a realisable, but suboptimal, transfer function has also been investigated to some extent.[76–79] The following paragraphs, as an example, give an outline of a causalising approximation as proposed by Perdigão and Sarmento.

Let us discuss the approximation where the transfer function $1/[2R_i(\omega)]$ in Eq. (6.44) has been replaced by a rational function $N(\omega)/D(\omega)$, where $N(\omega)$ and $D(\omega)$ are polynomials of order n and m, respectively. We want to ensure that the corresponding impulse-response function is causal, and that the control is stable. For this reason we impose the requirements that $n < m$, and that $D(\omega)$ has all its zeros in the upper half of the complex ω plane. In accordance with this choice, we replace Eq. (6.44) by

$$u(\omega) = F_e(\omega)N(\omega)/D(\omega) \tag{6.49}$$

and a corresponding replacement may be made to change the appropriate non-realisable block in Figure 6.7 by a realisable one. Inserting Eq. (6.49) into Eq. (6.42) gives

$$\alpha(\omega) = |F_e(\omega)|^2 \, |1 - 2R_i(\omega)N(\omega)/D(\omega)|^2. \tag{6.50}$$

If this is inserted into expression (6.35) for $E_{u,P}$, it becomes obvious that we, for a general $F_e(\omega)$, have $E_{u,P} > 0$. We now need to choose the unknown complex coefficients of the two polynomials in order to minimise $E_{u,P}$. For instance, if we had chosen $m = 3$ and $n = 2$, that is,

$$\frac{N(\omega)}{D(\omega)} = \frac{b_0 + b_1\omega + b_2\omega^2}{1 + a_1\omega + a_2\omega^2 + a_3\omega^3}, \tag{6.51}$$

then six complex coefficients a_1, a_2, a_3, b_0, b_1 and b_2 have to be determined. Thus $E_{u,P}$ becomes a function of the unknown coefficients, which are determined such as to minimise $E_{u,P}$ (observing the constraint that all zeros of $D(\omega)$ have positive imaginary parts). We see that the wave spectrum, represented by the quantity $|F_e(\omega)|^2$ in Eq. (6.50), will function as a weight factor in integral (6.35) which is to be minimised. Hence, we expect the approximation to be good in the frequency interval where $|F_e(\omega)|$ is large, in particular if the wave spectrum is narrow. We observe that for a different wave spectrum, that is, for a different function $F_e(\omega)$, a new determination of the best polynomials $N(\omega)$ and $D(\omega)$ has to be made. Note, however, that we are not, in real time, able to determine the Fourier transform $F_e(\omega)$ exactly, which would require integration over the interval $-\infty < t < \infty$ [see Eq. (2.137)]. In practice we have to determine $F_e(\omega)$ only approximately based on measurement during the passed last minutes or hours, only.

6.4 Absorption by a System of Several Oscillators

Absorption of ocean-wave-energy by a system of several oscillating bodies was first analysed by Budal,[58] and subsequently in a more systematic way by Evans[57] and Falnes.[21] Let us consider a system of $N(N \geq 1)$ bodies, each of them oscillating in up to six degrees of freedom. As in Section 5.5, to each motion mode $j(j = 1, 2, \ldots 6)$ of body $p(p = 1, 2, \ldots N)$ we associate an oscillator number $i = 6(p-1) + j$. If there is an incident wave, there is an excitation force $F_{e,i}$ as discussed in Subsections 5.5.1 and 5.5.5. As a response to the excitation force the oscillator attains a velocity u_i whereby an additional wave force, the radiation force $F_{r,i}$, is set up. In accordance with Eq. (5.154) the total wave force on oscillator i has a complex amplitude

$$\hat{F}_{t,i} = \hat{F}_{e,i} + \hat{F}_{r,i} = \hat{F}_{e,i} - \sum_{i'=1}^{6N} Z_{ii'} \hat{u}_{i'}, \tag{6.52}$$

where $Z_{ii'}$ is an element of the radiation-impedance matrix discussed in Subsections 5.5.1 and 5.5.3. In accordance with Eq. (2.78), the (time-average) power absorbed by oscillator i is

$$P_i = \tfrac{1}{2}\mathrm{Re}\{\hat{F}_{t,i}\hat{u}_i^*\} = \tfrac{1}{4}(\hat{F}_{t,i}\hat{u}_i^* + \hat{F}_{t,i}^*\hat{u}_i), \tag{6.53}$$

which in view of Eq. (6.52) may be written as

$$P_i = \frac{1}{4}(\hat{F}_{e,i}\hat{u}_i^* + \hat{F}_{e,i}^*\hat{u}_i) - \frac{1}{4}\sum_{i'=1}^{6N}(Z_{ii'}\hat{u}_{i'}\hat{u}_i^* + Z_{ii'}^*\hat{u}_{i'}^*\hat{u}_i). \tag{6.54}$$

By application of Eq. (5.25) we find the (time-average) power absorbed by the total system of oscillating bodies

$$P = \sum_{i=1}^{6N} P_i = P_e - P_r, \tag{6.55}$$

where

$$P_e = \frac{1}{4}\sum_i(\hat{F}_{e,i}\hat{u}_i^* + \hat{F}_{e,i}^*\hat{u}_i) = \frac{1}{4}(\hat{\mathbf{F}}_e^T\hat{\mathbf{u}}^* + \hat{\mathbf{F}}_e^\dagger\hat{\mathbf{u}}) \tag{6.56}$$

is the excitation power, and

$$P_r = \frac{1}{2}\sum_i\sum_{i'} R_{ii'}\hat{u}_{i'}\hat{u}_i^* = \frac{1}{2}\hat{\mathbf{u}}^\dagger\mathbf{R}\hat{\mathbf{u}} = \frac{1}{2}\hat{\mathbf{u}}^T\mathbf{R}\hat{\mathbf{u}}^* \tag{6.57}$$

is the radiated power. Here we have again applied the matrix notation introduced in Chapter 5. In Eq. (6.56) we may replace $\hat{\mathbf{F}}_e^\dagger\hat{\mathbf{u}}$ by $\hat{\mathbf{u}}^T\hat{\mathbf{F}}_e^*$. In deriving Eq. (6.57) we have utilised the fact that radiated power P_r and radiation-resistance matrix $\mathbf{R}$ are real. Moreover, we have made use of reciprocity relationship (5.173) and also Eq. (5.174), from which it follows that

$$\tfrac{1}{4}(Z_{ii'} + Z_{i'i}^*) = \tfrac{1}{4}(Z_{ii'} + Z_{ii'}^*) = \tfrac{1}{2}R_{ii'}. \tag{6.58}$$

Assuming that all $\hat{F}_{e,i}$ and all $\hat{u}_i$ are known, the total absorbed power P is given by Eqs. (6.55)–(6.57), provided that all matrix elements $R_{ii'}$ are also known. It is not necessary to know the imaginary part of the radiation-impedance matrix. However, in order to determine the partition of absorbed power among the individual oscillations, Eq. (6.54) shows that knowledge also of the imaginary part of all $Z_{ii'}$, or of the added-mass matrix, is required. This knowledge is also necessary in order to find from Eq. (6.52) the total wave force $\hat{F}_{t,i}$ acting on each oscillator.

In the case of no incident wave, we have $\mathbf{F}_e \equiv 0$ and, hence, $P_e = 0$. If, in spite of this, the bodies oscillate (as a result of external forcing), they would work as wave generators (that means $P < 0$), except in particular cases, (where $P = 0$, which is possible only if the radiation-resistance matrix is singular), when bodies may oscillate in an ideal fluid without the existence of non-zero radiated waves in the far-field region. In this situation it is impossible to have positive absorbed power P. Otherwise the principle of conservation of energy would be violated. Equation (6.55) then tells us that $P_r \geq 0$, or in view of Eq. (6.57),

$$\hat{\mathbf{u}}^\dagger \mathbf{R} \hat{\mathbf{u}} \geq 0, \tag{6.59}$$

for all possible values of $\hat{\mathbf{u}}$. Thus, according to this inequality, the radiation-resistance matrix $\mathbf{R}$ is positive semidefinite (Pease,[12] p. 50), because $\mathbf{R}$ is a real symmetric matrix ($\mathbf{R}^\dagger = \mathbf{R}^T = \mathbf{R} = \mathbf{R}^*$). If $\mathbf{R}$ is a non-singular matrix, it is positive definite. However, in many cases, as we have seen in Sections 5.7 and 5.8, the radiation-resistance matrix is singular (its rank is less than its order.) For instance, with a general axisymmetric body it is possible to choose non-zero complex velocity amplitudes for the surge and pitch modes in such a way that, by mutual cancellation, the resulting radiated wave vanishes. Hence, in the general case we can only state that the matrix is positive semidefinite.

In the examples referred to above, the singularity is due to geometrical symmetry, and the matrices are singular for all frequencies. We may note that the vanishing of a diagonal element at a particular frequency provides another possibility for a radiation-resistance matrix to become singular at that frequency. Such cases have been reported[80–83] already, many years ago (see also Problem 5.5. In this case the wave generation caused by the upper part of a pivoting flap is cancelled by the opposite action from the lower part below the pivot.)

Observe from Eqs. (6.54)–(6.57) that the net absorbed power results as a difference between two terms. The first term is linear in oscillator velocity, and linear also in the amplitude of the incident wave. The second term is bilinear (quadratic) in oscillator velocity. In absence of incident waves this is the only non-vanishing term. Furthermore, it should be observed that both terms vanish if all oscillators are kept immovable ($\mathbf{u} = 0$). Hence the second term does not represent reflected or diffracted power from fixed bodies, but only radiated power from oscillating bodies. The effect of wave diffraction on the bodies is included in excitation force $\mathbf{F}_e$.

If the oscillation amplitudes are relatively small while the amplitude of the incident wave is relatively large, the second, bilinear, term is negligible compared with the first term, which is linear in $\mathbf{u}$. Then, because $P_r \ll P_e$, we have from Eq. (6.55) that

$$P \approx P_e. \tag{6.60}$$

6.4.1 Maximum Absorbed Power and Useful Power

Let us now consider the situation in which the incident wave has a relatively small amplitude, while there is no constraint on how to choose the complex velocity amplitudes $\hat{u}_i$ for the oscillators. Then we wish to determine these amplitudes in such a way that the absorbed power P becomes a maximum. To this end we define the column vector

$$\boldsymbol{\delta} = \hat{\mathbf{u}} - \mathbf{U}, \tag{6.61}$$

where $\mathbf{U}$ is a solution of the linear algebraic equation

$$\mathbf{R}\mathbf{U} = \tfrac{1}{2}\hat{\mathbf{F}}_e. \tag{6.62}$$

We now utilise the fact that $\mathbf{R}$ is a matrix which is real and symmetric. Thus the transpose of Eq. (6.62) is $\tfrac{1}{2}\hat{\mathbf{F}}_e^T = \mathbf{U}^T\mathbf{R}^T = \mathbf{U}^T\mathbf{R}$. Inserting $\hat{\mathbf{u}} = \hat{\mathbf{U}} + \hat{\boldsymbol{\delta}}$ into Eqs. (6.56) and (6.57), we obtain from Eq. (6.55)

$$\begin{aligned}
P = P(\hat{\mathbf{u}}) &\equiv \tfrac{1}{4}\left(\hat{\mathbf{F}}_e^T\hat{\mathbf{u}}^* + \hat{\mathbf{F}}_e^\dagger\hat{\mathbf{u}}\right) - \tfrac{1}{2}\hat{\mathbf{u}}^\dagger\mathbf{R}\hat{\mathbf{u}} \\
&= \tfrac{1}{4}\hat{\mathbf{F}}_e^T\mathbf{U}^* + \tfrac{1}{4}\hat{\mathbf{F}}_e^\dagger\mathbf{U} + \tfrac{1}{4}\hat{\mathbf{F}}_e^T\boldsymbol{\delta}^* + \tfrac{1}{4}\hat{\mathbf{F}}_e^\dagger\boldsymbol{\delta} \\
&\quad - \tfrac{1}{2}\mathbf{U}^\dagger\mathbf{R}\mathbf{U} - \tfrac{1}{2}\boldsymbol{\delta}^\dagger\mathbf{R}\mathbf{U} - \tfrac{1}{2}\mathbf{U}^\dagger\mathbf{R}\boldsymbol{\delta} - \tfrac{1}{2}\boldsymbol{\delta}^\dagger\mathbf{R}\boldsymbol{\delta}.
\end{aligned} \tag{6.63}$$

Of these eight terms, only the first and eighth terms remain after mutual cancellation among the other terms. Because all terms are scalars, they equal their transposes. By transposing the third term we see that it is cancelled by the sixth term after utilising Eq. (6.62). Applying the transposed conjugate of Eq. (6.62) in the seventh term, we see that it cancels the fourth term, and similarly the second and the fifth terms cancel each other. Hence

$$P = P(\hat{\mathbf{u}}) = P(\mathbf{U} + \boldsymbol{\delta}) = \tfrac{1}{4}\hat{\mathbf{F}}_e^T\mathbf{U}^* - \tfrac{1}{2}\boldsymbol{\delta}^\dagger\mathbf{R}\boldsymbol{\delta}. \tag{6.64}$$

So far, this is just another version of Eqs. (6.55)–(6.57). However, observing inequality (6.59), which means that $\mathbf{R}$ is positive semidefinite, we see that the maximum absorbed power is

$$P_{\mathrm{MAX}} = P(\mathbf{U}) = \tfrac{1}{4}\hat{\mathbf{F}}_e^T\mathbf{U}^* \tag{6.65}$$

because the last term in Eq. (6.64) can only reduce and not increase P.

If radiation-resistance matrix $\mathbf{R}$ is non-singular, its inverse $\mathbf{R}^{-1}$ exists and Eq. (6.62) has a unique solution for the optimum velocity:

$$\hat{\mathbf{u}}_{\mathrm{OPT}} \equiv \mathbf{U} = \tfrac{1}{2}\mathbf{R}^{-1}\hat{\mathbf{F}}_e. \tag{6.66}$$

Then only $\boldsymbol{\delta} = 0$ can maximise Eq. (6.64). If $\mathbf{R}$ is singular, the algebraic equation (6.62) is indeterminate. We know that then the last term in Eq. (6.64) can vanish for some particular non-vanishing values of $\boldsymbol{\delta}$. (See, e.g., the last paragraph of Subsection 5.7.2.) Adding such a $\boldsymbol{\delta}$ to $\mathbf{U}$, we obtain another optimum velocity $\boldsymbol{\delta} + \mathbf{U}$, which gives the same maximum absorbed power as given by Eq. (6.65). Thus, even though the optimum velocity

$$\hat{\mathbf{u}}_{\text{OPT}} = \mathbf{U} \tag{6.67}$$

is ambiguous when $\mathbf{R}$ is singular, the maximum absorbed power is unique. This maximum necessarily results if the velocity is given as one of the possible solutions of Eq. (6.62) (also see Problem 7.5).

Because P_{MAX} is real and scalar, we may transpose and conjugate the right-hand side of Eq. (6.65). If we also use Eqs. (6.56), (6.57) and (6.62), we may choose among several expressions for the maximum absorbed power, such as

$$P_{\text{MAX}} = P_{r,\text{OPT}} = \tfrac{1}{2} P_{e,\text{OPT}} = \tfrac{1}{2}\hat{\mathbf{u}}_{\text{OPT}}^T \mathbf{R}\hat{\mathbf{u}}_{\text{OPT}}^* = \tfrac{1}{4}\hat{\mathbf{F}}_e^T \hat{\mathbf{u}}_{\text{OPT}}^* = \tfrac{1}{4}\hat{\mathbf{u}}_{\text{OPT}}^T \hat{\mathbf{F}}_e^*. \tag{6.68}$$

If $\mathbf{R}$ is non-singular, it is also possible to use Eq. (6.66) to obtain

$$P_{\text{MAX}} = \tfrac{1}{8}\hat{\mathbf{F}}_e^T \mathbf{R}^{-1}\hat{\mathbf{F}}_e^* = \tfrac{1}{8}\hat{\mathbf{F}}_e^\dagger \mathbf{R}^{-1}\hat{\mathbf{F}}_e \tag{6.69}$$

We now take into consideration energy loss caused by friction and viscosity by applying the simplified modelling proposed in Section 5.9. In analogy with Eq. (6.57) for the radiated power, the lost power is

$$P_{\text{loss}} = \tfrac{1}{2}\hat{\mathbf{u}}^T \mathbf{R}_f \hat{\mathbf{u}}^*, \tag{6.70}$$

where $\mathbf{R}_f$ is the friction-resistance matrix introduced in Eq. (5.312). The difference between the absorbed power and lost power is

$$P_u = P - P_{\text{loss}} = P_e - P_r - P_{\text{loss}}, \tag{6.71}$$

which we shall call useful power. Using Eqs. (6.56), (6.57) and (6.70), we get

$$P_u = \tfrac{1}{4}\big(\hat{\mathbf{F}}^T \hat{\mathbf{u}}^* + \hat{\mathbf{F}}_e^\dagger \hat{\mathbf{u}}\big) - \tfrac{1}{2}\hat{\mathbf{u}}^T (\mathbf{R} + \mathbf{R}_f)\hat{\mathbf{u}}^*. \tag{6.72}$$

If in Eq. (6.62) we replace $\mathbf{R}$ by $(\mathbf{R} + \mathbf{R}_f)$ we get the optimum condition for the oscillation velocities which maximise the useful power. If we use these optimum velocities in Eq. (6.68) and also there replace $\mathbf{R}$ by $(\mathbf{R} + \mathbf{R}_f)$, we obtain the maximum useful power. Even though $\mathbf{R}$ is in many cases a singular matrix, we expect that $\mathbf{R} + \mathbf{R}_f$ is usually non-singular. Then we have the maximum useful power

$$P_{u,\text{MAX}} = \tfrac{1}{8}\hat{\mathbf{F}}_e^\dagger (\mathbf{R} + \mathbf{R}_f)^{-1}\hat{\mathbf{F}}_e^* \tag{6.73}$$

and the corresponding vector

$$\hat{\mathbf{u}} = \tfrac{1}{2}(\mathbf{R} + \mathbf{R}_f)^{-1}\hat{\mathbf{F}}_e \tag{6.74}$$

for the optimum complex velocity-amplitude components.

Because of the off-diagonal elements of radiation-resistance matrix $\mathbf{R}$, condition (6.62) or (6.74) for the optimum tells us that, for oscillator number i, it is not necessarily best, in the multi-oscillator case, to have velocity u_i in phase with excitation force $F_{e,i}$.

6.4.2 Maximum Absorbed Power by an Axisymmetric Body

Let us now consider some examples of maximum power absorption with no constraints on the amplitudes.

With heave motion as the only mode of oscillation for a single axisymmetric body, radiation-resistance matrix $\mathbf{R}$ and excitation-force vector $\hat{\mathbf{F}}_e$ are simplified to the scalars R_{33} and $F_{e,3} = f_{30} A$, respectively. Then we have according to Eqs. (6.66) and (6.69) [see also Eq. (3.45)]

$$P_{\text{MAX}} = \frac{|\hat{F}_{e,3}|^2}{8 R_{33}} \quad \text{when } \hat{u}_3 = \hat{u}_{3,\text{OPT}} = \frac{\hat{F}_{e,3}}{2 R_{33}}. \tag{6.75}$$

Using reciprocity relation (5.279), we find that this gives

$$P_{\text{MAX}} = \frac{|f_3|^2}{8 R_{33}} |A|^2 = \frac{|f_{30}|^2}{8 R_{33}} |A|^2 = \frac{2\rho g^2 D(kh)}{8\omega k} |A|^2 = \frac{1}{k} J, \tag{6.76}$$

where $J = (\rho g^2 D/4\omega)|A|^2$ is the incident wave-energy transport per unit frontage of the incident wave, in accordance with Eq. (4.136). Thus the maximum absorption width (defined as the ratio between P and J) is

$$d_{a,\text{MAX}} = \frac{P_{\text{MAX}}}{J} = \frac{1}{k} = \frac{\lambda}{2\pi}. \tag{6.77}$$

This result was first derived independently by Budal and Falnes,[84] Evans[85] and Newman.[30]

Next let us consider the case of a single axisymmetric body oscillating in just the three translation modes: surge, sway and heave. The oscillation velocity components, the excitation-force components and the radiation-resistance matrix may be written as

$$\hat{\mathbf{u}} = \begin{bmatrix} \hat{u}_1 \\ \hat{u}_2 \\ \hat{u}_3 \end{bmatrix} \quad \hat{\mathbf{F}}_e = \begin{bmatrix} \hat{F}_{e,1} \\ \hat{F}_{e,2} \\ \hat{F}_{e,3} \end{bmatrix} \quad \mathbf{R} = \begin{bmatrix} R_{11} & 0 & 0 \\ 0 & R_{11} & 0 \\ 0 & 0 & R_{33} \end{bmatrix}, \tag{6.78}$$

where we have set $R_{22} = R_{11}$ for reasons of symmetry. The conditions for optimum (6.62) and for maximum power (6.68) give

$$\hat{\mathbf{u}} = \begin{bmatrix} \hat{F}_{e,1}/2 R_{11} \\ \hat{F}_{e,2}/2 R_{22} \\ \hat{F}_{e,3}/2 R_{33} \end{bmatrix} = \mathbf{U} \tag{6.79}$$

and

$$P_{MAX} = \frac{1}{4}\hat{\mathbf{F}}_e^T \mathbf{U}^* = \frac{1}{8}\left(\frac{|\hat{F}_{e1}|^2}{R_{11}} + \frac{|\hat{F}_{e2}|^2}{R_{22}} + \frac{|\hat{F}_{e3}|^2}{R_{33}} \right), \tag{6.80}$$

$$P_{MAX} = \frac{1}{8}\left(\frac{|f_1|^2}{R_{11}} + \frac{|f_2|^2}{R_{22}} + \frac{|f_3|^2}{R_{33}} \right)|A|^2. \tag{6.81}$$

From Eqs. (5.262) and (5.266) we have

$$f_1(\beta) = -f_{10}\cos\beta, \tag{6.82}$$

$$f_2(\beta) = -f_{20}\sin\beta, \tag{6.83}$$

$$f_3(\beta) = f_{30}, \tag{6.84}$$

and by using reciprocity relation (5.279) we finally arrive at

$$P_{MAX} = \frac{1}{8}\frac{4\rho g^2 D(kh)}{\omega k}\left(\cos^2\beta + \sin^2\beta + \frac{1}{2} \right)|A|^2$$

$$= \frac{3}{k}\frac{\rho g^2 D(kh)}{4\omega}|A|^2 = \frac{3}{k}J. \tag{6.85}$$

The maximum absorption width is

$$d_{a,MAX} = \frac{P_{MAX}}{J} = \frac{3}{k} = \frac{3}{2\pi}\lambda. \tag{6.86}$$

This result was first obtained by Newman.[30]
   For the case with oscillation in just the three modes surge, heave and pitch, the radiation-resistance matrix is singular. (It is a 3×3 matrix of rank 2.) As shown in Problem 6.3, the result is again

$$P_{MAX} = (3/k)J, \tag{6.87}$$

and

$$d_{a,MAX} = 3/k = (3/2\pi)\lambda. \tag{6.88}$$

   Let us finally consider an example in which two concentric axisymmetric buoys, $p = 1$ and $p = 2$, are restricted to oscillate in the heave mode only. Then the relevant oscillators are $i = 6(p-1) + 3$; that is, $i = 3$ and $i = 9$ [cf. Eq. (5.152)]. Then according to Eq. (5.279), the system's radiation-resistance matrix is

$$\mathbf{R} = \begin{bmatrix} R_{33} & R_{39} \\ R_{93} & R_{99} \end{bmatrix} = \frac{\omega k}{2\rho g^2 D(kh)}\begin{bmatrix} |f_{30}|^2 & f_{30}f_{90}^* \\ f_{90}f_{30}^* & |f_{90}|^2 \end{bmatrix}. \tag{6.89}$$

This result is also in agreement with Eq. (5.288). It is easy to verify that the corresponding determinant vanishes, which means that $\mathbf{R}$ is a singular matrix. Choosing an arbitrary value U_9, we find that the indeterminate equation (6.62) for optimum gives

$$U_3 = \frac{f_{30}A/2 - R_{39}U_9}{R_{33}} = \frac{f_{90}A/2 - R_{99}U_9}{R_{93}}. \tag{6.90}$$

From Eq. (6.65) the maximum power is

$$P_{\text{MAX}} = \frac{(f_{30}U_3^* + f_{90}U_9^*)A}{4} = \frac{|f_{30}A|^2}{8R_{33}} = \frac{f_{30}f_{90}^*|A|^2}{8R_{93}}, \tag{6.91}$$

which is independent of the choice of U_9. Inserting from Eqs. (6.89) and (4.136) we get

$$P_{\text{MAX}} = \frac{\rho g^2 D(kh)}{4\omega k}|A|^2 = \frac{J}{k}. \tag{6.92}$$

Thus the absorption width is the same as given by Eq. (6.77). This observation gives a clue to understanding why radiation-resistance matrices may turn out to be singular.

The maximum absorbed power as given by Eqs. (6.76) and (6.92) corresponds to optimum destructive interference between the incident plane wave and the iso-tropically radiated circular wave generated by the heaving oscillation of one body or two bodies [in the case of Eq. (6.76) or Eq. (6.92), respectively]; see Figure 6.2. As for this optimum interference, it does not matter from which of the two bodies the circular wave originates. It is not possible to increase the maximum absorbed power by adding another isotropically radiating body to the system. This is the physical explanation of the singularity of the radiation-resistance matrix given by Eq. (6.89). If only isotropically radiating heave modes are involved, the rank of the matrix cannot be more than one.

Let us now consider the more general case in which all six modes of motion are allowed. The maximum power as given by Eq. (6.85) or (6.87) corresponds to optimum interference of the incident wave with three radiated circular waves, which have three possible different θ variations, in accordance with Eq. (5.262). Thus, for instance, if surge is already involved and optimised, it is not possible to improve the optimum situation by also involving pitch, because both these modes produce radiated waves with the same θ variation. If three modes of motion (or three oscillators) corresponding to three different θ variations [see Eq. (5.262)] are already involved with their complex amplitudes chosen to yield optimum wave interference between the three radiated circular waves and the plane incident wave, then it is not possible to increase the absorbed power by increasing the number of degrees of freedom (or number of oscillators) beyond three. Hence the rank of the radiation-resistance matrix for an axisymmetric system [as given by Eq. (5.279)] cannot be more than three. Thus the matrix is necessarily singular if its size is larger than 3×3.

6.4.3 Maximum Absorbed Power in the Two-Dimensional Case

For the two-dimensional case (see Section 5.8), Eqs. (6.68) for maximum power and (6.62) for optimum oscillation are modified into

$$P_{\text{MAX}}' = \tfrac{1}{4}\hat{\mathbf{F}}_e'^T\mathbf{U}^* = \tfrac{1}{2}\mathbf{U}^T\mathbf{R}'\mathbf{U}^*, \tag{6.93}$$

where the complex velocity amplitude $\hat{u}_{OPT} = U$ has to satisfy the algebraic equation

$$\mathbf{R}'\mathbf{U} = \tfrac{1}{2}\hat{\mathbf{F}}_e'. \tag{6.94}$$

Here $\mathbf{R}'$ and $\mathbf{F}_e'$ represent the radiation resistance and the excitation force, respectively, per unit width. We now express the excitation force and the radiation resistance (both per unit width) in terms of the Kochin functions by using Eqs. (5.301)–(5.305). We shall assume that the given incident wave is propagating in the positive x direction ($\beta = 0$). Then the maximum absorbed power per unit width is rewritten as

$$P'_{MAX} = \frac{1}{4}\hat{\mathbf{F}}_e'^{T}\mathbf{U}^* = \frac{1}{4}\hat{\mathbf{F}}_e'^{\dagger}\mathbf{U} = \frac{\rho g\, D(kh)}{4k}\, A^*\, \mathbf{h}'^{\dagger}(\pi)\mathbf{U}, \tag{6.95}$$

where the optimum complex amplitudes are given by

$$\{\mathbf{h}'(0)[\mathbf{h}'(0)]^{\dagger} + \mathbf{h}'(\pi)[\mathbf{h}'(\pi)]^{\dagger}\}\mathbf{U} = \frac{gk}{\omega}\mathbf{h}'(\pi)A. \tag{6.96}$$

We introduce the relative absorbed power

$$\epsilon = \frac{P'}{J} = \frac{4\omega P'}{\rho g^2\, D(kh)|A|^2} \tag{6.97}$$

and the relative optimum oscillation amplitude

$$\zeta = \frac{\omega}{gk}\left(\frac{\mathbf{U}}{A}\right). \tag{6.98}$$

Note that ϵ is just the fraction of the incident wave-energy transport (4.136) being absorbed. Its maximum value

$$\epsilon_{MAX} = \frac{4\omega}{\rho g^2\, D(kh)|A|^2}\frac{\rho g\, D(kh)}{4k}\, A^*\, \mathbf{h}'^{\dagger}(\pi)\frac{gk}{\omega}A\zeta, \tag{6.99}$$

$$\epsilon_{MAX} = \mathbf{h}'^{\dagger}(\pi)\zeta \tag{6.100}$$

is obtained when the vector of relative amplitudes ζ satisfies the equation

$$\{\mathbf{h}'(0)[\mathbf{h}'(0)]^{\dagger} + \mathbf{h}'(\pi)[\mathbf{h}'(\pi)]^{\dagger}\}\zeta = \mathbf{h}'(\pi). \tag{6.101}$$

Note that the expression within the braces here is a normalised version of the radiation-resistance matrix, and hence it equals its own complex conjugate. [See the last expression of Eq. (5.301).] The optimum condition (6.101) may be written as

$$\mathbf{h}'(\pi)\{[\mathbf{h}'(\pi)]^{\dagger}\zeta - 1\} + \mathbf{h}'(0)[\mathbf{h}'(0)]^{\dagger}\zeta = 0. \tag{6.102}$$

A matrix of the type $\mathbf{h}\mathbf{h}^{\dagger}$ is of rank 1, or of rank zero in the trivial case when $\mathbf{h}^{\dagger}\mathbf{h} = 0$ (Pease,[12] p. 239). If vectors $\mathbf{h}'(0)$ and $\mathbf{h}'(\pi)$ are linearly independent, the radiation-resistance matrix is of rank 2. Otherwise, its rank is at most equal to 1. In general, the radiation resistance of an oscillating two-dimensional body is a

3×3 matrix. It follows that this matrix is necessarily singular, because its rank is 2 or less.

Let us assume that the two vectors $\mathbf{h}'(\pi)$ and $\mathbf{h}'(0)$ are linearly independent. Then the condition (6.102) for optimum cannot be satisfied unless

$$[\mathbf{h}'(\pi)]^{\dagger}\boldsymbol{\zeta} - 1 = 0, \tag{6.103}$$

$$[\mathbf{h}'(0)]^{\dagger}\boldsymbol{\zeta} = 0. \tag{6.104}$$

Thus we have two scalar equations which the components of the optimum amplitude vector $\boldsymbol{\zeta}$ have to satisfy. Hence, the system of equations is indeterminate if $\boldsymbol{\zeta}$ has more than two components, that is, for a case in which the radiation-resistance matrix is singular, as stated previously.

Combining Eqs. (6.100) and (6.103), we obtain the unambiguous value

$$\epsilon_{\text{MAX}} = [\mathbf{h}'(\pi)]^{\dagger}\boldsymbol{\zeta} = 1 \tag{6.105}$$

for the maximum relative absorbed wave power. This means that 100% of the incident wave-energy is absorbed by the oscillating body. It can be shown [Falnes,[86] Eqs. (53)–(59)] that when conditions (6.103) and (6.104) are satisfied, then radiated waves cancel waves diffracted toward the left ($x \rightarrow -\infty$) as well as the sum of the incident wave and waves diffracted toward the right ($x \rightarrow +\infty$).

A necessary, but not sufficient, condition for the vectors $[\mathbf{h}'(\pi)]^T$ and $[\mathbf{h}'(0)]^T$ to be linearly independent is that they are of order two or more. That is, at least two modes of oscillation have to be involved.

If the body oscillates in one mode, only, the above vectors simplify to scalars. Hence Eqs. (6.100) and (6.101), in this case, simplify to

$$\epsilon_{\text{MAX}} = h'^{*}(\pi)\zeta, \tag{6.106}$$

$$[|h'(0)|^2 + |h'(\pi)|^2]\zeta = h'(\pi). \tag{6.107}$$

This means that not more than a fraction

$$\epsilon_{\text{MAX}} = \frac{1}{1 + |h'(0)/h'(\pi)|^2} \tag{6.108}$$

of the incident wave energy can be absorbed by the body. If the oscillating body is able to radiate a wave only in the negative direction, then $h'(0) = 0$, and hence 100% absorption is possible. If the body is symmetric so that equally large waves are radiated in opposite directions, we have $|h'(0)| = |h'(\pi)|$, which means that not more than 50% of the incident wave energy can be absorbed.

It is necessary that the body has a non-symmetric radiation ability if it is to absorb more than 50% of the incident wave energy. Such a non-symmetric body is the Salter "duck", intended to oscillate in the pitch mode. Already in 1974 Salter reported[71,87] measured absorbed power corresponding to more than 80% of the incident wave power.

Next we consider an oscillating body which (in its time-average position) is symmetric with respect to the plane $x = 0$. It is obvious that the wave radiated by the heave motion is symmetric, whereas the waves radiated by the surge motion and by the pitch motion are antisymmetric. In terms of the Kochin functions this means that

$$h'_3(\pi) = h'_3(0), \tag{6.109}$$

$$h'_1(\pi) = -h'_1(0), \quad h'_5(\pi) = -h'_5(0). \tag{6.110}$$

It may be remarked that if the body is a horizontal circular cylinder, and if the pitch rotation is referred to the cylinder axis, then $|h'_5| = 0$. Furthermore, if the cylinder is completely submerged on deep water, we have

$$h'_1(0) = -ih'_3(0). \tag{6.111}$$

This means that if the cylinder axis is oscillating with equal amplitudes in heave and surge with phases differing by $\pi/2$, that is, if the centre of the cylinder is describing a circle in the xz plane, then the waves generated by the cylinder motion travel away from the cylinder along the free surface, but in one direction only. This was first shown in a theoretical work by Ogilvie,[88] and this caused Evans[89] to propose a wave-energy converter utilising a submerged horizontal cylinder, which has later been called the "Bristol cylinder".

Returning now to the case of a general symmetric body oscillating in the three modes, heave ($j = 3$), surge ($j = 1$) and pitch ($j = 5$), we see that the vectors $\mathbf{h}'(\pi)$ and $\mathbf{h}'(0)$ are linearly independent. If only two modes are involved, the vectors are linearly independent if the two modes are heave and surge or heave and pitch. Thus, in all these cases 100% absorption is possible if the optimum can be fulfilled.

However, if surge and pitch are the only two modes involved in the body's oscillation, the vectors $\mathbf{h}'(\pi)$ and $\mathbf{h}'(0)$ are linearly dependent; $\mathbf{h}'(\pi) = -\mathbf{h}'(0)$. In this case the optimum condition gives

$$[\mathbf{h}'(\pi) - \mathbf{h}'(0)]^\dagger \boldsymbol{\zeta} = 1 \tag{6.112}$$

as obtained by summing the two equation (6.103) and (6.104). Hence, from Eqs. (6.100) and (6.112) we have

$$\epsilon_{\text{MAX}} = [\mathbf{h}'(\pi)]^\dagger \boldsymbol{\zeta} = \tfrac{1}{2}, \tag{6.113}$$

which means that not more than 50% power absorption is possible in this case.

We noted above that the 3×3 matrix for the radiation resistance is singular. There is a good reason for this. It is possible to absorb 100% of the incident wave power by optimum oscillation in two modes, one symmetric mode (heave) and one antisymmetric mode (surge or pitch). Hence it is not possible to absorb more wave power by including a third mode. With all three modes involved in the optimisation problem, $\mathbf{R}$ is singular, and hence the system of equations for determining the optimum values of U_1, U_3 and U_5 is indeterminate. The optimum complex amplitude U_3 is determined. If U_1 is arbitrarily chosen, then U_5 is determined, and vice versa. In this way the antisymmetric wave, resulting from the combined surge-and-pitch oscillation, is optimum in the far-field region. If both the symmetric wave and the antisymmetric wave are optimum, all incident wave energy is absorbed by the oscillating body (also see Problem 6.4).

6.4.4 Maximum Absorbed Power with Amplitude Constraints

In practice there are certain limitations on the excursion, velocity and acceleration of oscillating bodies. Thus, for all designed bodies of the system there are upper bounds to amplitudes. For sufficiently low waves these limitations are not reached, and the optimisation without constraints in the previous subsections is applicable. For moderate and large wave heights, such limitations or amplitude constraints may be important.

Cases in which constraints come into play for only some of the bodies of a system are complicated to analyse. More complicated numerical optimisation has to be applied.[39,90] It is simpler when the wave height is so large that no optimum oscillation amplitude is less than the bound caused by the design specifications of the system.[58,41] It is also relatively simple to analyse a case in which only one single, but global constraint, is involved.[91,92] However, this case is not further pursued here.

Problems

Problem 6.1: Power Absorbed by Heaving Buoy

An axisymmetric buoy has a cylindrical part, of height $2l = 8$ m and of diameter $2a = 6$ m, above a hemispherical lower end. The buoy is arranged to have a draught of $l + a = 7$ m when in its equilibrium position. The heave amplitude $|\hat{s}|$ should be limited to l. Assume that this power buoy is equipped with a machinery by which the oscillatory motion can be controlled. Firstly, the heave speed is controlled to be in phase with the heave excitation force. Secondly, the damping of the oscillation should be adjusted for (optimum amplitude corresponding to) maximum absorbed power with small waves, and for necessary limitation of the heave amplitude in larger waves. Apart from the amplitude limitation we shall assume that linear theory is applicable. Moreover, we shall assume that the wave, as well as the heave oscillation, is sinusoidal. Deep water is assumed. The density of sea water is $\rho = 1020$ kg/m^3, and the acceleration of gravity $g = 9.81$ m/s^2.

(a) Using numerical results from Problem 5.12, calculate and draw graphs for the absorbed power P as function of the amplitude $|A|$ of the incident wave when the wave period T is 6.3 s, 9.0 s, 11.0 s and 15.5 s. For each period it is necessary to determine the wave amplitude $|A_c|$ above which the design limit of the heave amplitude comes into play.

(b) Let us next assume that the average absorbed power has to be limited to $P_{max} = 300$ kW because of the design capacity of the installed machinery. For this reason it is desirable (as an alternative to completely stopping the heave oscillation) to adjust the phase angle γ between the velocity and the excitation force to a certain value ($\gamma \neq 0$) such that the absorbed power becomes $P_{\max}$. For this situation of very large incident wave, calculate and draw a graph for γ as a function of $|A|$ for each of the four mentioned wave periods.

(c) Discuss (verbally) how this will be modified if we, instead of specifying $P_{\max} = 300$ kW for the machinery, specify a maximum external damping force amplitude $|F_u|_{\max} = 2.4 \times 10^5$ N or

(d) a maximum load resistance $R_u = 1.0 \times 10^5$ N s m^{-1}.

Problem 6.2: Ratio of Absorbed Power to Volume

Assume that an incident regular (sinusoidal) wave of amplitude $|A| = 1$ m is given (on deep water). Consider wave periods in the interval $5 \text{ s} < T < 16 \text{ s}$. A heaving buoy is optimally controlled for maximum absorption of wave energy (cf. Problem 6.1). The buoy is shaped as a cylinder with a hemispherical bottom, and it has a diameter of $2a$ and a total height of $a + 2l$, where $2l$ is the height of the cylindrical part. The equilibrium draught is $a + l$. Apart for the design limit l for the heave amplitude, we shall assume that linear theory is valid.

(a) Derive an expression for the ratio P/V between the absorbed power P and the volume V of the buoy, in terms of $\rho, g, a, l, |A|, \omega = 2\pi/T, |f_3|/S$, and $\omega R_{33}/S$. (See Problem 5.12 for a definition of some of these symbols.) With a fixed value for l, show that (P/V) has its largest value when $a \to 0$. Draw a curve for $(P/V)_{\max}$ versus T.

(b) Using results from Problem 6.1, draw a curve for P/V versus T, when $a = 3$ m and $l = 4$ m. In addition to the scale for P/V (in kW/m^3), include a scale for P (in kW). Also draw a curve for how P would have been, if there had been no limitation of the heave amplitude.

Problem 6.3: Maximum Absorbed Power by an Axisymmetric Body

The maximum power absorbed by an oscillating body in a plane incident wave is

$$P_{\max} = \tfrac{1}{2} \mathbf{U} \mathbf{R} \mathbf{U}^*,$$

where $\mathbf{R}$ is the radiation-resistance matrix, and where the optimum velocity vector $\mathbf{U}$ has to satisfy the algebraic equation

$$\mathbf{R} \mathbf{U} = \tfrac{1}{2} \hat{\mathbf{F}},$$

where $\hat{\mathbf{F}}$ is the vector composed of all (six) excitation-force components of the body. Let λ denote the wavelength and J the wave-power transport of the incident plane, harmonic wave.

Use the reciprocity relation between the radiation-resistance matrix and the excitation-force vector to show that an axisymmetric body, oscillating with optimum complex amplitudes, absorbs a power

(a) $P_{max} = (3\lambda/2\pi)J$
  if it oscillates in the surge and heave modes, only,
(b) $P_{max} = (\lambda/\pi)J$
  if it oscillates in the surge and pitch modes, only, and
(c) $P_{max} = (3\lambda/2\pi)J$
  if it oscillates in the surge, heave and pitch modes, only.

State conditions which the optimum complex velocity amplitudes have to satisfy in each of the three cases, (a)–(c). It is assumed that linear theory is applicable;
that is, the wave amplitude is sufficiently small to ensure that the oscillation amplitudes are not restricted.

Problem 6.4: Maximum Absorbed Power by a Symmetric Two-Dimensional Body

Show that if a plane wave is perpendicularly incident upon a two-dimensional
body which has a vertical symmetry plane, parallel to the wave front, then, at
optimum oscillation, the body absorbs

(a) half of the incident wave power, if it oscillates in the surge and pitch modes
  only, and
(b) all incident wave power, if it oscillates in surge and heave, only, or if it oscillates
  in surge, heave and pitch.

State in each case the conditions which the optimum complex velocity amplitudes have to satisfy.

Problem 6.5: Maximum Absorbed Power with Optimised Amplitude

The maximum absorbed power when the amplitude, but not the phase, of the
oscillation is optimised, is for two different situations given by Eqs. (3.41) and
(6.8). In the former case the oscillation is restricted by the dynamic equation
(3.30), whereas in the latter case the oscillation amplitude is chosen at will (which
may necessitate that a control device is part of the system).

(a) For the situation to which Eq. (3.41) pertains, determine the phase angle
  γ between the velocity and the excitation force, in terms of $x \equiv (\omega m_m + \omega m_r - S/\omega)/R_r$ and express $P_{a,max}$ in terms of $\hat{F}_e$, R_r and x.
(b) Setting $\gamma_j = \gamma$, $R_{jj} = R_r$ and $\hat{F}_{e,j} = \hat{F}_e$, apply Eq. (6.8) to express P_{max} in
  terms of $\hat{F}_e$, R_r and x.
(c) Then express $p \equiv P_{max}/P_{a,max}$ in terms of x, and show that $p \geq 1$. Finally, show
  that $p = 1$ for resonance ($x = 0$).

CHAPTER SEVEN

Wave Interaction with Oscillating Water Columns

By "oscillating water column", we understand the water contained below a water-air interface inside a hollow structure with a submerged opening where the OWC water is communicating with the water of the open sea (see Figures 4.2, 4.5 or 7.1). Two kinds of interaction are considered: the radiation problem and the excitation problem. The radiation problem concerns the radiation of waves caused by an oscillating dynamic air pressure above the interface. The excitation problem concerns the oscillation caused by an incident wave when the dynamic air pressure is zero. Comparisons are made with wave-body interactions. Wave-energy extraction by OWCs is also discussed. Finally, the case in which several OWCs and several oscillating bodies are interacting with waves is considered (see Figure 7.1).

Many of the wave-energy converters which have, so far, been investigated in several countries are of the type with an OWC. The power takeoff may be hydraulic machinery, but pneumatic power takeoff, using air turbines, is more common. In the latter case there is a dynamic air pressure above the water surface inside the OWC chamber. In such a case the OWC may be referred to as a "periodic surface pressure"[93] or an "oscillating surface-pressure distribution".[94] It is this kind of OWC which is the subject of study in the present chapter. When Eqs. (4.22)–(4.28) were discussed, one OWC as indicated in Figures 4.2 and 4.5 was considered. In Figure 7.1 two OWCs are indicated: one is in a floating structure, and the other is in a fixed structure. As with an oscillating body (cf. Chapter 5), with an OWC two kinds of interaction are considered.

The first kind is the *radiation problem*. A radiated wave may be generated by the OWC as a result of an established dynamic air pressure p_k. The wave may be generated on otherwise calm water if the dynamic air pressure is supplied to the system by external means.

The second kind is the *scattering* (or *excitation*) *problem*. An incident wave produces oscillation of the water column and hence air is pumped by the internal water surface (at $z = z_k$). This results in a dynamic air pressure if the air volume is not too large (infinite).

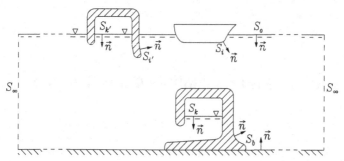

Figure 7.1: System of bodies and chambers for pressure distributions (OWCs) contained within an imaginary cylindrical control surface S_∞. Wetted surfaces of oscillating bodies are indicated by S_i and $S_{i'}$, whereas S_k and $S_{k'}$ denote internal water surfaces. Fixed surfaces, including the sea bed, are given as S_b, and S_0 denotes the external free water surface. The arrows indicate unit normals.[95]

If $z_k = 0$, then in accordance with Eq. (4.15), the static pressure of the internal air equals the ambient air pressure. Then we may – simply by opening the chamber to the air – ensure that the dynamic air pressure vanishes ($p_k = 0$).

In approximate theoretical studies of an OWC, one may think of the internal water surface S_k as an imaginary, weightless, rigid piston which is considered as an oscillating body. Such a theory does not correctly model the hydrodynamics, because the boundary condition (4.40)

$$\hat{\eta}_k = -\frac{i\omega}{g}[\hat{\phi}]_{z=z_k} - \frac{1}{\rho g}\hat{p}_k \tag{7.1}$$

is not exactly satisfied. It may, however, give good approximate results for low frequencies, when the wavelength is very long compared with the (characteristic) horizontal length of internal water surface S_k. Then, far apart from resonances of this surface, it moves approximately as a horizontal plane piston.

7.1 The Applied-Pressure Description for a Single OWC

As opposed to this rigid-piston approximation, general, and more correct, theoretical results are presented here which are based on the linearised hydrodynamic theory for an ideal irrotational fluid.[93,94]

The basic equations for the complex amplitude of the velocity potential are given by Eqs. (4.35), (4.36) and (4.37) as

$$\nabla^2\hat{\phi} = 0 \tag{7.2}$$

in the fluid region,

$$\left[\frac{\partial\hat{\phi}}{\partial n}\right]_{S_b} = 0 \tag{7.3}$$

on fixed solid surfaces S_b,

$$\left[-\omega^2\hat{\phi} + g\frac{\partial\hat{\phi}}{\partial z}\right]_{S_0} = 0 \tag{7.4}$$

on the free water surface S_0, and

$$\left[-\omega^2 \hat{\phi} + g \frac{\partial \hat{\phi}}{\partial z}\right]_{S_k} = -\frac{i\omega}{\rho} \hat{p}_k \tag{7.5}$$

on the internal water surface S_k. Of these four linear equations in $\hat{\phi}$, the first three are homogeneous, whereas Eq. (7.5) is inhomogeneous. The right-hand side of Eq. (7.5) is a driving term caused by the dynamic air pressure or "air-pressure fluctuation" p_k. This system of linear equations may be supplemented with a radiation condition at infinite distance, as discussed in Sections 4.1, 4.3 and 4.6. In the case of an incident wave, this may serve as the driving function.

Let there be an incident wave [cf. Eqs. (4.92) and (4.102)]

$$\hat{\phi}_0 = \frac{-g}{i\omega} e(kz) \hat{\eta}_0, \tag{7.6}$$

where

$$\hat{\eta}_0 = A \exp\{-ik(x \cos \beta + y \sin \beta)\}. \tag{7.7}$$

It is assumed that the sea bed is horizontal at a depth $z = -h$. The angle of incidence is β with respect to the x axis. We decompose the resulting velocity potential as

$$\phi = \phi_0 + \phi_d + \phi_r, \tag{7.8}$$

where the two last terms represent a diffracted wave and a radiated wave, respectively. We require

$$\nabla^2 \begin{bmatrix} \phi_0 \\ \phi_d \\ \phi_r \end{bmatrix} = 0 \quad \text{in the fluid region,} \tag{7.9}$$

$$\left(-\omega^2 + g \frac{\partial}{\partial z}\right) \begin{bmatrix} \hat{\phi}_0 \\ \hat{\phi}_d \\ \hat{\phi}_r \end{bmatrix} = 0 \quad \text{on } S_0, \tag{7.10}$$

and

$$\frac{\partial}{\partial n} \begin{bmatrix} \phi_0 + \phi_d \\ \phi_r \end{bmatrix} = 0 \quad \text{on } S_b. \tag{7.11}$$

(Note that $\partial \phi_0 / \partial n = 0$ and hence also $\partial \phi_d / \partial n = 0$ on the plane sea bed $z = -h$, which is a part of S_b.) Furthermore,

$$\left(-\omega^2 + g \frac{\partial}{\partial z}\right)(\hat{\phi}_0 + \hat{\phi}_d) = 0 \quad \text{on } S_k \tag{7.12}$$

$$\left(-\omega^2 + g \frac{\partial}{\partial z}\right)\hat{\phi}_r = -\frac{i\omega}{\rho} \hat{p}_k \quad \text{on } S_k. \tag{7.13}$$

With the use of Eq. (7.8), it is easy to verify that if the homogeneous

Eqs. (7.9)–(7.12) and the inhomogeneous boundary condition (7.13) are all satisfied, then the required Eqs. (7.2)–(7.5) are necessarily satisfied.

Writing

$$\hat{\phi}_r = \varphi_k \hat{p}_k, \tag{7.14}$$

we find that the proportionality coefficient φ_k must satisfy the Laplace equation,
the above homogeneous boundary conditions for $\hat{\phi}_r$ on S_0 and S_b, and the inhomogeneous boundary condition

$$\left(-\omega^2 + g\frac{\partial}{\partial z}\right)\varphi_k = -\frac{i\omega}{\rho} \quad \text{on } S_k. \tag{7.15}$$

The coefficient φ_k introduced in Eq. (7.14) is analogous to the coefficient φ_j in
Eq. (5.10) or (5.36).

The volume flow produced by the oscillating internal water surface is given by

$$Q_{t,k} = \iint\limits_{S_k} v_z \, dS = \iint\limits_{S_k} \frac{\partial \phi}{\partial z} \, dS. \tag{7.16}$$

The SI unit for volume flow is m³/s. It is convenient to decompose the total volume
flow into two terms, as follows:

$$Q_{t,k} = Q_{e,k} + Q_{r,k}, \tag{7.17}$$

where we have introduced an *excitation volume flow*,

$$Q_{e,k} = \iint\limits_{S_k} \frac{\partial}{\partial z}(\phi_0 + \phi_d) \, dS, \tag{7.18}$$

and a *radiation volume flow*,

$$Q_{r,k} = \iint\limits_{S_k} \frac{\partial \phi_r}{\partial z} \, dS. \tag{7.19}$$

Because ϕ_0 and ϕ_d are linear in A and ϕ_r is linear in p_k, we have in terms of
complex amplitudes that

$$\hat{Q}_{t,k} = \hat{Q}_{e,k} + \hat{Q}_{r,k} = q_{e,k}A - Y_{kk}\hat{p}_k, \tag{7.20}$$

where we have introduced the *excitation-volume-flow coefficient*

$$q_{e,k} = \hat{Q}_{e,k}/A = \iint\limits_{S_k} \frac{\partial}{\partial z}(\hat{\phi}_0 + \hat{\phi}_d)\frac{1}{A} \, dS, \tag{7.21}$$

and the *radiation admittance*

$$Y_{kk} = -\iint\limits_{S_k} \frac{\partial \varphi_k}{\partial z} \, dS. \tag{7.22}$$

Note that the excitation volume flow $Q_{e,k}$ is the volume flow when the air-pressure fluctuation is zero ($p_k = 0$). This is a consequence of decompositions (7.8) and (7.17) which we have chosen (following Evans[94]). Also observe that the larger $|Y_{kk}|$ is, the larger $|Q_{r,k}|$ is admitted for a given $|\hat{p}_k|$. The term "admittance" for the ratio between the complex amplitudes of volume flow and air-pressure fluctuation is adopted from the theory of electric circuits, where electric admittance is the inverse of electric impedance. Their real/imaginary parts are called conductance/susceptance and resistance/reactance, respectively. Thus we may decompose the radiation admittance into real and imaginary parts, $Y_{kk} = G_{kk} + i B_{kk}$, where

$$G_{kk} = \text{Re}\{Y_{kk}\} = \tfrac{1}{2}(Y_{kk} + Y_{kk}^*) \tag{7.23}$$

is the *radiation conductance*, and

$$B_{kk} = \text{Im}\{Y_{kk}\} \tag{7.24}$$

is the *radiation susceptance*. The SI unit for Y_{kk}, G_{kk} and B_{kk} is $\text{m}^3\text{s}^{-1}/\text{Pa} = \text{m}^5\,\text{s}^{-1}\,\text{N}^{-1}$. Thus in the present context, the product of mechanical impedance (cf. Chapters 2 and 5) and (pneumatic/hydraulic) admittance is not a dimensionless quantity. Its SI unit would be $(\text{N}\,\text{s}\,\text{m}^{-1})(\text{m}^5\,\text{s}^{-1}\,\text{N}^{-1}) = \text{m}^4$.

7.1.1 Absorbed Power and Radiation Conductance

Whereas the (pneumatic/hydraulic) admittance represents the ratio between volume flow and pressure fluctuation, the product of these two quantities has the dimension of power, in SI units: $(\text{m}^3\,\text{s}^{-1})(\text{N}\,\text{m}^{-2}) = \text{N}\,\text{m}\,\text{s}^{-1} = \text{W}$. Proceeding in analogy with Eqs. (6.1)–(6.6), we have for the (time-average) power P absorbed from the wave,

$$P = \overline{p_k(t)Q_{t,k}(t)} = \tfrac{1}{2}\text{Re}\{\hat{p}_k \hat{Q}_{t,k}^*\} = \tfrac{1}{2}\text{Re}\{\hat{p}_k(\hat{Q}_{e,k} - Y_{kk}\hat{p}_k)^*\}, \tag{7.25}$$

where we have used Eq. (7.20) and an analogue of Eq. (2.78). As in Eq. (6.2) we write the absorbed power as

$$P = P_e - P_r. \tag{7.26}$$

The excitation power is

$$P_e = \tfrac{1}{4}\hat{p}_k \hat{Q}_{e,k}^* + \tfrac{1}{4}\hat{p}_k^* \hat{Q}_{e,k} = \tfrac{1}{2}\text{Re}\{\hat{p}_k \hat{Q}_{e,k}^*\} = \tfrac{1}{2}\text{Re}\{\hat{p}_k q_{e,k}^* A^*\}, \tag{7.27}$$

and the radiated power is

$$P_r = \tfrac{1}{2}\hat{p}_k \hat{Y}_{kk}^* \hat{p}_k^* + \tfrac{1}{2}\hat{p}_k^* \hat{Y}_{kk}\hat{p}_k = \tfrac{1}{2}G_{kk}|\hat{p}_k|^2, \tag{7.28}$$

where G_{kk} is given by Eq. (7.23). Because $P_r \geq 0$, the radiation conductance cannot be negative.

7.1.2 Reactive Power and Radiation Susceptance

Thus, whereas radiation conductance G_{kk} is related to radiated power, radiation susceptance B_{kk} represents reactive power, and it may be related to the difference $W_p - W_k$ between potential energy and kinetic energy in the near-field region of the wave radiated as a result of air-pressure fluctuation p_k. To be more explicit, it can be shown[95] that

$$\frac{1}{4\omega} B_{kk} |\hat{p}_k|^2 = W_p - W_k, \tag{7.29}$$

which is analogous to Eq. (5.188) (but observe the sign on the right-hand side). Also compare Eqs. (2.87)–(2.90). As the frequency goes to zero, kinetic energy W_k tends to zero, whereas potential energy W_p reduces to hydrostatic energy $W_{p0} = \rho g S_k |\hat{\eta}_{k0}|^2/4$ associated with the vertical displacement $\eta_{k0} = p_k/\rho g$ of internal water surface S_k. Thus as $\omega \to 0$,

$$B_{kk} \to \frac{4\omega W_{p0}}{|\hat{p}_k|^2} = \frac{\omega S_k}{\rho g}. \tag{7.30}$$

Note that the effect of hydrostatic stiffness is included in radiation susceptance B_{kk}. This is in contrast to the case of a heaving body, where the effect is represented by a separate term – such as the term S_b in Eq. (5.318) – and not in the radiation reactance (or added mass). As the frequency increases from zero, there will be contributions to the potential energy from the wave elevation of the free water surface outside the OWC structure. Moreover, the kinetic energy will be significant. From the discussion in Subsection 5.9.1 on resonances for heaving bodies, it appears plausible that also OWCs may have resonances. Thus, from physical arguments we expect that Eq. (5.328) approximately represents the resonance frequency for an OWC contained in a surface-piercing vertical tube which is submerged to a depth l, and which has a diameter small in comparison with l. Because the effect of hydrostatic stiffness is included in radiation susceptance B_{kk}, resonance occurs for frequencies where $B_{kk} = 0$, that is, for frequencies at which the radiation admittance Y_{kk} is real.

7.1.3 An Axisymmetric Example

As an example, consider a circular vertical tube, the lower end of which just touches the water surface; see Figure 7.2. This is an OWC chamber with particularly simple geometry.

Because diffraction is negligible in this case, it is easy [cf. Problem 7.1 and Eq. (7.46)] to derive expressions for the excitation volume flow $Q_{e,k}$ and the radiation conductance G_{kk}, namely

$$\hat{Q}_{e,k} = q_{e,k} A \quad \text{with} \quad q_{e,k} = \frac{i\omega}{k} 2\pi a J_1(ka), \tag{7.31}$$

$$G_{kk} = \frac{k}{8J} |\hat{Q}_{e,k}|^2 = \frac{\omega k}{2\rho g^2 D} |q_{e,k}|^2 = \frac{2\omega}{\rho g} [\pi a J_1(ka)]^2. \tag{7.32}$$

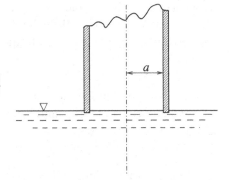

Figure 7.2: Vertical circular tube just penetrating
the free water surface represents a simple form of
an OWC chamber.

Here J_1 is the first-order Bessel function of the first kind. Because

$$(2/ka)J_1(ka) = 1 + \mathcal{O}\{k^2a^2\} \quad \text{as } ka \to 0, \tag{7.33}$$

we have

$$\hat{Q}_{e,k} \approx i\omega A\pi a^2 \tag{7.34}$$

in the long-wavelength limit, that is, for $ka \ll 1$. Because $i\omega A$ is the complex
amplitude of the vertical component of the fluid velocity corresponding to the
incident wave at the origin, this is a result to be expected in the long-wavelength
case.

We know that G_{kk} can never be negative. However, G_{kk} (as well as $\hat{Q}_{e,k}$) has
the same zeros as $J_1(ka)$, namely $ka = 3.832, 7.016, 10.173, \ldots$.

The derivation of the radiation susceptance

$$B_{kk} = \text{Im}\{Y_{kk}\} \tag{7.35}$$

is more complicated. Results for deep water are given by Evans.[94] The graph in
Figure 7.3 demonstrates that B_{kk} is positive for low frequencies (when potential
energy dominates kinetic energy), whereas negative values are most typically
found for high frequencies. The zero crossings in the graph show the first seven

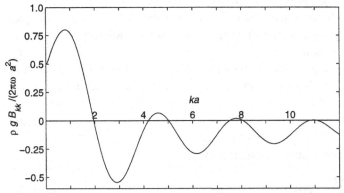

Figure 7.3: Radiation susceptance $B_{kk} = \text{Im}\{Y_{kk}\}$ for a circular OWC in a circular tube with
negligible submergence.[94] Here $(\rho g/2\omega\pi a^2)B_{kk} \to \frac{1}{2}$ as $ka \to 0$.

resonances. The first of them is at $ka = 1.96$, corresponding to a disc of radius of approximately three tenths of a wavelength ($a/\lambda = ka/2\pi = 1.96/2\pi = 0.31$). The corresponding angular eigenfrequency is

$$\omega_0 = \sqrt{k_0 g} = \sqrt{1.96\, g/a} = 1.4\,\sqrt{g/a}. \tag{7.36}$$

If we associate this lowest eigenfrequency with that of a heaving, rigid, weightless, circular disc of radius a, and hence with hydrostatic stiffness $S_b = \rho g \pi a^2$, its added mass at ω_0 would be

$$m_{33} = \frac{S_b}{\omega_0^2} = \frac{g\pi a^2}{1.96\, g/a} = 0.51\,\pi a^3 \rho = 0.76\,\frac{2\pi}{3} a^3 \rho. \tag{7.37}$$

This equals approximately three quarters of the mass of water displaced by a hemisphere of radius a. Note that the effect of hydrostatic stiffness is included in the radiation susceptance B_{kk} (as opposed to the radiation reactance of a floating body, as for instance the envisaged rigid disk). In the long-wavelength limit ($ka \to 0$) this hydrostatic effect is the sole contribution to B_{kk}. Thus in accordance with Eq. (7.30),

$$B_{kk} \to \omega \pi a^2/(\rho g) \quad \text{as } ka \to 0. \tag{7.38}$$

Note that the considered circular tube with negligible submergence can sustain only infinitesimal amplitudes. A more practical case would require a finite submergence of the tube.

7.1.4 Maximum Absorbed Power

Below let us make a discussion analogous to that of Subsection 6.2.1 for the case of an oscillating body. The excitation power as given by Eq. (7.27) may be written as

$$P_e = \tfrac{1}{2}|\hat{p}_k \hat{Q}_{e,k}|\cos(\gamma_k) = \tfrac{1}{2}|\hat{p}_k q_{e,k} A|\cos(\gamma_k), \tag{7.39}$$

where γ_k is the phase angle between the air-pressure fluctuation $\hat{p}_k$ and the excitation volume flow $\hat{Q}_{e,k}$. Note that, for a given incident wave and a given phase angle, P_e is linear in the air-pressure amplitude $|\hat{p}_k|$ whereas the radiated power P_r, as given by Eq. (7.28), is quadratic in $|\hat{p}_k|$. Hence, the absorbed power $P = P_e - P_r$ may be represented by a parabola, as indicated in the diagram of Figure 7.4. It is easy to show that the parabola has a maximum when $\hat{p}_k$ has the optimum value

$$|\hat{p}_k|_{\text{opt}} = \frac{|\hat{Q}_{e,k}|}{2G_{kk}}\cos(\gamma_k), \tag{7.40}$$

corresponding to the maximum absorbed power

$$P = P_{\max} \equiv \frac{|\hat{Q}_{e,k}|^2}{8G_{kk}}\cos^2(\gamma_k). \tag{7.41}$$

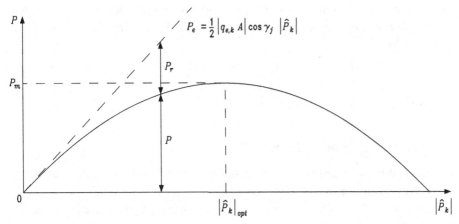

Figure 7.4: Absorbed power P (solid curve) versus air-chamber pressure amplitude $|\hat{p}_k|$. Radiated power P_r is the difference between excitation power P_e (dashed line) and the absorbed power.

Note also that

$$(P_r)_{\text{opt}} = \tfrac{1}{2}(P_e)_{\text{opt}} = P_{\max} \tag{7.42}$$

and that $P = 0$ for $|\hat{p}_k| = 2|\hat{p}_k|_{\text{opt}}$.

Furthermore, if such a condition may be accomplished (by resonance or by phase control) that the air-chamber pressure fluctuation p_k is in phase with the excitation volume flow $Q_{e,k}$ then $\gamma_k = 0$, and the absorbed power takes the maximum value

$$P_{\text{MAX}} = \frac{|\hat{Q}_{e,k}|^2}{8G_{kk}} \tag{7.43}$$

when

$$\hat{p}_k = \hat{p}_{k,\text{OPT}} = \frac{\hat{Q}_{e,k}}{2G_{kk}}. \tag{7.44}$$

This maximum absorbed power must correspond to maximum destructive interference as discussed in Section 6.1.

For instance, for an axisymmetric system, we know from Eq. (6.76) that

$$P_{\text{MAX}} = \frac{1}{k}J = \frac{\rho g^2 D(kh)}{4\omega k}|A|^2, \tag{7.45}$$

which must also hold if the optimally absorbing system is an axisymmetric OWC. With this in mind it follows from Eqs. (7.21), (7.43) and (7.45) that

$$G_{kk} = \frac{k}{8J}|\hat{Q}_{e,k}|^2 = \frac{\omega k}{2\rho g^2 D(kh)}|q_{e,k}|^2. \tag{7.46}$$

This reciprocity relation for the axisymmetric case, where $\hat{Q}_{e,k} = q_{e,k}A$ is independent of the angle β of wave incidence, is analogous to the reciprocity relation

(5.151) for a heaving axisymmetric body. In other words, the maximum power that may be absorbed by the axisymmetric OWC equals the power transport on a wave front of width

$$da = \frac{1}{k} = \frac{\lambda}{2\pi}. \tag{7.47}$$

This is the maximum absorption width.

7.1.5 Reciprocity Relations for an OWC

For the OWC there are reciprocity relations similar to those of the oscillating body. At present let us just state some of the relations and defer the proofs to Subsections 7.2.6 and 7.2.7. Relations for the three-dimensional case as well as for the two-dimensional case are given below.

In presenting the reciprocity relations we need the far-field coefficients for radiated waves. In the three-dimensional case this coefficient $a_k(\theta)$ is, in analogy with Eq. (5.129), given by the asymptotic expression for the radiated wave:

$$\hat{\phi}_r \sim \hat{p}_k a_k(\theta) e(kz)(kr)^{-1/2} e^{-ikr}, \tag{7.48}$$

as $kr \to \infty$. The two-dimensional far-field coefficient $a_k^\pm$ is, in analogy with Eqs. (4.92) and (5.292), defined by the asymptotic expression

$$\hat{\eta}_r \sim \hat{p}_k a_k^\pm e^{-ik|x|}, \tag{7.49}$$

where the plus and minus signs refer to radiation in the positive x direction and negative x direction, respectively.

For the three-dimensional case the relations analogous to Eqs. (5.140) and (5.150) are as follows. The radiation conductance $G_{kk} = \mathrm{Re}\{Y_{kk}\}$ is related to the far-field coefficient $a_k(\theta)$ by

$$G_{kk} = \frac{\omega \rho\, D(kh)}{2k} \int_0^{2\pi} |a_k(\theta)|^2\, d\theta \tag{7.50}$$

and to the excitation volume flow $\hat{Q}_{e,k}(\beta)$ by

$$G_{kk} = \frac{k}{16\pi J} \int_{-\pi}^{\pi} |\hat{Q}_{e,k}(\beta)|^2\, d\beta. \tag{7.51}$$

We may observe that Eq. (7.46) follows directly from Eq. (7.51) when $\hat{Q}_{e,k}$ is independent of β. The analogue of the Haskind relation is

$$q_{e,k}(\beta) = \hat{Q}_{e,k}(\beta)/A = -\rho g[D(kh)/k]\sqrt{2\pi}\, a_k(\beta \pm \pi) e^{i\pi/4}. \tag{7.52}$$

The minus sign is not a misprint, as the reader might suspect when comparing it with Eq. (5.147). The following comparison may serve as an explanation for this minus sign. For the heave mode ($j = 3$), body velocity u_j and force F_j are chosen positive when they are directed upward. However, if the air-pressure fluctuation

is positive, then this acts as a downward force on the OWC, whereas the volume
flow is positive when the motion is upward.

For the case of propagation in a wave channel of width d, under conditions of
no cross waves, that is, for $kd < \pi$ (see Problem 4.6), the reciprocity relations for
the radiation conductance are

$$G'_{kk} = \frac{1}{d}G_{kk} = \frac{\rho g^2 D(kh)}{2\omega}(|a_k^+|^2 + |a_k^-|^2), \tag{7.53}$$

$$G'_{kk} = \frac{1}{d}G_{kk} = \frac{1}{8J}\left(\left|\frac{\hat{Q}_{e,k}(0)}{d}\right|^2 + \left|\frac{\hat{Q}_{e,k}(\pi)}{d}\right|^2\right)$$

$$= \frac{1}{8J}(|\hat{Q}'_{e,k}(0)|^2 + |\hat{Q}'_{e,k}(\pi)|^2), \tag{7.54}$$

whereas the excitation volume flow may be expressed as

$$\hat{Q}_{e,k}(0) = A\frac{\rho g^2 D(kh) d}{\omega}a_k^- \tag{7.55}$$

if the incident wave originates from $x = -\infty$ (i.e., if $\beta = 0$), and

$$\hat{Q}_{e,k}(\pi) = A\frac{\rho g^2 D(kh) d}{\omega}a_k^+ \tag{7.56}$$

if it originates from $x = +\infty(\beta = \pi)$. These relations apply, of course, to the two-
dimensional case, because it was assumed that no cross wave exists in the wave
channel. For the two-dimensional case we have introduced quantities per unit
width (in the y direction), denoted by $\hat{Q}'_{e,k}$ for the excitation volume flow and G'_{kk}
for the radiation conductance. These reciprocity relations for a two-dimensional
OWC are analogous to similar reciprocity relations for two-dimensional bodies.
See Eqs. (5.301)–(5.305).

Let us, as an application, consider the two-dimensional situation with an incident wave (in the positive x direction) in a wave channel of width $d < \pi/k$. A
combination of Eqs. (7.54) and (7.43) gives for this case

$$P_{\max} = \frac{Jd|\hat{Q}_{e,k}(0)|^2}{|\hat{Q}_{e,k}(0)|^2 + |\hat{Q}_{e,k}(\pi)|^2} \tag{7.57}$$

or alternatively

$$P_{\max} = \frac{Jd}{1 + |a_k^+/a_k^-|^2}, \tag{7.58}$$

where Eqs. (7.55) and (7.56) have been used. Note that the OWC result, Eq. (7.58),
is analogous to the oscillating-body result, Eq. (6.108). [To see this, one may find
it helpful to use Eqs. (5.296) and (5.297) to express the Kochin functions in terms
of far-field coefficients.] For an OWC which radiates symmetrically (equal waves
in both directions, i.e., $a_k^+ = a_k^-$) we have $P_{\max} = Jd/2$, which corresponds to 50%
absorption of the incident wave power. In contrast, if the OWC is unable to radiate in the positive direction ($a_k^+ = 0$), then $P_{\max} = Jd$, which means 100% wave

absorption. The case of $a_k^+ = 0$ may be realised, for instance, by an OWC span-
ning a wave channel at the downstream end where the absorbing beach has been
removed. With an open air chamber ($\hat{p}_k = 0$) no power is absorbed ($P = 0$) and
the incident wave will be totally reflected. With optimum air-chamber pressure
according to Eq. (7.44), we have 100% absorption, which means that the optimum
radiated wave in the negative direction just cancels the reflected one.

7.1.6 OWC with Pneumatic Power Takeoff

Next we discuss the practical possibility of achieving an optimum phase and optimum amplitude of the air-chamber pressure. We assume that the air chamber
has a volume V_a and that an air turbine is placed in a duct between the chamber and the outer atmosphere. For simplicity, we represent the turbine by a pneumatic admittance Λ_t, which we assume to be a constant at our disposal. (To a
reasonable approximation a Wells turbine[96] is linear and its Λ_t is real and, within
certain limits, inversely proportional to the speed of rotation.) We assume that
Λ_t is independent of the air-chamber pressure fluctuation $\hat{p}_k$, which requires
that

$$|\hat{p}_k| \ll p_a, \tag{7.59}$$

where p_a is the ambient absolute air pressure.

If we neglect air compressibility, the volume flow at the turbine equals that at
the internal water surface, which means that the air pressure is given by

$$\hat{p}_k = \hat{Q}_{t,k}/\Lambda_t. \tag{7.60}$$

If air compressibility is taken into consideration, it can be shown (see Problem 7.2)
that

$$\hat{p}_k = \hat{Q}_{t,k}/\Lambda, \tag{7.61}$$

where

$$\Lambda = \Lambda_t + i\omega(V_a/\kappa p_a) \tag{7.62}$$

and κ is the exponent in the gas law of adiabatic compression ($\kappa = 1.4$ for air).
The imaginary part of Λ may be of some importance in a full-scale OWC, but it is
usually negligible in down-scaled laboratory model experiments.

Thus the dynamics of the OWC determines the air-chamber pressure by the
relation

$$\hat{p}_k = q_{e,k}A/(Y_{kk} + \Lambda), \tag{7.63}$$

which is obtained by combining Eqs. (7.20) and (7.61). Hence, the optimum phase
condition (that $\hat{p}_k$ is in phase with the excitation volume flow $\hat{Q}_{e,k} = q_{e,k}A$ or

that $\gamma_k = 0$) is fulfilled if

$$B_{kk} + \frac{\omega V_a}{\kappa p_a} = \text{Im}\{Y_{kk} + \Lambda\} = 0. \tag{7.64}$$

Compare Eqs. (7.24) and (7.62). If air compressibility is negligible, this corresponds to the hydrodynamic resonance condition $B_{kk} = 0$. Moreover, a comparison of Eqs. (7.63) and (7.44) shows that amplitude $\hat{p}_k$ is optimum if the additional condition

$$\text{Re}\{\Lambda\} = G_{kk} \equiv \text{Re}(Y_{kk}) \tag{7.65}$$

is satisfied. That is, we have to choose a (real) turbine admittance Λ_t which is equal to the radiation conductance G_{kk}.

As a numerical example, let us consider an axisymmetric OWC as shown in Figure 7.2 and let the diameter be $2a = 8$ m. Thus the internal water surface is $S_k = \pi a^2 = 50\,\text{m}^2$. Further assume that the average air-chamber volume is $V_a \approx 300$ m^3 and the ambient air pressure is $p_a = 10^5$ Pa. For a typical wave period $T = 2\pi/\omega = 9$ s, the (deep-water) wavelength is $\lambda = 2\pi/k = 126$ m, that is $ka = 0.20$. Thus $J_1(ka) = 0.099$, and Eq. (7.32) gives $G_{kk} = 2.2 \times 10^{-4}$ m^5/(s N). From the curve in Figure 7.3 we see that $(\rho g/2\pi\omega a^2)B_{kk} \approx 0.5$, that is $B_{kk} \approx 0.0018$ m^5/(s N), which is larger than G_{kk} by one order of magnitude. Taking into consideration the effect of the air compressibility, we have $\text{Im}\{\Lambda\} = \omega V_a/(\kappa p_a) = 0.0015$ m^5/(s N), which is also positive and of the same order of magnitude as B_{kk}. Hence the optimum phase condition is far from being satisfied in this example. If, however, the phase could be optimised by some artificial means, for instance, by control of air valves in the system, then the optimum turbine admittance would be $\Lambda_t \approx G_{kk} \approx 0.0002$ m^5/(s N).

Finally, let us add some remarks of practical relevance. We have used linear analysis and neglected viscous effects. This is a reasonable approximation to the reality only if the oscillation amplitude $\hat{s}_b$ of the water at the barrier (the inlet mouth) of the OWC does not exceed the radius of curvature ρ_b at the barrier. Then the so-called Keulegan-Carpenter number $N = \pi|\hat{v}_b|/(\omega\rho_b)$ is small (v_b is the water velocity at the barrier). This number is related to the occurrence of vortices around a body in an oscillating flow.[97] Assuming harmonic oscillations yields $|\hat{v}_b| = \omega|\hat{s}_b|$, where $|\hat{s}_b|$ is the excursion amplitude of the water at the barrier. Then the Keulegan-Carpenter number reduces to $N = \pi|\hat{s}_b|/\rho_b$. Then $|\hat{s}_b| < \rho_b$ means $N < \pi$, a condition which ensures laminar flow and absence of vortices. With $N \geq \pi$, vortex shedding will occur and appreciable viscous loss results. In practice the circular tube of the OWC has to have a finite wall thickness and a finite submergence, which is in contrast to the case discussed in connection with Figure 7.2 when both these dimensions were assumed to approach zero. If the wall thickness is, say, 1 m, the radius of curvature at the lower end could be 0.5 m. Then we may expect that the present linear theory is applicable if the volume flow does not exceed $0.5\,\text{m} \times 50\,\text{m}^2 \times 2\pi/(9\,\text{s}) \approx 17\,\text{m}^3/\text{s} \sim 20\,\text{m}^3/\text{s}$, which, according to Eq. (7.43), corresponds to a P_{MAX} that does not exceed

$(20 \, \mathrm{m^3/s})^2/(8 \times 0.0002 \, \mathrm{m^5 \, s^{-1} \, N^{-1}}) = 0.25 \times 10^6 \, \mathrm{W} = 0.25 \, \mathrm{MW}$. This limit for linear behaviour may be increased by increasing the radius of curvature at the lower barrier end. This may be achieved, for instance, by making the lower end of the vertical tube somewhat horn shaped.

7.2 Systems of OWCs and Oscillating Bodies

(The text in this section is partly taken from *Applied Ocean Research*, Vol. 7, J. Falnes and P. McIver "Surface wave interactions with systems of oscillating bodies and pressure distributions" pp. 225–234, 1985, with permission from Elsevier Science.) Wave interaction with an interacting system containing oscillating bodies as well as OWCs was first studied by Falnes and McIver[95] and independently by Fernandes.[98] Here let us consider a system of N_i bodies which can oscillate about a mean equilibrium position and of N_k chambers containing air with a pressure which can oscillate about a mean value. At the lower end, each air chamber is closed by an OWC (see Figure 7.1). If a chamber structure can oscillate, it belongs to the set of oscillating bodies. The oscillating bodies are partly or completely submerged (also see Figure 5.13). The equilibrium level of an internal water surface may differ from the mean level of the external water surface, provided the equilibrium internal air pressure is adjusted correspondingly.

If each body is free to oscillate in all its six modes of motion, the system has

$$N = 6N_i + N_k \tag{7.66}$$

independent oscillators. We assume that all oscillations are harmonic with a common angular frequency ω.

The states of the first $6N_i$ oscillators are characterised by the velocity components u_{ij} ($i = 1, 2, 3, \ldots, N_i$) of the oscillating bodies. The second subscript j ($j = 1, 2, 3, 4, 5, 6$) denotes the mode of motion (surge, sway, heave, roll, pitch and yaw, respectively). For the remaining oscillators the states are given by the dynamic pressures p_k ($k = 1, 2, 3, \ldots, N_k$) of the air in the chambers. With a given frequency the state of each oscillator is then given by a single complex amplitude ($\hat{u}_{ij}$ or $\hat{p}_k$).

Consider, for the moment, the case in which the oscillation amplitude is zero for each oscillator and let a plane incident wave

$$\eta_0(x, y) = A e^{-ikr(\beta)} \tag{7.67}$$

be given as in Eq. (4.102). Here A is the complex elevation amplitude at the origin $r(\beta) = 0$, and

$$r(\beta) = x \cos \beta + y \sin \beta, \tag{7.68}$$

where (x, y) are horizontal Cartesian coordinates and β is the angle of incidence. We choose vertical reference lines (x_i, y_i) for the bodies, taken, for instance, through the centres of mass, and (x_k, y_k) for the internal water surfaces. We further define the undisturbed surface elevation of the incident wave at the reference line of the oscillators:

$$A_i = \eta_0(x_i, y_i), \qquad A_k = \eta_0(x_k, y_k). \tag{7.69}$$

7.2.1 Phenomenological Theory

The incident wave produces a hydrodynamic force on the body surface S_i and a volume flow which is due to induced motion of the internal water surface S_k. When $u_{ij} = 0$ for all i and j, and $p_k = 0$ for all k, we term them the excitation force F_{ij} and the excitation volume flow Q_k, respectively. In linear theory their complex amplitudes are proportional to the incident wave amplitude A. We write

$$\hat{F}_{ij} = f_{ij} A_i, \qquad \hat{Q}_k = q_k A_k. \tag{7.70}$$

The complex coefficients of proportionality are functions of ω and β. We term them excitation coefficients, the excitation-force coefficient f_{ij} and the excitation volume flow coefficient q_k. (Note that we in this section omit the subscript e on excitation parameters F_{ij} and Q_k.)

Next consider the case in which the amplitudes of the oscillators are not zero, $u_{ij} \neq 0$ and $p_k \neq 0$. Then these will contribute to the j component of the force on body i and to the volume flow in chamber k. Because of our assumption of linearity we have proportionality between input and output. Below let us introduce additional complex coefficients of proportionality ($Z_{ij,i'j'}$, $H_{ij,k}$, $Y_{k,k'}$ and $H_{k,ij}$). Further we may use the principle of superposition. Thus we write the j component of the total force acting on body i as

$$\hat{F}_{t,ij} = f_{ij} A_i - \sum_{i'j'} Z_{ij,i'j'} \hat{u}_{i'j'} - \sum_k H_{ij,k} \hat{p}_k, \tag{7.71}$$

where the second sum runs from $k = 1$ to $k = N_k$. In the first sum i' runs from 1 to N_i and j' from 1 to 6. [Instead of using a single index $l = 6(i' - 1) + j'$, we denote a body oscillation by an apparent double index $i'j'$ to distinguish it from a pressure oscillator, denoted by a single index k.]

Similarly, the total volume flow which is due to the oscillation of the internal water surface S_k is

$$\hat{Q}_{t,k} = q_k A_k - \sum_{k'} Y_{k,k'} \hat{p}_{k'} - \sum_{ij} H_{k,ij} \hat{u}_{ij}. \tag{7.72}$$

We shall find it convenient to express the above relations in vector and matrix formulation. Let us first, for the oscillators, the body modes and the OWCs, define the column vector of complex amplitudes of incident wave elevation at the reference positions (or more precisely, the vertical reference lines)

$$\mathbf{A}_u = (A_{11}, A_{12}, \ldots, A_{ij}, \ldots, A_{N_i6})^T, \tag{7.73}$$

$$\mathbf{A}_p = (A_1, A_2, \ldots, A_k, \ldots, A_{N_k})^T, \tag{7.74}$$

where (for $j = 1, 2, \ldots, 6$) $A_{ij} \equiv A_i$ with A_k and A_i given by Eq. (7.69).

To characterise the state of the oscillators we use the vectors

$$\mathbf{u} = (u_{11}, u_{12}, \ldots, u_{ij}, \ldots, u_{N_i 6})^T, \tag{7.75}$$

$$\mathbf{p} = (p_1, p_2, \ldots, p_k, \ldots, p_{N_k})^T. \tag{7.76}$$

Moreover, we define the excitation-force vector $\mathbf{F}$ and the excitation volume flow vector $\mathbf{Q}$ by

$$\mathbf{F} = (F_{11}, F_{12}, \ldots, F_{ij}, \ldots, F_{N_i 6})^T, \tag{7.77}$$

$$\mathbf{Q} = (Q_1, Q_2, \ldots, Q_k, \ldots, Q_{N_k})^T. \tag{7.78}$$

Then we have

$$\hat{\mathbf{F}} = \mathbf{f A}_u, \qquad \hat{\mathbf{Q}} = \mathbf{q A}_p, \tag{7.79}$$

where the excitation coefficient matrices are diagonal and given by

$$\mathbf{f} = \mathrm{diag}(f_{11}, f_{12}, \ldots, f_{ij}, \ldots, f_{N_i 6}), \tag{7.80}$$

$$\mathbf{q} = \mathrm{diag}(q_1, q_2, \ldots, q_k, \ldots, q_{N_k}). \tag{7.81}$$

The set of coefficients $Z_{ij,i'j'}$ in Eq. (7.71) is the radiation-impedance matrix $\mathbf{Z}$ for the oscillating bodies (cf. Section 5.5). Furthermore, the set of complex coefficients $Y_{k,k'}$ in Eq. (7.72) is just the radiation-admittance matrix $\mathbf{Y}$ for the oscillating surface pressure distribution, as introduced first by Evans.[94] It will be shown later (see Section 7.2.5) that both of these matrices are symmetric, which means that they do not change by transposition

$$\mathbf{Z}^T = \mathbf{Z}, \qquad \mathbf{Y}^T = \mathbf{Y}. \tag{7.82}$$

The new complex coefficients $H_{ij,k}$ and $H_{k,ij}$, which have the dimension of length squared (for $j = 1, 2, 3$ and length cubed for $j = 4, 5, 6$), represent the hydrodynamic coupling between the oscillating bodies and the oscillating pressure distributions. From analogy with reciprocal electric circuits (see Problem 7.3) we expect that

$$H_{ij,k} = -H_{k,ij}, \tag{7.83}$$

a relation which we shall prove later [see Eq. (7.160)]. We define the radiation coupling matrix $\mathbf{H}$ by

$$\mathbf{H} = \{H_{ij,k}\} \equiv \mathbf{H}_{up}. \tag{7.84}$$

The matrices $\mathbf{Z}$, $\mathbf{Y}$ and $\mathbf{H}$ are of order $(N - N_k) \times (N - N_k)$, $N_k \times N_k$ and $(N - N_k) \times N_k$, respectively.

Mechanical impedance has the dimension of force divided by velocity. Contrary to the common usage in electric circuit theory, we have here for notational convenience defined an admittance which is not dimensionally inverse to impedance. Note that $Y'_{k,k'} \equiv Y_{k,k'}/S_k S_{k'}$ is dimensionally inverse to $Z_{ij,i'j'}$. However, it is more convenient to write just $Y_{k,k'}$ instead of $S_k S_{k'} Y'_{k,k'}$ in Eq. (7.72).

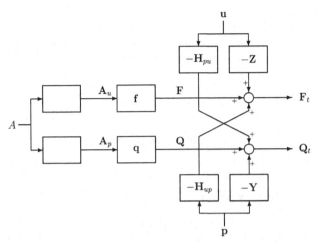

Figure 7.5: Block diagram of the system of oscillating bodies and OWCs. The incident wave, the oscillating bodies' velocities and the oscillating air-chamber pressures, (represented by A, $\mathbf{u}$ and $\mathbf{p}$, respectively) are considered as inputs to the system. The body forces and the volume flows (represented by $\mathbf{F}_t$ and $\mathbf{Q}_t$) are considered as outputs here. The empty boxes correspond to Eqs. (7.67) and (7.69); otherwise the diagram illustrates Eqs. (7.85) and (7.86).

Now Eqs. (7.71) and (7.72) for the total body force and the total volume flow may be written in matrix notation as

$$\hat{\mathbf{F}}_t = \mathbf{f}\mathbf{A}_u - \mathbf{Z}\hat{\mathbf{u}} - \mathbf{H}_{up}\hat{\mathbf{p}}, \tag{7.85}$$

$$\hat{\mathbf{Q}}_t = \mathbf{q}\mathbf{A}_p - \mathbf{Y}\hat{\mathbf{p}} - \mathbf{H}_{pu}\hat{\mathbf{u}}. \tag{7.86}$$

These equations represent a linear system which may be illustrated in a block diagram, as shown in Figure 7.5. Because of the antisymmetry relation (7.83) we have $\mathbf{H}_{up} = -\mathbf{H}_{pu}^T = \mathbf{H}$. Hence we may write Eqs. (7.85) and (7.86) as

$$\begin{bmatrix} \hat{\mathbf{F}}_t \\ \hat{\mathbf{Q}}_t \end{bmatrix} = \begin{bmatrix} \hat{\mathbf{F}} \\ \hat{\mathbf{Q}} \end{bmatrix} - \begin{bmatrix} \mathbf{Z} & \mathbf{H} \\ -\mathbf{H}^T & \mathbf{Y} \end{bmatrix} \begin{bmatrix} \hat{\mathbf{u}} \\ \hat{\mathbf{p}} \end{bmatrix}, \tag{7.87}$$

where the first term represents the excitation quantities and the last term the radiation problem.

For real ω it is convenient to decompose the complex radiation matrices into their real and imaginary parts:

$$\mathbf{Z} = \mathbf{R} + i\mathbf{X}, \tag{7.88}$$

$$\mathbf{Y} = \mathbf{G} + i\mathbf{B}, \tag{7.89}$$

$$\mathbf{H} = \mathbf{C} + i\mathbf{J}. \tag{7.90}$$

In analogy with electric-circuit usage, we may term $\mathbf{R}$, $\mathbf{G}$, $\mathbf{X}$ and $\mathbf{B}$ the radiation resistance, conductance, reactance and susceptance matrices, respectively. Note that in hydrodynamic texts, $\mathbf{R}$ is by some authors termed the radiation damping matrix; here, however, we want to reserve this term for the composite $N \times N$

(complex, but hermitian) matrix

$$\Delta = \begin{bmatrix} \mathbf{R} & -i\mathbf{J} \\ i\mathbf{J}^T & \mathbf{G} \end{bmatrix}. \tag{7.91}$$

7.2.2 Absorbed Power

The time-averaged power absorbed by the system from the waves is

$$P = \mathrm{Re}\{\mathcal{P}\}, \tag{7.92}$$

where we have introduced the complex power

$$\mathcal{P} = \sum_{ij} \mathcal{P}_{ij} + \sum_k \mathcal{P}_k = \sum_{ij} \frac{1}{2} \hat{F}_{t,ij} \hat{u}_{ij}^* + \sum_k \frac{1}{2} \hat{p}_k \hat{Q}_{t,k}^*. \tag{7.93}$$

Note that $P_{ij} = \mathrm{Re}\{\mathcal{P}_{ij}\}$ is the power absorbed by body i as a result of oscillation in mode j and that $P_k = \mathrm{Re}\{\mathcal{P}_k\}$ is the power absorbed through the oscillating internal water surface S_k. Using Eqs. (7.70)–(7.72), we get

$$\mathcal{P}_{ij} = \frac{1}{2}\left(\hat{F}_{ij} \hat{u}_{ij}^* - \sum_{i'j'} Z_{ij,i'j'} \hat{u}_{i'j'} \hat{u}_{ij}^* - \sum_k H_{ij,k} \hat{p}_k \hat{u}_{ij}^* \right), \tag{7.94}$$

$$\mathcal{P}_k = \frac{1}{2}\left(\hat{Q}_k^* \hat{p}_k - \sum_{k'} Y_{k,k'}^* \hat{p}_{k'}^* \hat{p}_k - \sum_{ij} H_{k,ij}^* \hat{u}_{ij}^* \hat{p}_k \right). \tag{7.95}$$

When summing over all oscillators, we find it convenient to use matrix notation. We find the total absorbed power

$$P = P_e - P_r, \tag{7.96}$$

where the excitation power is

$$P_e = \mathrm{Re}\{\mathcal{P}_e\} = \tfrac{1}{2}\mathrm{Re}\{\hat{\mathbf{F}}^T \hat{\mathbf{u}}^* + \hat{\mathbf{p}}^T \hat{\mathbf{Q}}^*\} \tag{7.97}$$

and the radiated power is

$$P_r = \mathrm{Re}\{\mathcal{P}_r\} \tag{7.98}$$

with

$$\begin{aligned} 2\mathcal{P}_r &= \hat{\mathbf{u}}^\dagger(\mathbf{Z}\hat{\mathbf{u}} + \mathbf{H}\hat{\mathbf{p}}) + \hat{\mathbf{p}}^T \mathbf{Y}^* \hat{\mathbf{p}}^* - \mathbf{H}^\dagger \hat{\mathbf{u}}^*) \\ &= \hat{\mathbf{u}}^\dagger \mathbf{Z}\hat{\mathbf{u}} + \hat{\mathbf{p}}^T \mathbf{Y}^* \hat{\mathbf{p}}^* + 2i\hat{\mathbf{u}}^\dagger \mathbf{J}\hat{\mathbf{p}}. \end{aligned} \tag{7.99}$$

To obtain this expression we have used Eqs. (7.82), (7.87) and (7.90). Using also Eqs. (7.88) and (7.89), we find the radiated power

$$P_r = \tfrac{1}{2}(\mathcal{P}_r + \mathcal{P}_r^*) = \tfrac{1}{2}\hat{\mathbf{u}}^\dagger \mathbf{R}\hat{\mathbf{u}} + \tfrac{1}{2}\hat{\mathbf{p}}^\dagger \mathbf{G}\hat{\mathbf{p}} - \mathrm{Im}\{\hat{\mathbf{u}}^\dagger \mathbf{J}\hat{\mathbf{p}}\}. \tag{7.100}$$

Introducing the N-dimensional column vectors

$$\hat{\boldsymbol{v}} = \begin{bmatrix} \hat{\mathbf{u}} \\ -\hat{\mathbf{p}} \end{bmatrix}, \quad \hat{\boldsymbol{\kappa}} = \begin{pmatrix} \hat{\mathbf{F}} \\ -\hat{\mathbf{Q}} \end{pmatrix}, \tag{7.101}$$

we can write the radiated power as

$$\begin{aligned} P_r &= \tfrac{1}{2}\hat{\mathbf{u}}^\dagger \mathbf{R}\hat{\mathbf{u}} + \tfrac{1}{2}\hat{\mathbf{p}}^\dagger \mathbf{G}\hat{\mathbf{p}} - \tfrac{1}{2i}(\hat{\mathbf{u}}^\dagger \mathbf{J}\hat{\mathbf{p}} - \hat{\mathbf{u}}^T \mathbf{J}^* \hat{\mathbf{p}}^*) \\ &= \tfrac{1}{2}(\hat{\mathbf{u}}^\dagger \mathbf{R}\hat{\mathbf{u}} + \hat{\mathbf{p}}^\dagger \mathbf{G}\hat{\mathbf{p}} + \hat{\mathbf{u}}^\dagger i\mathbf{J}\hat{\mathbf{p}} - \hat{\mathbf{p}}^\dagger i\mathbf{J}^T \hat{\mathbf{u}}) \\ &= \tfrac{1}{2}\mathbf{u}^\dagger (\mathbf{R}\hat{\mathbf{u}} + i\mathbf{J}\hat{\mathbf{p}}) + \tfrac{1}{2}\mathbf{p}^\dagger (\mathbf{G}\hat{\mathbf{p}} - i\mathbf{J}^T \hat{\mathbf{u}}) \\ &= \tfrac{1}{2}\begin{pmatrix} \hat{\mathbf{u}}^\dagger & -\hat{\mathbf{p}}^\dagger \end{pmatrix} \begin{pmatrix} \mathbf{R}\hat{\mathbf{u}} + i\mathbf{J}\hat{\mathbf{p}} \\ i\mathbf{J}^T \hat{\mathbf{u}} - \mathbf{G}\hat{\mathbf{p}} \end{pmatrix} \\ &= \tfrac{1}{2}\begin{pmatrix} \hat{\mathbf{u}}^\dagger & -\hat{\mathbf{p}}^\dagger \end{pmatrix} \begin{pmatrix} \mathbf{R} & -i\mathbf{J} \\ i\mathbf{J}^T & \mathbf{G} \end{pmatrix} \begin{pmatrix} \hat{\mathbf{u}} \\ -\hat{\mathbf{p}} \end{pmatrix} = \tfrac{1}{2}\hat{\boldsymbol{v}}^\dagger \boldsymbol{\Delta}\hat{\boldsymbol{v}}, \end{aligned} \tag{7.102}$$

where we have used definition (7.91). We observe that because P_r is real (and also a scalar), it follows that it equals its own complex conjugate (and also its own transpose). Thus,

$$P_r = \tfrac{1}{2}\hat{\boldsymbol{v}}^\dagger \boldsymbol{\Delta}\hat{\boldsymbol{v}} = \tfrac{1}{2}\hat{\boldsymbol{v}}^T \boldsymbol{\Delta}^* \hat{\boldsymbol{v}}^* = \tfrac{1}{2}\hat{\boldsymbol{v}}^\dagger \boldsymbol{\Delta}^\dagger \hat{\boldsymbol{v}}. \tag{7.103}$$

Because this is true for arbitrary vectors $\hat{\boldsymbol{v}}$, it follows that $\boldsymbol{\Delta} = \boldsymbol{\Delta}^\dagger$ or

$$\boldsymbol{\Delta}^T = \boldsymbol{\Delta}^*. \tag{7.104}$$

The radiation damping matrix $\boldsymbol{\Delta}$ is thus a Hermitian matrix because it equals its own adjoint (see, e.g., [Pease[12], p. 50]). Statement (7.83) is consistent with this fact, which is still to be proved; see Eq. (7.160).

In the absence of an incident wave the excitation parameters vanish ($\mathbf{F} \equiv 0$ and $\mathbf{Q} \equiv 0$), and then it follows from the principle of conservation of energy that $P \le 0$. Hence, from Eq. (7.96) we see that $P_r \ge 0$ and from Eq. (7.102) that

$$\hat{\boldsymbol{v}}^\dagger \boldsymbol{\Delta}\hat{\boldsymbol{v}} \ge 0 \tag{7.105}$$

for all possible values of vector $\hat{\boldsymbol{v}}$. This shows that the Hermitian radiation damping matrix $\boldsymbol{\Delta}$ is positive semidefinite.

Inserting Eqs. (7.97), (7.101) and (7.103) into Eq. (7.96), we find that the total absorbed power may be written as

$$P = P(\hat{\boldsymbol{v}}) = \tfrac{1}{4}(\hat{\boldsymbol{\kappa}}^T \hat{\boldsymbol{v}}^* + \hat{\boldsymbol{\kappa}}^\dagger \hat{\boldsymbol{v}}) - \tfrac{1}{2}\hat{\boldsymbol{v}}^\dagger \boldsymbol{\Delta}\hat{\boldsymbol{v}}. \tag{7.106}$$

If there is no constraint on the complex amplitudes of the components (velocities or air pressures) of column vector $\hat{\boldsymbol{v}}$, it can be shown (cf. Problem 7.5) that the maximum value of the absorbed power P is

$$P_{\text{MAX}} = P(\mathbf{U}) = \tfrac{1}{4}\hat{\boldsymbol{\kappa}}^\dagger \boldsymbol{U} = \tfrac{1}{2}\mathbf{U}^\dagger \boldsymbol{\Delta}\mathbf{U}, \tag{7.107}$$

corresponding to an optimum oscillation amplitude vector

$$\hat{\boldsymbol{v}}_{\text{OPT}} = \mathbf{U}, \tag{7.108}$$

where $\boldsymbol{U}$ is a solution of the algebraic equation

$$\boldsymbol{\Delta}\mathbf{U} = \tfrac{1}{2}\hat{\boldsymbol{\kappa}}. \tag{7.109}$$

Note that the derivation of Eq. (7.107) is an extension of the derivation of Eq. (6.65), which corresponds to a case with a real and symmetric radiation damping matrix instead of a complex and Hermitian one. In both cases the matrices are positive semidefinite.

If radiation damping matrix $\boldsymbol{\Delta}$ is non-singular, we have the unique solution

$$\hat{\boldsymbol{v}}_{\text{OPT}} = \boldsymbol{U} = \tfrac{1}{2}\boldsymbol{\Delta}^{-1}\hat{\boldsymbol{\kappa}}, \tag{7.110}$$

$$P_{\text{MAX}} = \tfrac{1}{8}\hat{\boldsymbol{\kappa}}^\dagger \boldsymbol{\Delta}^{-1}\hat{\boldsymbol{\kappa}}. \tag{7.111}$$

For the situation of a singular matrix $\boldsymbol{\Delta}$, we may argue as in Subsection 6.4.1, but we do not repeat this discussion here. See, however, Problem 7.5.

7.2.3 Hydrodynamic Formulation

As in Eq. (5.158) we shall again decompose the velocity potential into three terms corresponding to the incident wave, the diffracted wave and the radiated wave:

$$\phi = \phi_0 + \phi_d + \phi_r. \tag{7.112}$$

We need, however, to specify in more detail the requirements placed upon the terms ϕ_d and ϕ_r, caused by diffraction and radiation, respectively. Firstly, we require that ϕ_d and ϕ_r satisfy the radiation condition of outgoing waves at infinite distance. Compare Eqs. (4.246) and (4.256). Secondly, if we compare Figures 7.1 and 5.13, we see that we now (in contrast to Subsection 5.5.2) have to apply the inner surface boundary condition (4.37) with a non-vanishing right-hand term. We shall decompose the radiated wave into components corresponding to the individual oscillators,

$$\hat{\phi}_r = \sum_{ij}\varphi_{ij}\hat{u}_{ij} + \sum_k \varphi_k \hat{p}_k = \boldsymbol{\varphi}_u^T\hat{\mathbf{u}} + \boldsymbol{\varphi}_p^T\hat{\mathbf{p}}, \tag{7.113}$$

where $\boldsymbol{\varphi}_u$ is the column vector composed of all the complex potential coefficients φ_{ij} which depend on x, y, z and ω. The transpose $\boldsymbol{\varphi}_u^T$ is, of course, the corresponding line vector. Similarly, $\boldsymbol{\varphi}_p$ is the column vector composed of all the coefficients φ_k.

Referring to the surfaces defined in Figure 7.1, we shall require each of φ_{ij}, φ_k and $(\phi_0 + \phi_d)$ to satisfy the homogeneous boundary conditions

$$\frac{\partial \phi}{\partial n} = 0 \quad \text{on the surface } S_b, \tag{7.114}$$

$$-\omega^2 \hat{\phi} + g \frac{\partial \hat{\phi}}{\partial z} = 0 \quad \text{on the surface } S_0, \tag{7.115}$$

as well as the following conditions on S_i and S_k. On the wet surface S_i of oscillating body i ($i = 1, 2, \ldots, N_i$), we require

$$\frac{\partial \varphi_{i'j}}{\partial n} = n_{ij} \delta_{ii'} = \begin{cases} n_{ij} & (i' = i) \\ 0 & (i' \neq i) \end{cases}, \tag{7.116}$$

$$\frac{\partial \varphi_k}{\partial n} = 0, \tag{7.117}$$

$$\frac{\partial}{\partial n}(\phi_0 + \phi_d) = 0, \tag{7.118}$$

where n_{ij} is, in accordance with Eqs. (5.5) and (5.6), the j component of the unit normal $\vec{n}_i$ of the wet surface S_i directed into the fluid (Figure 7.1). On internal water surface S_k ($z = z_k$) in chamber k ($k = 1, 2, \ldots, N_k$), we require

$$\left(\frac{\partial}{\partial z} - \frac{\omega^2}{g}\right)\varphi_{k'} = -\frac{i\omega}{\rho g}\delta_{kk'} = \begin{cases} -i\omega/\rho g & (k' = k) \\ 0 & (k' \neq k) \end{cases}, \tag{7.119}$$

$$\left(\frac{\partial}{\partial z} - \frac{\omega^2}{g}\right)\varphi_{ij} = 0, \tag{7.120}$$

$$\left(\frac{\partial}{\partial z} - \frac{\omega^2}{g}\right)(\hat{\phi}_0 + \hat{\phi}_d) = 0. \tag{7.121}$$

Using this we prove as follows that $\hat{\phi} = \hat{\phi}_0 + \hat{\phi}_d + \sum_{i'} \sum_j \varphi_{i'j} \hat{u}_{i'j} + \sum_{k'} \varphi_{k'} \hat{p}_{k'}$ satisfies the inhomogeneous boundary conditions on S_k and on S_i. Firstly, using boundary conditions (7.116)–(7.118) on S_i gives

$$\frac{\partial \hat{\phi}}{\partial n} = \frac{\partial}{\partial n}\left(\hat{\phi}_0 + \hat{\phi}_d + \sum_{i'} \sum_j \varphi_{i'j} \hat{u}_{i'j} + \sum_{k'} \varphi_{k'} \hat{p}_{k'}\right)$$

$$= 0 + \frac{\partial}{\partial n} \sum_{i'} \sum_j \varphi_{i'j} \hat{u}_{i'j} + 0$$

$$= \sum_{i'} \sum_j \frac{\partial}{\partial n}\varphi_{i'j} \hat{u}_{i'j} = \sum_{i'} \sum_j n_{ij} \delta_{ii'} \hat{u}_{i'j}$$

$$= \sum_j n_{ij} \hat{u}_{ij} = \hat{\vec{U}}_i \cdot \vec{n}_i + \hat{\vec{\Omega}}_i \cdot \vec{s}_i \times \vec{n}_i = \hat{\vec{u}}_i \cdot \vec{n}_i = \hat{u}_{i,n}, \tag{7.122}$$

where we in the last steps have used a generalisation of Eq. (5.7). Hence the inhomogeneous boundary condition (4.36)

$$\frac{\partial \phi}{\partial n} = \hat{u}_{i,n} \tag{7.123}$$

is fulfilled on each wet body surface S_i. Secondly, on S_k we use conditions (7.119)–(7.121), which give

$$\left(\frac{\partial}{\partial z} - \frac{\omega^2}{g}\right)\hat{\phi} = \left(\frac{\partial}{\partial z} - \frac{\omega^2}{g}\right)\left(\hat{\phi}_0 + \hat{\phi}_d + \sum_{i'}\sum_{j}\varphi_{i'j}\hat{u}_{i'j} + \sum_{k'}\varphi_{k'}\hat{p}_{k'}\right)$$

$$= 0 + 0 + \left(\frac{\partial}{\partial z} - \frac{\omega^2}{g}\right)\sum_{k'}\varphi_{k'}\hat{p}_{k'}$$

$$= \sum_{k'}\left(\frac{\partial}{\partial z} - \frac{\omega^2}{g}\right)\varphi_{k'}\hat{p}_{k'}$$

$$= \sum_{k'}-\frac{i\omega}{\rho g}\delta_{kk'}\hat{p}_{k'}$$

$$= -\frac{i\omega}{\rho g}\hat{p}_k. \tag{7.124}$$

Hence the inhomogeneous boundary condition (4.37)

$$\left(-\omega^2 + g\frac{\partial}{\partial z}\right)\hat{\phi} = -\frac{i\omega}{\rho}\hat{p}_k \tag{7.125}$$

is satisfied on each of the surfaces S_k.

7.2.4 Hydrodynamic Parameters

Next let us consider the excitation parameters and radiation parameters. The j component of the total force on body i, resulting from the hydrodynamic pressure $\hat{p} = -i\omega\rho\hat{\phi}$, is obtained by integration,

$$\hat{F}_{t,ij} = -\iint_{S_i} \hat{p}n_{ij}\, dS = i\omega\rho \iint_{S_i} n_{ij}\hat{\phi}\, dS, \tag{7.126}$$

in accordance with Eq. (5.23). Using decompositions (7.112) and (7.113), we easily see that $F_{t,ij}$ is as given by Eq. (7.71) with

$$\hat{F}_{ij} = f_{ij}A_i = i\omega\rho \iint_{S_i} n_{ij}(\hat{\phi}_0 + \hat{\phi}_d)\, dS, \tag{7.127}$$

$$Z_{ij,i'j'} = -i\omega\rho \iint_{S_i} n_{ij}\varphi_{i'j'}\, dS, \tag{7.128}$$

$$H_{ij,k} = -i\omega\rho \iint_{S_i} n_{ij}\varphi_k\, dS. \tag{7.129}$$

The total volume flow through the mean water surface S_k is

$$\hat{Q}_{t,k} = \iint_{S_k} \hat{v}_z\, dS = \iint_{S_k} \frac{\partial\hat{\phi}}{\partial z}\, dS. \tag{7.130}$$

Using decompositions (7.112) and (7.113), we immediately see that $\hat{Q}_{t,k}$ is as given by Eq. (7.72) with

$$\hat{Q}_k = q_k A_k = \iint_{S_k} \frac{\partial}{\partial z}(\hat{\phi}_0 + \hat{\phi}_d)\, dS, \tag{7.131}$$

$$Y_{k,k'} = -\iint_{S_k} \frac{\partial}{\partial z}\varphi_{k'}\, dS, \tag{7.132}$$

$$H_{k,ij} = -\iint_{S_k} \frac{\partial}{\partial z}\varphi_{ij}\, dS. \tag{7.133}$$

Our next task will be to prove symmetry relations (7.82) and (7.83) and some additional reciprocity relations, including extensions of the Haskind relation.

7.2.5 Reciprocity Relations for Radiation Parameters

Because of boundary condition (7.116) on the wet surfaces S_i, we may rewrite expression (7.128) for the radiation-impedance matrix as

$$Z_{ij,i'j'} = -i\omega\rho \iint_{S_i} \frac{\partial\varphi_{ij}}{\partial n}\varphi_{i'j'}\, dS. \tag{7.134}$$

Because of the same boundary conditions we may extend the region of integration from S_i to include also the wet surfaces of all the other bodies. Furthermore, we observe from boundary condition (7.120) on S_k that

$$0 = -i\omega\rho \iint_{S_k} \left(\frac{\partial}{\partial n} + \frac{\omega^2}{g}\right)\varphi_{ij}\varphi_{i'j'}\, dS, \tag{7.135}$$

because $\partial/\partial n = -\partial/\partial z$ on S_k. We may also note that because $\partial\varphi_{ij}/\partial n$ is real on S_i, we may, if we wish, replace φ_{ij} by φ_{ij}^* in Eq. (7.134) as well as in Eq. (7.135). We are now in the position to extend the region of integration to the totality of wave-generating surfaces:

$$S = \sum_{i=1}^{N_i} S_i + \sum_{k=1}^{N_k} S_k. \tag{7.136}$$

Also compare Section 4.7. After reverting to matrix notation, by summation we obtain

$$\mathbf{Z} = -i\omega\rho \iint_S \frac{\partial\boldsymbol{\varphi}_u}{\partial n}\boldsymbol{\varphi}_u^T\, dS - \sum_k i\omega\rho \iint_{S_k} \frac{\omega^2}{g}\boldsymbol{\varphi}_u\boldsymbol{\varphi}_u^T\, dS \tag{7.137}$$

or alternatively

$$\mathbf{Z} = -i\omega\rho \iint_S \frac{\partial\boldsymbol{\varphi}_u^*}{\partial n}\boldsymbol{\varphi}_u^T\, dS - \sum_k i\omega\rho \iint_{S_k} \frac{\omega^2}{g}\boldsymbol{\varphi}_u^*\boldsymbol{\varphi}_u^T\, dS. \tag{7.138}$$

Note that these expressions for the radiation-impedance matrix $\mathbf{Z}$ are extensions of Eqs. (5.168) and (5.169) to the case in which wave-generating OWCs are also included in the system of oscillators.

When subtracting from Eq. (7.137) its own transpose, we find that the last term (the sum) does not contribute because of cancellations, and hence $\mathbf{Z} - \mathbf{Z}^T$ vanishes as in Eq. (5.171). That is,

$$\mathbf{Z}^T = \mathbf{Z}, \tag{7.139}$$

which means that $\mathbf{Z}$ is a symmetric matrix.

Furthermore we add to Eq. (7.138) its own adjoint (transpose and complex conjugate). This gives, for real ω, the radiation-resistance matrix

$$\mathbf{R} = \mathrm{Re}(\mathbf{Z}) = \frac{1}{2}(\mathbf{Z} + \mathbf{Z}^*) = \frac{1}{2}(\mathbf{Z} + \mathbf{Z}^\dagger)$$

$$= -\frac{1}{2}i\omega\rho \iint\limits_{S} \left(\frac{\partial \boldsymbol{\varphi}_u^*}{\partial n} \boldsymbol{\varphi}_u^T - \boldsymbol{\varphi}_u^* \frac{\partial \boldsymbol{\varphi}_u^T}{\partial n} \right) dS$$

$$= \frac{i}{2}\omega\rho\, \mathbf{I}(\boldsymbol{\varphi}_u^*, \boldsymbol{\varphi}_u^T) = -\frac{i}{2}\omega\rho\, \mathbf{I}(\boldsymbol{\varphi}_u, \boldsymbol{\varphi}_u^\dagger). \tag{7.140}$$

Because here the last term (the sum) in Eq. (7.138) is also cancelled out, Eq. (7.140) differs from Eq. (5.174) only by the subscript u. An alternative expression

$$\mathbf{R} = -i\omega\rho \lim_{r \to \infty} \iint\limits_{S_\infty} \frac{\partial \boldsymbol{\varphi}_u^*}{\partial r}\, \boldsymbol{\varphi}_u^T\, dS \tag{7.141}$$

may be derived from Eq. (7.140) by using Eq. (4.255).

Next we consider radiation-admittance matrix $\mathbf{Y}$ as given, in component version, by Eq. (7.132). Boundary condition (7.119) on $S_{k''}$ may be written as $\delta_{k'k''} = -[i\omega\rho + (\rho g / i\omega)\partial/\partial z]\varphi_{k'}$. Inserting this into Eq. (7.132) gives

$$Y_{k,k'} = \iint\limits_{S_k} \frac{\partial \varphi_{k'}}{\partial z}\left(i\omega\rho + \frac{\rho g}{i\omega}\frac{\partial}{\partial z} \right) \varphi_k\, dS$$

$$= \sum_{k''} \iint\limits_{S_{k''}} \frac{\partial \varphi_{k'}}{\partial z}\left(i\omega\rho + \frac{\rho g}{i\omega}\frac{\partial}{\partial z} \right) \varphi_k\, dS. \tag{7.142}$$

Because $\partial/\partial n = -\partial/\partial z$ on $S_{k''}$ and $\partial\varphi_{k'}/\partial n = 0$ on S_i according to boundary condition (7.117), we have

$$Y_{k,k'} = \sum_{k''} \iint\limits_{S_{k''}} \frac{\partial \varphi_{k'}}{\partial n}\left(-i\omega\rho + \frac{\rho g}{i\omega}\frac{\partial}{\partial n} \right) \varphi_k\, dS$$

$$= \iint\limits_{S} \frac{\partial \varphi_{k'}}{\partial n}\left(-i\omega\rho + \frac{\rho g}{i\omega}\frac{\partial}{\partial n} \right) \varphi_k\, dS$$

$$= -i\omega\rho \iint\limits_{S} \varphi_k \frac{\partial \varphi_{k'}}{\partial n}\, dS + \frac{\rho g}{i\omega} \iint\limits_{S} \frac{\partial \varphi_{k'}}{\partial n}\frac{\partial \varphi_k}{\partial n}\, dS. \tag{7.143}$$

Because we, for real ω, have

$$1 = -\left(i\omega\rho + \frac{\rho g}{i\omega}\frac{\partial}{\partial z}\right)\varphi_k = \left(i\omega\rho + \frac{\rho g}{i\omega}\frac{\partial}{\partial z}\right)\varphi_k^* \quad \text{on } S_k, \tag{7.144}$$

we also have

$$Y_{k,k'} = i\omega\rho \iint_S \varphi_k^* \frac{\partial\varphi_{k'}}{\partial n}\,dS - \frac{\rho g}{i\omega} \iint_S \frac{\partial\varphi_{k'}}{\partial n}\frac{\partial\varphi_k^*}{\partial n}\,dS \tag{7.145}$$

and hence

$$Y_{k,k'}^* = -i\omega\rho \iint_S \varphi_k \frac{\partial\varphi_{k'}^*}{\partial n}\,dS + \frac{\rho g}{i\omega} \iint_S \frac{\partial\varphi_{k'}^*}{\partial n}\frac{\partial\varphi_k}{\partial n}\,dS. \tag{7.146}$$

Using these expressions, we derive the following results. Firstly, in agreement with Eqs. (4.240) and (4.249), we have

$$Y_{k,k'} - Y_{k',k} = -i\omega\rho \iint_S \left(\varphi_k \frac{\partial\varphi_{k'}}{\partial n} - \varphi_{k'}\frac{\partial\varphi_k}{\partial n}\right)dS$$

$$= -i\omega\rho I(\varphi_k, \varphi_{k'}) = 0, \tag{7.147}$$

because φ_k and $\varphi_{k'}$ satisfy the same radiation condition. Hence

$$Y_{k,k'} = Y_{k',k} \tag{7.148}$$

or, in matrix notation as stated in Eqs. (7.82),

$$\mathbf{Y}^T = \mathbf{Y}. \tag{7.149}$$

This proves that the radiation-admittance matrix is symmetric. Secondly, the real part of this matrix, the so-called radiation-conductance matrix, is

$$G_{k,k'} = \frac{1}{2}(Y_{k,k'} + Y_{k,k'}^*) = \frac{1}{2}(Y_{k,k'} + Y_{k',k}^*)$$

$$= \frac{i\omega\rho}{2} \iint_S \left(\varphi_k^* \frac{\partial\varphi_{k'}}{\partial n} - \varphi_{k'}\frac{\partial\varphi_k^*}{\partial n}\right)dS, \tag{7.150}$$

or, using Eq. (4.240),

$$G_{k',k} = G_{k,k'} = \text{Re}\{Y_{k,k'}\} = \frac{i\omega\rho}{2} I(\varphi_k^*, \varphi_{k'}). \tag{7.151}$$

An alternative expression

$$G_{k,k'} = i\omega\rho \lim_{r\to\infty} \iint_{S_\infty} \varphi_k^* \frac{\partial\varphi_{k'}}{\partial n}\,dS \tag{7.152}$$

may be obtained by using Eq. (4.255). These expressions may be rewritten in matrix notation as

$$\mathbf{G} = \text{Re}\{\mathbf{Y}\} = \frac{i\omega\rho}{2}\mathbf{I}(\boldsymbol{\varphi}_p^*, \boldsymbol{\varphi}_p^T) = -\frac{i\omega\rho}{2}\mathbf{I}(\boldsymbol{\varphi}_p, \boldsymbol{\varphi}_p^\dagger), \tag{7.153}$$

$$\mathbf{G} = i\omega\rho \lim_{r\to\infty} \iint_{S_\infty} \boldsymbol{\varphi}_p^* \frac{\partial\boldsymbol{\varphi}_p^T}{\partial r}\,dS = -i\omega\rho \lim_{r\to\infty} \iint_{S_\infty} \frac{\partial\boldsymbol{\varphi}_p^*}{\partial r}\boldsymbol{\varphi}_p^T\,dS. \tag{7.154}$$

Here we have assumed that ω is real, and we have observed the fact that $\mathbf{G}$ is a symmetric real matrix.

Next let us present a proof of reciprocity relation (7.83), that is, $\mathbf{H}^T = -\mathbf{H}$, where the coupling matrix $\mathbf{H}$ is defined by Eq. (7.84). As a first step we use the inhomogeneous boundary condition (7.116) for φ_{ij} in Eq. (7.129), giving

$$H_{ij,k} = -i\omega\rho \sum_{i'} \iint_{S_{i'}} \varphi_k \frac{\partial \varphi_{ij}}{\partial n} \, dS. \tag{7.155}$$

In view of boundary condition (7.120) and also noting that $\partial/\partial z = -\partial/\partial n$ on $S_{k'}$, we can rewrite this as

$$H_{ij,k} = -i\omega\rho \iint_{S} \varphi_k \frac{\partial \varphi_{ij}}{\partial n} \, dS - \frac{i\omega^3\rho}{g} \sum_{k'} \iint_{S_{k'}} \varphi_k \varphi_{ij} \, dS \tag{7.156}$$

The integration surface S is defined by Eq. (7.136).

Moreover, using the inhomogeneous boundary condition (7.119) for φ_k in Eq. (7.133), we find

$$H_{k,ij} = \frac{\rho g}{i\omega} \sum_{k'} \iint_{S_{k'}} \frac{\partial \varphi_{ij}}{\partial z} \left(\frac{\partial \varphi_k}{\partial z} - \frac{\omega^2}{g} \varphi_k \right) dS. \tag{7.157}$$

Furthermore, using the homogeneous boundary conditions (7.117) for φ_k and (7.120) for φ_{ij}, we obtain

$$H_{k,ij} = i\omega\rho \iint_{S} \varphi_{ij} \frac{\partial \varphi_k}{\partial n} \, dS + \frac{i\omega^3\rho}{g} \sum_{k'} \iint_{S_{k'}} \varphi_{ij} \varphi_k \, dS. \tag{7.158}$$

Adding this to Eq. (7.156), we now find

$$H_{ij,k} + H_{k,ij} = -i\omega\rho \iint_{S} \left(\varphi_k \frac{\partial \varphi_{ij}}{\partial n} - \varphi_{ij} \frac{\partial \varphi_k}{\partial n} \right) dS + 0$$

$$= -i\omega\rho \, I(\varphi_k, \varphi_{ij}) = 0 \tag{7.159}$$

because of Eq. (4.249). Thus we have

$$H_{ij,k} = -H_{k,ij}, \tag{7.160}$$

which is in agreement with statement (7.83).

Note that for real ω the right-hand side of Eq. (7.119) is purely imaginary. Hence, in the integrand of Eq. (7.157) we may replace φ_k by $-\varphi_k^*$. The same replacement may be made in Eq. (7.158). Furthermore, because the right-hand side of Eq. (7.116) is real, we may replace φ_{ij} by φ_{ij}^* in Eq. (7.155) and, hence, also in Eq. (7.156). We have

$$J_{ij,k} = \mathrm{Im}\{H_{ij,k}\} = \frac{1}{2i}(H_{ij,k} - H_{ij,k}^*) = \frac{1}{2i}(H_{ij,k} + H_{k,ij}^*). \tag{7.161}$$

Taking the sum of Eq. (7.156) with φ_{ij} replaced by φ_{ij}^* and of the complex conjugate of Eq. (7.158) with φ_k replaced by $-\varphi_k^*$, we obtain

$$J_{ij,k} = -\frac{\omega\rho}{2}I(\varphi_k, \varphi_{ij}^*) = \frac{\omega\rho}{2}I(\varphi_{ij}^*, \varphi_k), \tag{7.162}$$

where we have used Eqs. (4.240) and (4.244). Using Eq. (4.255), we also have

$$J_{ij,k} = -\omega\rho \lim_{r\to\infty} \iint_{S_\infty} \varphi_k \frac{\partial \varphi_{ij}^*}{\partial r} dS = \omega\rho \lim_{r\to\infty} \iint_{S_\infty} \frac{\partial \varphi_k}{\partial r} \varphi_{ij}^* dS. \tag{7.163}$$

In matrix notation we have

$$\mathbf{J} = \frac{\omega\rho}{2}\mathbf{I}(\boldsymbol{\varphi}_u^*, \boldsymbol{\varphi}_p^T) = \omega\rho \lim_{r\to\infty} \iint_{S_\infty} \boldsymbol{\varphi}_u^* \frac{\partial \boldsymbol{\varphi}_p^T}{\partial r} dS. \tag{7.164}$$

On the basis of Eqs. (7.140), (7.151) and (7.162) we shall later, in Subsection 7.2.7, express the matrices $\mathbf{G}$, $\mathbf{R}$ and $\mathbf{J}$ in terms of far-field coefficients, Kochin functions, or excitation parameters.

7.2.6 Extension of the Haskind Relation

According to Eqs. (7.127) and (7.131) the excitation force is

$$\hat{F}_{ij} = f_{ij}A_i = i\omega\rho \iint_{S_i} n_{ij}(\hat{\phi}_0 + \hat{\phi}_d) dS \tag{7.165}$$

and the excitation volume flow is

$$\hat{Q}_k = q_k A_k = \iint_{S_k} \frac{\partial}{\partial z}(\hat{\phi}_0 + \hat{\phi}_d) dS. \tag{7.166}$$

(Note that we here, as elsewhere in Section 7.2, omit subscript e on excitation parameters F_{ij} and Q_k.) With Haskind's formula[55] for the excitation force on one single body, as well as with the following extension of the Haskind relation, the diffraction potential $\hat{\phi}_d$ is eliminated. Instead, the radiation potential coefficient φ_{ij} enters into the integral. For the OWC case the radiation potential coefficient φ_k enters instead of φ_{ij} into the Haskind relation for the excitation volume flow.

Using boundary conditions (7.116) and (7.118), namely $\partial\varphi_{i'j}/\partial n = n_{ij}\delta_{ii'}$ and $\partial/\partial n(\hat{\phi}_0 + \hat{\phi}_d) = 0$ on S_i, we may write Eq. (7.165) as

$$\hat{F}_{ij} = i\omega\rho \iint_{S_i} \left[(\hat{\phi}_0 + \hat{\phi}_d)\frac{\partial\varphi_{ij}}{\partial n} - \varphi_{ij}\frac{\partial}{\partial n}(\hat{\phi}_0 + \hat{\phi}_d)\right] dS. \tag{7.167}$$

Note that both terms of the integrand are zero on $S_{i'}$ when $i' \neq i$. On S_k where $\partial/\partial n = -\partial/\partial z$, the two terms cancel each other because of the homogeneous boundary conditions (7.120) and (7.121). Hence the integration area may be extended from S_i to S; that is,

$$\hat{F}_{ij} = i\omega\rho I[(\hat{\phi}_0 + \hat{\phi}_d), \varphi_{ij}] = i\omega\rho I(\hat{\phi}_0, \varphi_{ij}) + i\omega\rho I(\hat{\phi}_d, \varphi_{ij}), \tag{7.168}$$

where use has been made of Eq. (4.240). Now, because $\hat{\phi}_d$ and φ_{ij} satisfy the same radiation condition, Eq. (4.249) gives $I(\hat{\phi}_d, \varphi_{ij}) = 0$ and, hence,

$$\hat{F}_{ij} = i\omega\rho\, I(\hat{\phi}_0, \varphi_{ij}) \tag{7.169}$$

results. Note that if φ_{ij} is known in the far-field only, and not on S, we calculate the integral on S_∞ in accordance with Eq. (4.243).

Using boundary conditions (7.119) and (7.121), namely $-[i\omega\rho + (\rho g/i\omega)\partial/\partial z]$ $\varphi_k = \delta_{kk'}$ and $[i\omega\rho + (\rho g/i\omega)\partial/\partial z](\hat{\phi}_0 + \hat{\phi}_d) = 0$ on $S_{k'}$ in Eq. (7.166), we rewrite the excitation volume flow as

$$\begin{aligned}
\hat{Q}_k &= -\iint_{S_k} \left\{ \frac{\partial}{\partial z}(\hat{\phi}_0 + \hat{\phi}_d)\left(i\omega\rho + \frac{\rho g}{i\omega}\frac{\partial}{\partial z}\right)\varphi_k \right. \\
&\qquad \left. - \left(i\omega\rho + \frac{\rho g}{i\omega}\frac{\partial}{\partial z}\right)(\hat{\phi}_0 + \hat{\phi}_d)\frac{\partial\varphi_k}{\partial z} \right\} dS \\
&= -i\omega\rho \iint_{S_k} \left[\varphi_k\frac{\partial}{\partial z}(\hat{\phi}_0 + \hat{\phi}_d) - (\hat{\phi}_0 + \hat{\phi}_d)\frac{\partial\varphi_k}{\partial z} \right] dS \\
&= i\omega\rho \iint_{S_k} \left[\varphi_k\frac{\partial}{\partial n}(\hat{\phi}_0 + \hat{\phi}_d) - (\hat{\phi}_0 + \hat{\phi}_d)\frac{\partial\varphi_k}{\partial n} \right] dS.
\end{aligned} \tag{7.170}$$

Note that in the first of these three integrals, both terms of the integrand vanish on $S_{k'}$ because of boundary conditions (7.119) and (7.121). Hence the integration area may be extended from S_k to $\sum_{k'} S_{k'}$. Furthermore, both terms of the integrand in the last integral vanish on all S_i according to boundary conditions (7.117) and (7.118). Consequently, the integration area may be extended to S. Then Eq. (4.240) may be applied; that is,

$$\hat{Q}_k = -i\omega\rho\, I[(\hat{\phi}_0 + \hat{\phi}_d), \varphi_k] = -i\omega\rho\, I(\hat{\phi}_0, \varphi_k) - i\omega\rho\, I(\hat{\phi}_d, \varphi_k). \tag{7.171}$$

In accordance with Eq. (4.249), the last term vanishes because $\hat{\phi}_d$ and φ_k satisfy the same radiation condition. Hence, the excitation volume flow is given by the formula

$$\hat{Q}_k = -i\omega\rho\, I(\hat{\phi}_0, \varphi_k), \tag{7.172}$$

which is an extension of Haskind's formula.

To summarise, we may rewrite Eqs. (7.169) and (7.172) for the excitation parameters in vector notation as

$$\hat{\mathbf{F}} = i\omega\rho\,\mathbf{I}(\hat{\phi}_0, \boldsymbol{\varphi}_u), \tag{7.173}$$

$$\hat{\mathbf{Q}} = -i\omega\rho\,\mathbf{I}(\hat{\phi}_0, \boldsymbol{\varphi}_p), \tag{7.174}$$

where $\boldsymbol{\varphi}_u$ is the column vector composed of all φ_{ij} and $\boldsymbol{\varphi}_p$ is the column vector composed of all φ_k. From definition (4.278) of the Kochin function $h_{ij}(\beta)$, it follows that

$$h_{ij}(\beta \pm \pi) = -\frac{k}{D(kh)} I[e(kz)e^{-ik(x\cos\beta + y\sin\beta)}, \varphi_{ij}]. \tag{7.175}$$

Thus, with an incident wave

$$\hat{\phi}_0 = -\frac{g}{i\omega} Ae(kz)e^{-ik(x\cos\beta + y\sin\beta)} \tag{7.176}$$

we have for the excitation force

$$\hat{F}_{ij} = \hat{F}_{ij}(\beta) = i\omega\rho I(\hat{\phi}_0, \varphi_{ij}) = \frac{\rho g D(kh)}{k} h_{ij}(\beta \pm \pi)A. \tag{7.177}$$

Similarly, we have for the excitation volume flow

$$\hat{Q}_k = \hat{Q}_k(\beta) = -i\omega\rho I(\hat{\phi}_0, \varphi_k) = -\frac{\rho g D(kh)}{k} h_k(\beta \pm \pi)A. \tag{7.178}$$

The Kochin functions $h_{ij}(\theta)$ and $h_k(\theta)$ for the radiated waves may, according to Eq. (4.282), be expressed by the global far-field coefficients $a_{ij}(\theta)$ and $a_k(\theta)$. Alternatively, they may be expressed by the local far-field coefficients $b_{ij}(\theta)$ and $b_k(\theta)$, using Eq. (4.271). Thus, in terms of the far-field coefficients we have

$$\begin{aligned} h_{ij}(\beta \pm \pi) &= e^{i\pi/4}\sqrt{2\pi} a_{ij}(\beta \pm \pi) \\ &= e^{i\pi/4}\sqrt{2\pi} b_{ij}(\beta \pm \pi) e^{ikx_i \cos(\beta\pm\pi) + iky_i \sin(\beta\pm\pi)} \\ &= e^{i\pi/4}\sqrt{2\pi} b_{ij}(\beta \pm \pi) e^{-ik(x_i \cos\beta + y_i \sin\beta)}. \end{aligned} \tag{7.179}$$

Hence, in terms of the local far-field coefficient for the radiated wave, we may rewrite the excitation force as [also see Eq. (5.203)]

$$\hat{F}_{ij}(\beta) = \frac{\rho g D(kh)}{k} \sqrt{2\pi} b_{ij}(\beta \pm \pi) A_i e^{i\pi/4}, \tag{7.180}$$

with

$$A_i = \eta_0(x_i, y_i) = Ae^{-ik(x_i \cos\beta + y_i \sin\beta)}. \tag{7.181}$$

Thus, we have

$$\hat{F}_{ij}(\beta) = f_{ij}(\beta)A_i, \tag{7.182}$$

where the excitation-force coefficient is

$$\begin{aligned} f_{ij}(\beta) &= \frac{\rho g D(kh)}{k} \sqrt{2\pi} b_{ij}(\beta \pm \pi) e^{i\pi/4} \\ &= \frac{\rho g D(kh)}{k} h_{ij}(\beta \pm \pi) e^{ik(x_i \cos\beta + y_i \sin\beta)}. \end{aligned} \tag{7.183}$$

Similarly we have for the excitation volume flow

$$\hat{Q}_k(\beta) = q_k(\beta)A_k, \tag{7.184}$$

with

$$A_k = \eta_0(x_k, y_k). \tag{7.185}$$

The excitation-volume-flow coefficient is

$$q_k(\beta) = -\frac{\rho g D(kh)}{k}\sqrt{2\pi}\,b_k(\beta \pm \pi)e^{i\pi/4}. \tag{7.186}$$

Note that by measuring the excitation coefficients with a certain direction of wave incidence β, we also obtain the corresponding far-field coefficients for waves radiated in the opposite direction (see Figure 5.12).

To summarise our generalisation of the Haskind relation, we define a $6N_i$-dimensional and an N_k-dimensional Kochin-function (column) vector $\mathbf{h}_u(\theta)$ and $\mathbf{h}_p(\theta)$ composed of all $h_{ij}(\theta)$ and all $h_k(\theta)$, respectively. In vector notation, Eqs. (7.177) and (7.178) for the excitation parameters may then be written as

$$\hat{\boldsymbol{\kappa}} = \hat{\boldsymbol{\kappa}}(\beta) = \begin{bmatrix} \hat{\mathbf{F}}(\beta) \\ -\hat{\mathbf{Q}}(\beta) \end{bmatrix} = \frac{\rho g D(kh)}{k} A \begin{bmatrix} \mathbf{h}_u(\beta \pm \pi) \\ \mathbf{h}_p(\beta \pm \pi) \end{bmatrix}$$

$$\equiv \frac{\rho g D(kh)}{k} A \mathbf{h}(\beta \pm \pi). \tag{7.187}$$

Here we have introduced the N-dimensional Kochin function vector $\mathbf{h}(\theta)$. According to Eq. (4.282) it is related to a global far-field coefficient vector

$$\mathbf{a}(\theta) = \begin{bmatrix} \mathbf{a}_u(\theta) \\ \mathbf{a}_p(\theta) \end{bmatrix} \tag{7.188}$$

as

$$\mathbf{h}(\theta) = \sqrt{2\pi}\,\mathbf{a}(\theta)e^{i\pi/4}. \tag{7.189}$$

The proportionality coefficient vectors φ_u and φ_p, which were introduced in connection with Eq. (7.113), are related to the global far-field coefficient vector through the asymptotic approximation

$$\boldsymbol{\varphi} \equiv \begin{bmatrix} \boldsymbol{\varphi}_u \\ \boldsymbol{\varphi}_p \end{bmatrix} \sim \mathbf{a}(\theta)e(kz)(kr)^{-1/2}e^{-ikr} \tag{7.190}$$

as $kr \to \infty$. Compare. Eq. (4.246).

7.2.7 Reciprocity Relations for the Radiation Damping Matrix

The radiation damping matrix $\boldsymbol{\Delta}$ is expressed in terms of matrices $\mathbf{G}$, $\mathbf{R}$ and $\mathbf{J}$ according to definition (7.91). Using the above results, we shall derive reciprocity relations which express these matrices in terms of far-field coefficients, Kochin functions, or excitation parameters.

By using the general relation (4.254) – as well as its complex conjugate – for waves $\psi_{i,j}$ satisfying the radiation condition, we find that Eqs. (7.140), (7.153) and

(7.164) give

$$R = \frac{\omega\rho\,D\,(kh)}{2k} \int_0^{2\pi} \mathbf{a}_u(\theta)\,\mathbf{a}_u^\dagger(\theta)\,d\theta = \frac{\omega\rho\,D\,(kh)}{2k} \int_0^{2\pi} \mathbf{a}_u^*(\theta)\,\mathbf{a}_u^T(\theta)\,d\theta,$$

(7.191)

$$G = \frac{\omega\rho\,D\,(kh)}{2k} \int_0^{2\pi} \mathbf{a}_p(\theta)\,\mathbf{a}_p^\dagger(\theta)\,d\theta = \frac{\omega\rho\,D\,(kh)}{2k} \int_0^{2\pi} \mathbf{a}_p^*(\theta)\,\mathbf{a}_p^T(\theta)\,d\theta,$$

(7.192)

$$i\mathbf{J} = \frac{\omega\rho\,D\,(kh)}{2k} \int_0^{2\pi} \mathbf{a}_u^*(\theta)\,\mathbf{a}_p^T(\theta)\,d\theta = -\frac{\omega\rho\,D\,(kh)}{2k} \int_0^{2\pi} \mathbf{a}_u(\theta)\,\mathbf{a}_p^\dagger(\theta)\,d\theta,$$

(7.193)

respectively. Observe that we have assumed ω and k to be real, and that $\mathbf{R}$, $\mathbf{G}$ and $\mathbf{J}$, by definition, are real. Furthermore, by transposing we get

$$i\mathbf{J}^T = \frac{\omega\rho\,D\,(kh)}{2k} \int_0^{2\pi} \mathbf{a}_p(\theta)\,\mathbf{a}_u^\dagger(\theta)\,d\theta = -\frac{\omega\rho\,D\,(kh)}{2k} \int_0^{2\pi} \mathbf{a}_p^*(\theta)\,\mathbf{a}_u^T(\theta)\,d\theta.$$

(7.194)

By using the above expression in definition (7.91) of the radiation damping matrix, we have similarly

$$\Delta = \begin{bmatrix} \mathbf{R} & -i\mathbf{J} \\ i\mathbf{J}^T & \mathbf{G} \end{bmatrix} = \frac{\omega\rho\,D\,(kh)}{2k} \int_0^{2\pi} \begin{bmatrix} \mathbf{a}_u(\theta)\,\mathbf{a}_u^\dagger(\theta) & \mathbf{a}_u(\theta)\,\mathbf{a}_p^\dagger(\theta) \\ \mathbf{a}_p(\theta)\,\mathbf{a}_u^\dagger(\theta) & \mathbf{a}_p(\theta)\,\mathbf{a}_p^\dagger(\theta) \end{bmatrix} d\theta.$$

(7.195)

Using Eqs. (7.188), (7.189) and (7.187), we may write this as

$$\Delta = \frac{\omega\rho\,D\,(kh)}{2k} \int_0^{2\pi} \mathbf{a}(\theta)\,\mathbf{a}^\dagger(\theta)\,d\theta$$

$$= \frac{\omega\rho\,D\,(kh)}{4\pi k} \int_0^{2\pi} \mathbf{h}(\theta)\,\mathbf{h}^\dagger(\theta)\,d\theta$$

$$= \frac{\omega k}{4\pi\rho g^2 D\,(kh)|A|^2} \int_{-\pi}^{\pi} \hat{\boldsymbol{\kappa}}(\beta)\,\hat{\boldsymbol{\kappa}}^\dagger(\beta)\,d\beta.$$

(7.196)

The factor in front of the last integral may also be written as

$$\frac{\omega k}{4\pi\rho g^2 D\,(kh)|A|^2} = \frac{k}{8\pi\rho g v_g|A|^2} = \frac{k}{16\pi\,J}$$

(7.197)

according to Eqs. (4.110) and (4.136). Here v_g is the group velocity and J is the wave-energy transport (wave-power level).

For the case with only oscillating bodies, and no OWC, Eq. (7.196) specialises to Eqs. (5.140) and (5.149), and then $\Delta = \mathbf{R}$ is a real symmetrical matrix. Also Eqs. (7.32), (7.50) and (7.51) are special cases of Eq. (7.196) when the system contains just one OWC.

However, for systems which contain at least one oscillating body and one OWC, the radiation damping matrix $\mathbf{\Delta}$ is complex. That it is, in fact, a Hermitian matrix (i.e., $\mathbf{\Delta} = \mathbf{\Delta}^\dagger$) is easily demonstrated by transposing and conjugating Eq. (7.196).

All diagonal elements of $\mathbf{\Delta}$ are real, and they are also non-negative. In particular,

$$R_{ij,ij} = \frac{\omega\rho\, D\,(kh)}{2k} \int_0^{2\pi} |a_{ij}(\theta)|^2\, d\theta = \frac{k}{16\pi J} \int_{-\pi}^{\pi} |\hat{F}_{ij}(\beta)|^2\, d\beta \geq 0, \quad (7.198)$$

$$G_{kk} = \frac{\omega\rho\, D\,(kh)}{2k} \int_0^{2\pi} |a_k(\theta)|^2\, d\theta = \frac{k}{16\pi J} \int_{-\pi}^{\pi} |\hat{Q}_k(\beta)|^2\, d\beta \geq 0. \quad (7.199)$$

Hence, the sum of all diagonal elements, that is, the so-called trace of the radiation damping matrix $\mathbf{\Delta}$, is necessarily real and non-negative. The trace of the matrix equals also the sum of all eigenvalues of the matrix (cf. Pease,[12] p. 86). This is consistent with inequality (7.105), which means that Hermitian matrix $\mathbf{\Delta}$ is positive semidefinite. Such a matrix has only real non-negative eigenvalues (cf. Pease,[12] p. 109).

7.2.8 Axisymmetric Systems

In Section 5.7 we considered axisymmetric bodies interacting with radiated or incident waves. Let us now discuss a system of concentric axisymmetric bodies and OWCs. We shall, however, restrict the discussion to the case of isotropic radiation. That is, we exclude body oscillations apart from the heave mode 2 (i.e., $j = 3$). For this axisymmetric case, the far-field coefficients and Kochin functions are independent of direction θ. Moreover, excitation parameters are independent of direction β of wave incidence. Hence we have

$$\mathbf{\Delta} = \frac{\pi\omega\rho\, D\,(kh)}{k}\mathbf{aa}^\dagger = \frac{\omega\rho\, D\,(kh)}{2k}\mathbf{hh}^\dagger = \frac{k}{8J}\hat{\boldsymbol{\kappa}}\hat{\boldsymbol{\kappa}}^\dagger \quad (7.200)$$

according to Eqs. (7.196) and (7.197). In particular we have

$$R_{i3,i'3} = \frac{k}{8J}\hat{F}_{i3}\hat{F}_{i'3}^*, \quad (7.201)$$

$$G_{k,k'} = \frac{k}{8J}\hat{Q}_k\hat{Q}_{k'}^*, \quad (7.202)$$

$$iJ_{i3,k} = -iJ_{k,i3} = -\frac{k}{8J}\hat{F}_{i3}\hat{Q}_k^*. \quad (7.203)$$

A floating structure containing two concentric OWCs is indicated in Figure 7.6. If the system in addition contains, for example, a concentric, toroidally shaped, heaving body (not shown in the figure), then indices i and k in Eqs. (7.201)–(7.203) may both take the values 1 or 2 ($N_i = 2$ and $N_k = 2$).

However, let us for a while consider a simpler case, that of two concentric OWCs in a fixed structure. (Assume, e.g., that the structure of Figure 7.6 is supplied with a concentric cylindrical strut – not shown – which holds the structure

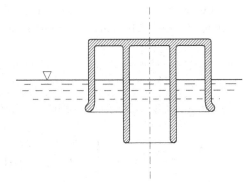

Figure 7.6: Axisymmetric floating structure containing two concentric OWCs.

fixed relative to the sea bed.) According to Eq. (7.202), the radiation-conductance matrix is

$$\mathbf{G} = \frac{k}{8J} \begin{bmatrix} |\hat{Q}_1|^2 & \hat{Q}_1 \hat{Q}_2^* \\ \hat{Q}_1^* \hat{Q}_2 & |\hat{Q}_2|^2 \end{bmatrix} \tag{7.204}$$

in this case. The determinant vanishes because

$$\frac{8J}{k}|\mathbf{G}| = \begin{vmatrix} |\hat{Q}_1|^2 & \hat{Q}_1 \hat{Q}_2^* \\ \hat{Q}_1^* \hat{Q}_2 & |\hat{Q}_2|^2 \end{vmatrix} = |\hat{Q}_1|^2 |\hat{Q}_2|^2 - \hat{Q}_1 \hat{Q}_2^* \hat{Q}_1^* \hat{Q}_2 = 0. \tag{7.205}$$

Hence $\mathbf{G}^{-1}$ does not exist and $\mathbf{G}$ is singular. In general if there is more than one OWC, $\mathbf{G}$ is singular in the axisymmetric case. Thus we have

$$G_{12} = \pm\sqrt{G_{11} G_{22}}. \tag{7.206}$$

Let us consider the ratio between the complex amplitudes of the excitation volume flows of two concentric OWCs. Using Eq. (7.202) we find

$$\frac{\hat{Q}_k}{\hat{Q}_{k'}} = \frac{\hat{Q}_k \hat{Q}_{k'}^*}{|\hat{Q}_{k'}|^2} = \frac{8J}{k} \frac{G_{k,k'}}{|\hat{Q}_{k'}|^2}. \tag{7.207}$$

It thus appears that $\hat{Q}_k/\hat{Q}_{k'}$ is real. If it is positive ($G_{k,k'} > 0$), the two excitation volume flows $\hat{Q}_k$ and $\hat{Q}_{k'}$ have the same phase. If $G_{k,k'} < 0$, they are in opposite phases.

Returning now to Eq. (7.201), we find it easy to verify that the heave excitation forces $\hat{F}_{i3}$ and $\hat{F}_{i'3}$ are in equal or opposite phases according to the sign of the non-diagonal element $R_{i3,i'3}$ of the radiation-resistance matrix. Note that $R_{i3,i'3} = \pm(R_{i3,i3} R_{i'3,i'3})^{1/2}$. Furthermore, using Eq. (7.203), we see that $\hat{Q}_k/\hat{F}_{i3}$ is purely imaginary, which means that the excitation heave force and the excitation volume flow have phases differing $\pi/2$ from each other.

Let us finally consider an example with $N_i = N_k = 1$, namely an OWC contained in a heaving axisymmetric body (such as the system shown in Figure 7.6, except with one of the two vertical "tubes" removed). For this case, vectors $\boldsymbol{\kappa}$ and υ defined by Eq. (7.101) are two dimensional:

$$\boldsymbol{\kappa} = \begin{bmatrix} \hat{F}_{i3} \\ -\hat{Q}_k \end{bmatrix}, \qquad \hat{\boldsymbol{v}} = \begin{bmatrix} \hat{u}_{i3} \\ -\hat{p}_k \end{bmatrix}. \tag{7.208}$$

According to Eq. (7.200), the radiation damping matrix $\boldsymbol{\Delta}$ is

$$\boldsymbol{\Delta} = \frac{k}{8J}\boldsymbol{\kappa}\boldsymbol{\kappa}^{\dagger} = \frac{k}{8J}\begin{bmatrix} |\hat{F}_{i3}|^2 & -\hat{F}_{i3}\hat{Q}_k^* \\ -\hat{Q}_k\hat{F}_{i3}^* & |\hat{Q}_k|^2 \end{bmatrix} = \begin{bmatrix} R_{i3,i3} & -iJ_{i3,k} \\ iJ_{i3,k} & G_{kk} \end{bmatrix}. \tag{7.209}$$

Note that the radiation damping matrix is singular because its determinant vanishes:

$$\begin{vmatrix} |\hat{F}_{i3}|^2 & -\hat{F}_{i3}\hat{Q}_k^* \\ -\hat{Q}_k\hat{F}_{i3}^* & |\hat{Q}_k|^2 \end{vmatrix} = |\hat{F}_{i3}|^2|\hat{Q}_k|^2 - \hat{F}_{i3}\hat{Q}_k^*\hat{Q}_k\hat{F}_{i3}^* = 0. \tag{7.210}$$

Hence we have

$$J_{i3,k}^2 = R_{i3,i3}G_{kk}, \quad J_{i3,k} = \pm\sqrt{R_{i3,i3}G_{kk}}. \tag{7.211}$$

For a system with dimension small compared with the wavelength, we expect heave force $\hat{F}_{i3}$ and internal water surface displacement $\hat{Q}_k/i\omega$ to be closely in phase with the elevation of the incident wave. Then the real quantity $(\hat{Q}_k/i\omega)/\hat{F}_{i3}$ is positive. Hence, in this case

$$J_{i3,k} = -i\frac{k}{8J}\hat{Q}_k^*\hat{F}_{i3} = -i\frac{k}{8J}\frac{|\hat{Q}_k|^2\hat{F}_{i3}}{\hat{Q}_k} = \frac{k}{8J}|\hat{Q}_k|^2\frac{\hat{F}_{i3}}{\hat{Q}_k/i} \tag{7.212}$$

is positive, and then

$$J_{i3,k} = \sqrt{R_{i3,i3}G_{kk}}. \tag{7.213}$$

We have observed that the radiation damping matrix $\boldsymbol{\Delta}$ is singular for a system in which there is more than one mode of oscillation, for which the wave radiation is isotropic. We find a reasonable explanation for this fact if we consider maximum wave-power absorption from a plane incident wave, as discussed in Subsection 7.2.2, which shows that the optimum condition (7.109) represents a system of algebraic equations which is indeterminate if $\boldsymbol{\Delta}$ is singular. Physically, maximum power absorption is related to the optimum destructive interference between the incident plane wave and an isotropically radiated wave from the wave-power-absorbing system. A certain resulting radiated wave is required. It is, however, quite arbitrary how the various oscillator modes of the absorbing system contribute individually to this required isotropically radiated wave.

Problems

Problem 7.1: Circular Cylindrical OWC

The lower end of a vertical thin-walled tube of radius a just penetrates the free water surface. The tube constitutes the air chamber above an OWC. Assume the fluid to be ideal and neglect diffraction. Derive an expression for the excitation volume flow $\hat{Q}_e$ when there is an incident plane wave

$$\hat{\phi}_0 = \frac{-g}{i\omega} e(kz)\hat{\eta}_0,$$

$$\hat{\eta}_0 = Ae^{-ikx}.$$

Let the z axis coincide with the axis of the tube.

Furthermore, find an expression for the radiation conductance G_{kk} of the OWC. Simplify the expression for G_{kk} for the case of deep water ($kh \ll 1$).

[Hint: use reciprocity relation (7.199):

$$G_{kk} = \frac{k}{16\pi J} \int_{-\pi}^{\pi} |\hat{Q}_{e,k}(\beta)|^2 \, d\beta.$$

We may also use the integrals

$$\int_0^x t J_0(t) \, dt = x J_1(x), \qquad \int_0^{2\pi} e^{ix \cos t} \, dt = 2\pi J_0(x),$$

where $J_0(x)$ and $J_1(x)$ are Bessel functions of the first kind, and of order zero and one, respectively.]

Problem 7.2: Air Compressibility

Use the expression

$$p_a V_a^\kappa = (p_a + \Delta p_a)(V_a + \Delta V_a)^\kappa$$

for adiabatic compression to derive the last term in Eq. (7.62). Assume that Δp_a and ΔV_a are so small that small terms of higher order may be neglected.

Problem 7.3: Linear Electric 2-Port

Consider an electric "2-port" (also termed "4-pole"). The input voltage, input current, output voltage and output current have complex amplitudes U_1, I_1, U_2

and I_2, respectively. Assuming that the electric circuit is linear, we may express the voltages in terms of the currents as

$$\begin{bmatrix} U_1 \\ U_2 \end{bmatrix} = \mathbf{Z} \begin{bmatrix} I_1 \\ I_2 \end{bmatrix},$$

where

$$\mathbf{Z} = \begin{bmatrix} Z_{11} & Z_{12} \\ Z_{21} & Z_{22} \end{bmatrix}$$

is the impedance matrix of the 2-port.

By solving the linear algebraic equation we find it possible to write this relation in some other ways, such as

$$\begin{bmatrix} I_1 \\ I_2 \end{bmatrix} = \mathbf{Y} \begin{bmatrix} U_1 \\ U_2 \end{bmatrix}, \quad \begin{bmatrix} U_2 \\ I_2 \end{bmatrix} = \mathbf{K} \begin{bmatrix} U_1 \\ I_1 \end{bmatrix}, \quad \text{or} \quad \begin{bmatrix} U_1 \\ I_2 \end{bmatrix} = \mathbf{H} \begin{bmatrix} I_1 \\ U_2 \end{bmatrix},$$

where $\mathbf{Y}$ is the "admittance" matrix, $\mathbf{K}$ the "chain" matrix and $\mathbf{H}$ the "hybrid" matrix. Express the four matrix elements of $\mathbf{Y}$ and of $\mathbf{H}$ in terms of the four matrix elements of $\mathbf{Z}$.

If the impedance matrix is symmetrical, that is, if $Z_{21} = Z_{12}$, then the linear electric circuit is said to be reciprocal. What is the corresponding condition on the hybrid matrix?

One possible electric analogy of a mechanical system is to consider force and velocity as analogous quantities to voltage and current, respectively. For a mechanical system consisting of one OWC and a single one-mode oscillating body, the above hybrid matrix may be considered as the electric analogue of the square radiation matrix in Eq. (7.87). Discuss this analogy.

Problem 7.4: Maximum Absorbed Power

For a system of OWCs and oscillating bodies, the absorbed power may be written as [cf. Eq. (7.106)]

$$P = P(\hat{\boldsymbol{v}}) = \tfrac{1}{4}(\hat{\boldsymbol{\kappa}}^T \hat{\boldsymbol{v}}^* + \hat{\boldsymbol{\kappa}}^\dagger \hat{\boldsymbol{v}}) - \tfrac{1}{2}\hat{\boldsymbol{v}}^\dagger \mathbf{\Delta} \hat{\boldsymbol{v}}.$$

Assuming that the radiation damping matrix $\mathbf{\Delta}$ is non-singular (i.e., $\mathbf{\Delta}^{-1}$ exists), prove that

$$P_{\mathrm{MAX}} = \tfrac{1}{8} \hat{\boldsymbol{\kappa}}^\dagger \mathbf{\Delta}^{-1} \hat{\boldsymbol{\kappa}}$$

and derive an explicit expression for the optimum oscillation vector $\hat{\boldsymbol{v}}_0$ in terms of the excitation vector $\hat{\boldsymbol{\kappa}}$. Observe that $\mathbf{\Delta}$ is a complex, Hermitian, and positive semidefinite matrix. [Hint: introduce the vector $\hat{\boldsymbol{\delta}} = \hat{\boldsymbol{v}} - \tfrac{1}{2}(\mathbf{\Delta}^*)^{-1} \hat{\boldsymbol{\kappa}}$ and consider the non-negative quantity $\tfrac{1}{2}\hat{\boldsymbol{\delta}}^T \mathbf{\Delta} \hat{\boldsymbol{\delta}}$, which would have been the radiated power if $\hat{\boldsymbol{\delta}}$ had been the complex oscillation amplitude vector.]

Problem 7.5: Maximum Absorbed Wave Power

(a) On the basis of Eq. (7.106), show that the maximum absorbed power is as given by Eq. (7.107) where the optimum oscillation corresponds to U satisfying Eq. (7.109). [Hint: make the necessary generalisation of the derivation of Eq. (6.64) from Eq. (6.63).]

(b) If the radiation damping matrix Δ is singular, Eq. (7.109) has an infinity of possible solutions for U. Assume that U_1 and U_2 are two different solutions. Show that $P(U_1) = P(U_2)$. Thus, in spite of the indeterminateness of Eq. (7.109), the maximum absorbed power P_{MAX} is unambiguous. [Hint: use Eqs. (7.104), (7.107) and (7.109).]

Bibliography

1. C. C. Mei. *The Applied Dynamics of Ocean Surface Waves*. World Scientific Publishing, Singapore, 1983. Second printing, 1989 (ISBN 9971-50-789-7 or 9971-50-773-0).
2. O. M. Faltinsen. *Sea Loads on Ships and Offshore Structures*. Cambridge University Press, Cambridge, UK, 1990.
3. T. Sarpkaya and M. Isaacson. *Mechanics of Wave Forces on Offshore Structures*. Van Nostrand Reinhold, New York, 1981.
4. S. K. Chakrabarti. *Hydrodynamics of Offshore Structures*. Springer-Verlag, Berlin, 1987.
5. S. H. Salter. World progress in wave energy – 1988. *International Journal of Ambient Energy*, 10(1):3–24, 1989.
6. A. D. Carmichael and J. Falnes. State of the art in wave power recovery. In Richard J. Seymour, editor, *Ocean Energy Recovery*, Chap. 8, pp. 182–212. American Society of Civil Engineers, New York, 1992. (ISBN 0-87262-894-9).
7. Engineering Committee on Oceanic Resources, Working Group on Wave Energy Conversion. *Waves: A Vast Renewable Energy Resource*. Publication in preparation.
8. H. W. Bode. *Network Analysis and Feedback Amplifier Design*. Van Nostrand, New York, 1945.
9. B. Friedland. *Control System Design: An Introduction to State-Space Methods*. McGraw-Hill, New York, 1986.
10. C. Moler and C. Van Loan. Nineteen dubious ways to compute the exponential of a matrix. *Society for Industrial and Applied Mathematics Review*, 20:801–836, 1978.
11. P. Hr. Petkov, N. D. Christov, and M. M. Konstantinov. *Computational Methods for Linear Control Systems*. Prentice-Hall, Englewood Cliffs, NJ, 1991.
12. M. C. Pease. *Methods of Matrix Algebra*. Academic Press, New York, 1965.
13. A. Papoulis. *The Fourier Integral and Its Applications*. McGraw-Hill, New York, 1962.
14. R. N. Bracewell. *The Fourier Transform and Its Applications*. McGraw Hill, New York, 1986.
15. H. A. Kramers. La diffusion de la lumiere par les atomes. In *Atti del Congresso Internazionale dei Fisici*, Vol. II, pp. 545–557, Como, settembre 1927. Nicola Zanichelli, Bologna, 1928.
16. R. de L. Kronig. On the theory of dispersion of X-rays. *Journal of the Optical Society of America*, 12(6):547–557, 1926.

17. L. E. Kinsler and A. R. Frey. *Fundamentals of Acoustics*. Wiley, New York, third edition, 1982.

18. K. H. Panofsky and M. Phillips. *Classical Electricity and Magnetism*. Addison-Wesley, Reading, MA, 1955.

19. P. McIver and D. V. Evans. The occurrence of negative added mass in free-surface problems involving submerged oscillating bodies. *Journal of Engineering Mathematics*, 18:7–22, 1984.

20. J. N. Miles. Resonant response of harbours: an equivalent circuit analysis. *Journal of Fluid Mechanics*, 46:241–265, 1971.

21. J. Falnes. Radiation impedance matrix and optimum power absorption for interacting oscillators in surface waves. *Applied Ocean Research*, 2(2):75–80, 1980.

22. E. Meyer and E. G. Neumann. *Physikalische und technische Akustik*, pp. 180–182. Vierweg, Braunschweig, 1967.

23. E. Titchmarsh. *Eigenfunction Expansion Associated with Second-order Differential Equations*. Oxford University Press, New York, 1946.

24. J. N. Newman. *Marine Hydrodynamics*. MIT Press, Cambridge, MA, 1977.

25. M. S. Longuet-Higgins. The mean forces exerted by waves on floating or submerged bodies, with application to sand bars and wave power machines. *Proceedings of the Royal Society of London*, A352:463–480, 1977.

26. Y. Goda. *Random Seas and Design of Maritime Structures*. University of Tokyo Press, Tokyo, Japan, 1985. (ISBN 4-13-068110-9).

27. M. J. Tucker. *Waves in Ocean Engineering: Measurements, Analysis, Interpretation*. Ellis Horwood, New York, 1991. (ISBN 0-13-932955-2).

28. M. Abramowitz and I. A. Stegun. *Handbook of Mathematical Functions*. Dover Publications, New York, 1965.

29. J. V. Wehausen and E. V. Laitone. Surface waves. In S. Flügge, editor, *Encyclopedia of Physics*, Vol. IX, pp. 446–778. Springer-Verlag, Berlin, 1960.

30. J. N. Newman. The interaction of stationary vessels with regular waves. In *Proceedings of the 11th Symposium on Naval Hydrodynamics*, pp. 491–501. Mechanical Engineering Pub., London, 1976.

31. J. Falnes. On non-causal impulse response functions related to propagating water waves. *Applied Ocean Research*, 17(6):379–389, 1995.

32. B. King. *Time-Domain Analysis of Wave Exciting Forces on Ships and Bodies*. Technical Report 306. The Department of Naval Architecture and Marine Engineering, University of Michigan, Ann Arbor, MI, 1987.

33. F. T. Korsmeyer. The time domain diffraction problem. In *The Sixth International Workshop on Water Waves and Floating Bodies*. Woods Hole, MA, 1991.

34. S. Naito and S. Nakamura. Wave energy absorption in irregular waves by feed-forward control system. In D. V. Evans and A. F. de O. Falcão, editors, *Hydrodynamics of Ocean Wave-Energy Utilization*, pp. 169–280. Springer-Verlag, Berlin, 1986. IUTAM Symposium, Lisbon, 1985.

35. E. L. Morris, H. K. Zienkiewich, M. M. A. Pourzanjani, J. O. Flower, and M. R. Belmont. Techniques for sea state prediction. In *Second International Conference on Manoeuvring and Control of Marine Craft*, pp. 547–569. Computational Mechanics, Southampton, UK, 1992.

36. S. E. Sand. *Three-Dimensional Deterministic Structure of Ocean Waves*. Technical

Report 24. Institute of Hydrodynamics and Hydraulic Engineering (ISVA), Technical University of Denmark (DTH), DK-2800, Lyngby, 1979.

37. A. Erdélyi, editor. *Tables of Integral Transforms*. McGraw-Hill, New York, 1954.

38. M. J. L. Greenhow. The hydrodynamic interactions of spherical wave-power devices in surface waves. In B. Count, editor, *Power from Sea Waves*, pp. 287–343. Academic Press, London, 1980.

39. Å. Kyllingstad. *Approximate Analysis Concerning Wave-Power Absorption by Hydrodynamically Interacting Buoys*. Ph.D. thesis. Institutt for eksperimentalfysikk, NTH, Trondheim, Norway, 1982.

40. M. A. Srokosz. Some relations for bodies in a canal, with application to wave-power absorption. *Journal of Fluid Mechanics*, 99:145–162, 1980.

41. J. Falnes and K. Budal. Wave-power absorption by parallel rows of interacting oscillating bodies. *Applied Ocean Research*, 4(4):194–207, 1982.

42. T. Havelock. Waves due to a floating sphere making periodic heaving oscillations. *Proceedings of the Royal Society of London*, 231A:1–7, 1955.

43. A. Hulme. The wave forces acting on a floating hemisphere undergoing forced periodic oscillations. *Journal of Fluid Mechanics*, 121:443–463, 1982.

44. Håvard Eidsmoen. Hydrodynamic parameters for a two-body axisymmetric system. *Applied Ocean Research*, 17(2):103–115, 1995.

45. R. Eatock Taylor and E. R. Jeffreys. Variability of hydrodynamic load predictions for a tension leg platform. *Ocean Engineering*, 13(5):449–490, 1986.

46. *WAMIT User Manual*. (http://www.wamit.com).

47. F. T. Korsmeyer, C. H. Lee, J. N. Newman, and P. D. Sclavounos. The analysis of wave effects on tension-leg platforms. In *Proceedings of the Seventh International Conference on Offshore Mechanics and Arctic Engineering*, Vol. 2, pp. 1–14. American Society of Mechanical Engineers, New York, 1988.

48. W. E. Cummins. The impulse response function and ship motions. *Schiffstechnik*, 9:101–109, 1962.

49. J. Kotik and V. Mangulis. On the Kramers-Kronig relations for ship motions. *International Shipbuilding Progress*, 9(97): 9, 1962.

50. Martin Greenhow. A note on the high-frequency limits of a floating body. *Journal of Ship Research*, 28:226–228, 1984.

51. B. M. Count and E. R. Jefferys. Wave power, the primary interface. In *Proceedings of the 13th Symposium on Naval Hydrodynamics*, Paper 8, pp. 1–10. The Shipbuilding Research Association of Japan, Tokyo, 1980.

52. L. J. Tick. Differential equations with frequency-dependent coefficients. *Journal of Ship Research*, 3(3):45–46, 1959.

53. S. Goldman. *Transformation Calculus and Electric Transients*. Constable and Co., London, 1949.

54. S. P. Timoshenko and J. M. Gere. *Mechanics of Materials*. Van Nostrand, New York, 1973.

55. M. D. Haskind. The exciting forces and wetting of ships (in Russian). *Izvestiya Akademii Nauk SSSR, Otdelenie Tekhnicheskikh Nauk*, 7:65–79, 1957.

56. J. N. Newman. The exciting forces on fixed bodies in waves. *Journal of Ship Research*, 6(3):10–17, 1962.

57. D. V. Evans. Some analytic results for two and three dimensional wave-energy

absorbers. In B. Count, editor, *Power from Sea Waves*, pp. 213–249. Academic Press, London, 1980.

58. K. Budal. Theory of absorption of wave power by a system of interacting bodies. *Journal of Ship Research*, 21:248–253, 1977.

59. Å. Kyllingstad. A low-scattering approximation for the hydrodynamic interactions of small wave-power devices. *Applied Ocean Research*, 6:132–139, 1984.

60. T. Arzel, T. Bjarte-Larsson, and J. Falnes. Hydrodynamic parameters for a floating WEC force-reacting against a submerged body. Fourth European Wave Energy Conference *Proceedings of an International Conference held at Aalborg University in Denmark 4–6 December 2000*, pp. 265–272. Energy Centre Denmark, Danish Technological Institute, Copenhagen, 2001 (ISBN 87-90074-09-2).

61. K. Budal. Floating structure with heave motion reduced by force compensation. In *Proceedings of the Fourth International Offshore Mechanics and Arctic Engineering Symposium*, pp. 92–101. American Society of Mechanical Engineers, New York, 1985.

62. J. Falnes. Wave-energy conversion through relative motion between two single-mode oscillating bodies. *Journal of Offshore Mechanics and Arctic Engineering (ASME Transactions)*, 121:32–38, 1999.

63. M. McCormick. *Ocean Wave Energy Conversion*. Wiley, New York, 1981.

64. J. V. Wehausen. Causality and the radiation condition. *Journal of Engineering Mathematics*, 26:153–158, 1992.

65. D. V. Evans. Power from water waves. *Annual Review of Fluid Mechanics*, 13:157–187, 1981.

66. J. Falnes and K. Budal. Wave-power absorption by point absorbers. *Norwegian Maritime Research*, 6(4):2–11, 1978.

67. K. Budal and J. Falnes. Interacting point absorbers with controlled motion. In B. Count, editor, *Power from Sea Waves*, pp. 381–399. Academic Press, London, 1980. (ISBN 0-12-193550-7).

68. J. Falnes. Small is beautiful: how to make wave energy economic. In *1993 European Wave Energy Symposium*, pp. 367–372, NEL-Renewable Energy, East Kilbride, Scotland, UK, 1994 (ISBN 0-903640-84-8).

69. S. H. Salter. Power conversion systems for ducks. In *Proceedings of the International Conference on Future Energy Concepts*, Publication 171, pp. 100–108, Institution of Electrical Engineers, London, 1979.

70. P. Nebel. Maximizing the efficiency of wave-energy plants using complex-conjugate control. *Journal of Systems and Control Engineering*, 206(4):225–236, 1992.

71. S. H. Salter, D. C. Jeffery, and J. R. M. Taylor. The architecture of nodding duck wave power generators. *The Naval Architect*, pp. 1:21–24, 1976.

72. K. Budal and J. Falnes. Optimum operation of improved wave-power converter. *Marine Science Communications*, 3:133–159, 1977.

73. J. H. Milgram. Active water-wave absorbers. *Journal of Fluid Mechanics*, 43:845–859, 1970.

74. K. Budal, J. Falnes, T. Hals, L. C. Iversen, and T. Onshus. Model experiment with a phase controlled point absorber. In *Proceedings of the Second International Symposium on Wave and Tidal Energy*, pp. 191–206. BHRA Fluid Engineering, Cranford, Bedford, UK, 1981 (ISBN 0-906085-43-9).

75. K. Budal, J. Falnes, L. C. Iversen, P. M. Lillebekken, G. Oltedal, T. Hals, T. Onshus, and

A. S. Høy. The Norwegian wave-power buoy project. In H. Berge, editor, *Proceedings of the Second International Symposium on Wave Energy Utilization*, pp. 323–344. Tapir, Trondheim, Norway, 1982. (ISBN 82-519-0478-1).

76. J. N. B. A. Perdigão and A. J. N. A. Sarmento. A phase control strategy for OWC devices in irregular seas. In J. Grue, editor, *The Fourth International Workshop on Water Waves and Floating Bodies*, pp. 205–209. University of Oslo, Oslo, 1989.

77. J. N. B. A. Perdigão. *Reactive-Control Strategies for an Oscillating-Water-Column Device*. Ph.D. thesis. Universidade Técnica de Lisboa, Instituto Superior Técnico, 1998.

78. A. Clément and C. Maisondieu. Comparison of time-domain control laws for a piston wave absorber. In *European Wave Energy Symposium*, pp. 117–122. NEL-Renewable Energy, East Kilbride, Scotland, UK, 1993.

79. G. Chatry, A. Clément, and T. Gouraud. Self-adaptive control of a piston wave absorber. In *The Proceedings of the Eighth (1998) International Offshore and Polar Engineering Conference*, Vol. 1, pp. 127–133. Golden, Colo., ISOPE (International Society of Offshore and Polar Engineering), 1998.

80. J. L. Black, C. C. Mei, and M. C. G. Bray. Radiation and scattering of water waves by rigid bodies. *Journal of Fluid Mechanics*, 46:151–164, 1971.

81. C. C. Mei. Power extraction from water waves. *Journal of Ship Research*, 20:63–66, 1976.

82. J. R. Thomas. The absorption of wave energy by a three-dimensional submerged duct. *Journal of Fluid Mechanics*, 104:189–215, 1981.

83. G. P. Thomas. A hydrodynamic model of a submerged lenticular wave energy device. *Applied Ocean Research*, 5:69–79, 1983.

84. K. Budal and J. Falnes. A resonant point absorber of ocean waves. *Nature*, 256:478–479, 1975. With Corrigendum in Vol. 257, p. 626.

85. D. V. Evans. A theory for wave-power absorption by oscillating bodies. *Journal of Fluid Mechanics*, 77:1–25, 1976.

86. J. Falnes. Wave-power absorption by an array of attenuators oscillating with unconstrained amplitudes. *Applied Ocean Research*, 6:16–22, 1984.

87. S. H. Salter. Wave power. *Nature*, 249:720–724, 1974.

88. T. F. Ogilvie. First- and second-order forces on a cylinder submerged under a free surface. *Journal of Fluid Mechanics*, 16:451–472, 1963.

89. Clare, R., Evans, D. V., and Shaw, T. L. Harnessing sea wave energy by a submerged cylinder device. *Journal of Institution of Civil Engineers*. 73:356–385, 1982.

90. G. P. Thomas and D. V. Evans. Arrays of three-dimensional wave-energy absorbers. *Journal of Fluid Mechanics*, 108:67–88, 1981.

91. D. V. Evans. Maximum wave-power absorption under motion constraints. *Applied Ocean Research*, 3(4):200–203, 1981.

92. D. J. Pizer. Maximum wave-power absorption of point absorbers under motion constraints. *Applied Ocean Research*, 15:227–234, 1993.

93. A. J. N. A. Sarmento and A. F. de O. Falcão. Wave generation by an oscillating surface pressure and its application in wave-energy extraction. *Journal of Fluid Mechanics*, 150:467–485, 1985. (Presented previously at the *15th International Congress of Theoretical Applied Mechanics*, University of Toronto, Canada, 1980.)

94. D. V. Evans. Wave-power absorption by systems of oscillating surface pressure distributions. *Journal of Fluid Mechanics*, 114:481–499, 1982.

95. J. Falnes and P. McIver. Surface wave interactions with systems of oscillating bodies and pressure distributions. *Applied Ocean Research*, 7:225–234, 1985.

96. R. G. Alcorn, W. C. Beattie and R. Douglas. Turbine modelling and analysis using data obtained from the Islay wave-power plant. In *Proceedings of the Ninth (1999) International Offshore and Polar Engineering Conference*. Vol. 1, pp. 204–209. ISOPE. Cupertino, CA, 1999.

97. G. H. Keulegan and L. H. Carpenter. Forces on cylinders and plates in an oscillating fluid. *Journal of Research of the National Bureau of Standards*, 60:423–440, 1958.

98. A. C. Fernandes. Reciprocity relations for the analysis of floating pneumatic bodies with application to wave power absorption. In *Proceedings of the Fourth International Offshore Mechanics and Arctic Engineering Symposium*, Vol. 1, pp. 725–730. American Society of Mechanical Engineers, New York, 1985.

Index

The index contains **bold**, <u>underline</u>, and *italic* page numbers. **Bold** numbers refer to places where an idea or a quantity is explained or defined. References to Problems are <u>underline</u>; references to Figures are typed in *italic*.